Functions of the basal ganglia

The Ciba Foundation is an international scientific and educational charity. It was established in 1947 by the Swiss chemical and pharmaceutical company of CIBA Limited—now CIBA-GEIGY Limited. The Foundation operates independently in London under English trust law.

The Ciba Foundation exists to promote international cooperation in biological, medical and chemical research. It organizes about eight international multidisciplinary symposia each year on topics that seem ready for discussion by a small group of research workers. The papers and discussions are published in the Ciba Foundation symposium series. The Foundation also holds many shorter meetings (not published), organized by the Foundation itself or by outside scientific organizations. The staff always welcome suggestions for future meetings.

The Foundation's house at 41 Portland Place, London, W1N 4BN, provides facilities for all the meetings. Its library, open seven days a week to any graduate in science or medicine, also provides information on scientific meetings throughout the world and answers general enquiries on biomedical and chemical subjects. Scientists from any part of the world may stay in the house during working visits to London.

Functions of the basal ganglia

Ciba Foundation symposium 107

1984

Pitman
London

© Ciba Foundation 1984

ISBN 0 272 79777 4

Published in August 1984 by Pitman Publishing Ltd, 128 Long Acre, London WC2E 9AN, UK
Distributed in North America by CIBA Pharmaceutical Company (Medical Education Division), P.O. Box 12832, Newark, NJ 07101, USA

Suggested series entry for library catalogues:
Ciba Foundation symposia

Ciba Foundation symposium 107
viii + 281 pages, 42 figures, 8 tables

British Library Cataloguing in Publication Data

Functions of the basal ganglia.—(Ciba
  Foundation symposium; 107)
  1. Basal ganglia
  I. Series
  612'.825   QP383.3

Typeset and printed in Great Britain at The Pitman Press, Bath

Contents

Symposium on Functions of the basal ganglia, held at the Ciba Foundation, 22–24 November 1983
The original suggestion for this symposium came from Professor C. D. Marsden

Editors: David Evered (Organizer) and Maeve O'Connor

E. V. Evarts Opening remarks 1

W. J. H. Nauta and **V. B. Domesick** Afferent and efferent relationships of the basal ganglia 3
Discussion 23

J. P. Bolam Synapses of identified neurons in the neostriatum 30
Discussion 42

J. M. Deniau and **G. Chevalier** Synaptic organization of output pathways of the basal ganglia: an electroanatomical approach in the rat 48
Discussion 58

M. R. DeLong, A. P. Georgopoulos, M. D. Crutcher, S. J. Mitchell, R. T. Richardson and **G. E. Alexander** Functional organization of the basal ganglia: contributions of single-cell recording studies 64
Discussion 78

E. V. Evarts and **S. P. Wise** Basal ganglia outputs and motor control 83
Discussion 96

General Discussion 1 Basal ganglia links for movement, mood and memory 103 The limbic striatum and pallidum 107

A. M. Graybiel Neurochemically specified subsystems in the basal ganglia 114
Discussion 144

J. Glowinski, M. J. Besson and **A. Chéramy** Role of the thalamus in the

bilateral regulation of dopaminergic and GABAergic neurons in the basal ganglia   150
Discussion   160

C. Reavill, P. Jenner and **C. D. Marsden**   γ-Aminobutyric acid and basal ganglia outflow pathways   164
Discussion   171

General Discussion 2   The striatal mosaic and local control of dopamine   177

S. D. Iversen   Behavioural effects of manipulation of basal ganglia neurotransmitters   183
Discussion   195

I. Divac   The neostriatum viewed orthogonally   201
Discussion   210

General Discussion 3   Dopamine depletion and replacement   216   Models for basal ganglia disease   221

C. D. Marsden   Which motor disorder in Parkinson's disease indicates the true motor function of the basal ganglia?   225
Discussion   237

A. M. Wing and **E. Miller**   Basal ganglia lesions and psychological analyses of the control of voluntary movement   242
Discussion   253

Final general discussion   Local regulation of transmitter release in the basal ganglia   258   The subthalamic nucleus   261

E. V. Evarts   Closing remarks   269

Index of contributors   273

Subject index   274

Participants

G. W. Arbuthnott MRC Brain Metabolism Unit, University Department of Pharmacology, 1 George Square, Edinburgh EH8 9JZ, UK

J. P. Bolam University Department of Pharmacology, South Parks Road, Oxford OX1 3QT, UK

D. B. Calne Division of Medicine, Health Sciences Centre Hospital, 2211 Wesbrook Mall, The University of British Columbia, Vancouver, British Columbia V6T 1W5, Canada

M. B. Carpenter Department of Anatomy, Uniformed Services University of the Health Sciences, School of Medicine, 4301 Jones Bridge Road, Bethesda, Maryland 20814, USA

M. R. DeLong Department of Neurology and Neurosciences, The Johns Hopkins Hospital and School of Medicine, c/o The Baltimore City Hospitals, 4940 Eastern Avenue, Baltimore, Maryland 21224, USA

J. M. Deniau Laboratoire de Physiologie des Centres Nerveux, Université Pierre et Marie Curie, 4 Place Jussieu, 75230 Paris Cedex 05, France

G. Di Chiara Istituto di Farmacologia e Tossicologia Sperimental, Università di Cagliari, Viale Diaz 182, I-09100 Cagliari, Italy

M. DiFiglia Department of Neurology, Harvard Medical School, Massachusetts General Hospital, Warren Building 370, Fruit Street, Boston, Massachusetts 02114, USA

I. Divac Institute of Neurophysiology, University of Copenhagen, Blegdamsvej 3 C, DK-2200 Copenhagen N, Denmark

E. V. Evarts (*Chairman*) Laboratory of Neurophysiology, National Institute of Mental Health, Bldg 36, Room 2D10, Bethesda, Maryland 20205, USA

J. Glowinski Groupe NB, INSERM U114, Collège de France, 11 Place Marcelin Berthelot, 75231 Paris Cedex 05, France

A. M. Graybiel Department of Psychology and Brain Sciences, E25-618, Massachusetts Institute of Technology, Cambridge, Massachusetts 02139, USA

O. Hornykiewicz Institute of Biochemical Pharmacology, Faculty of Medicine, University of Vienna, Borschkegasse 8a, A-1090 Vienna, Austria

S. D. Iversen Neuroscience Research Centre, Merck Sharp and Dohme Limited, Hertford Road, Hoddesdon, Hertfordshire EN11 3BU, UK

P. G. Jenner Department of Neurology, Institute of Psychiatry, De Crespigny Park, London SE5 8AF, UK

S. T. Kitai Department of Anatomy, The University of Tennessee Center for the Health Sciences, 875 Monroe Avenue, Memphis, Tennessee 38163, USA

C. D. Marsden Department of Neurology, Institute of Psychiatry, De Crespigny Park, London SE5 8AF, UK

W. J. H. Nauta Department of Psychology and Brain Sciences, E25-618, Massachusetts Institute of Technology, Cambridge, Massachusetts 02139, USA

R. Porter The John Curtin School of Medical Research, The Australian National University, G.P.O. Box 334, Canberra City 2601, Australia

G. Rizzolatti Institute of Human Physiology, Faculty of Medicine and Surgery, University of Parma, Via A. Gramsci 14, 43100 Parma, Italy

E. T. Rolls Department of Experimental Psychology, University of Oxford, South Parks Road, Oxford OX1 3UD, UK

A. D. Smith University Department of Pharmacology, South Parks Road, Oxford OX1 3QT, UK

P. Somogyi (*Ciba Foundation Bursar*) Department of Human Physiology, School of Medicine, Flinders Medical Centre, Bedford Park, South Australia 5042

A. M. Wing MRC Applied Psychology Unit, 15 Chaucer Road, Cambridge CB2 2EF, UK

M. Yoshida Department of Neurology, Jichi Medical School, Minamika-wachi-machi, Kawachi-gun, Tochigi-Ken, Japan 329-04

Opening remarks

E.V. EVARTS

Laboratory of Neurophysiology, National Institute of Mental Health, Bethesda, Maryland 20205, USA

1984 Functions of the basal ganglia. Pitman, London (Ciba Foundation symposium 107) p 1–2

At the Annual Meeting of the Society for Neuroscience in 1976, Patrick McGeer delivered a lecture on the basal ganglia and called it 'Mood and movement: twin galaxies of the inner universe'. He was a smash hit, not only because of his gifts as an orator, but also because he had something of interest to say to each and every one of the neuroscience disciplines whose members were in attendance: then as now, the basal ganglia had something for everyone.

Neurologists have been interested in the basal ganglia since the substantia nigra was discovered to be the site of the lesion in Parkinson's disease, and pharmacologists entered the field *en masse* when Carlsson et al (1957) showed that reserpine-induced akinesia was reversed by L-dopa. The modern era of research in biochemical neuropathology was ushered in by Hornykiewicz's discovery of striatal dopamine depletion in Parkinson's disease, and neurochemists and neuroanatomists entered the field at about the same time, for, as Graybiel & Ragsdale (1983, p 427) have noted, 'the chemoarchitecture of the striatum became a focus for work on neurotransmitter distributions in the forebrain in the 1950s and 1960s when the caudate-putamen complex was found to have the highest content of acetylcholine and dopamine in the forebrain'.

Psychiatrists have been interested in the basal ganglia since the neuroleptic agents introduced in the 1950s were shown to be dopamine antagonists, and their interest was further stimulated by recognition that whereas some parts of the basal ganglia are primarily motor, other parts (e.g. the nucleus accumbens) project to centres for mood control rather than for muscle control. It was, indeed, this fact that enabled Patrick McGeer to link mood and movement in his lecture on the basal ganglia.

Psychologists became increasingly interested in the basal ganglia when Teuber (1966) called attention to the parallels between effects of prefrontal and basal ganglia lesions on perceptual–mnemonic processes in men and monkeys, and the neurophysiological approach to basal ganglia circuitry was reborn when Yoshida & Precht (1971) and then Uno & Yoshida (1975) discovered that both striatal and pallidal outputs act by inhibition and disinhibition rather than by excitation and disexcitation.

These few highlights from the recent decades of research on the basal ganglia show the remarkable success of interdisciplinary efforts: no other set of brain structures has been the subject of such a mutuality of research interest by scientists using different techniques. It is the mutuality of interest each of us has in the research of our colleagues here assembled that will provide one of the essential ingredients for the success of this Ciba Foundation symposium. There are many other ingredients, one of which is that the 'speakers' and the 'discussants' could easily have changed places. There was absolutely no need for me to prepare a list of who might discuss each paper, because many of us gathered in this room could discuss each of the presentations, and in any case there is a magic in 41 Portland Place that precludes the need to *plan* a discussion for these symposia. As chairman I shall occasionally propose one or two questions touching on the unsolved problems of the basal ganglia, since there are still some terrific riddles at the physiological and clinical levels; perhaps as a result of this meeting we shall be able to formulate the riddles more clearly, and in my summing up I shall try to list some of these unsolved problems and highlight some of the ideas that will have emerged from this meeting on the best approaches to them.

REFERENCES

Carlsson A, Lindqvist M, Magnusson T 1957 3,4-Dihydroxyphenylalanine and 5-hydroxytryptophan as reserpine antagonists. Nature (Lond) 160:1200

Graybiel AM, Ragsdale CW Jr 1983 Biochemical anatomy of the striatum. In: Emson PC (ed) Chemical neuroanatomy. Raven Press, New York, p 427-504

Teuber HL 1966 Alterations of perception after brain injury. In: Eccles JC (ed) The brain and conscious experience. Springer, New York, p 182-216

Uno M, Yoshida M 1975 Monosynaptic inhibition of thalamic neurons produced by stimulation of the pallidal nucleus in cats. Brain Res 99:377-380

Yoshida M, Precht W 1971 Monosynaptic inhibition of neurons of the substantia nigra by caudatonigral fibers. Brain Res 15:225-228

Afferent and efferent relationships of the basal ganglia

W. J. H. NAUTA and V. B. DOMESICK*

*Massachusetts Institute of Technology, Cambridge, MA 02139, and *Mailman Research Centre, McLean Hospital, Belmont, MA 02178, USA*

Abstract. A survey of the known circuitry of the basal ganglia leads to the following conclusions. (1) No complete account can yet be given of the neural pathways by which the basal ganglia affect the bulbospinal motor apparatus. Channels of exit from the basal ganglia originate from the internal pallidal segment, the pars reticulata of the substantia nigra, and the subthalamic nucleus, and each of these is directed in part rostrally to the cerebral cortex by way of the thalamus, in part caudally to the midbrain. The postsynaptic extension of the mesencephalic channels to bulbar and spinal motor neurons is largely unknown. Since the ascending channels are collectively of greatest volume, the notion remains plausible that the basal ganglia act in considerable part by modulating motor mechanisms of the cortex. (2) Recent findings in the rat suggest that the striatum is subdivided into a ventromedial, limbic system-afferented region and a dorsolateral, 'non-limbic' region largely corresponding to the main distribution of corticostriatal fibres from the motor cortex. These two subdivisions appear to give rise to different striatofugal lines, the outflow from the limbic-afferented sector partly re-entering the circuitry of the limbic system. (3) The limbic-afferented striatal sector suggests itself as an interface between the motivational and the more strictly motor aspects of movement. This suggestion is strengthened by evidence that the 'limbic striatum' seems enabled by its striatonigral efferents to modulate not only the source of its own dopamine innervation but also that of a large additional striatal region.

1984 Functions of the basal ganglia. Pitman, London (Ciba Foundation symposium 107) p 3-29

The term, extrapyramidal system, introduced by S. A. K. Wilson (1912), has never been adequately defined anatomically. Although it can be interpreted literally as denoting all of the brain's effector mechanisms that do not involve the pyramidal tract, convention over the years has made the term very nearly synonymous with the basal ganglia and their efferent connections. The much older term, basal ganglia, originally referred to all of the grey masses at the base of the cerebral hemisphere, including even the thalamus, but over time became restricted to the corpus striatum (striatum and pallidum) and two smaller structures, the subthalamic nucleus and the substantia nigra, both linked to the corpus striatum by reciprocal fibre connections.

In the following sections of this brief account the known neural circuitry of the basal ganglia is reviewed with no more than occasional reference to the neurophysiological and neurochemical evidence discussed in other papers in this volume.

Striatum

In all primates and many non-primate mammals the striatum is subdivided by a plate-like internal capsule into two districts, the dorsomedial caudate nucleus and the ventrolateral putamen. In other mammalian forms, including the rat, the anterior part of the internal capsule passes through the striatum in the form of a brush rather than a plate; in such forms the striatum lacks clear subdivision and is therefore often referred to as caudato-putamen.

All districts of the striatum exhibit basically the same cytoarchitecture: throughout its extent more than 98% of its neurons are small to medium-sized while the remaining 1–2% is made up of large multipolar cells with well-developed Nissl bodies (Namba 1957). Most of the smaller cells probably correspond to the so-called spiny neurons distinguished in Golgi material and the electron microscope (Kemp 1968a,b, Fox et al 1971) by numerous spines on all but the most proximal dendritic segments. A much smaller number of small cells, the so-called spidery neurons (Fox et al 1971/1972), have smooth dendrites but are distinct, by their size, from the large 'aspiny' neurons that correspond to the large multipolar cells seen in Nissl material.

It was long assumed that efferent connections of the striatum originate exclusively from the large neurons (Vogt & Vogt 1920), whereas the smaller cells were thought to be intrinsic striatal neurons. The results of retrograde-labelling experiments, however, suggest instead that most striatofugal fibres to the globus pallidus and substantia nigra originate from medium-sized, densely spiny, rather than large neurons (DiFiglia et al 1976, Somogyi et al 1981a, Bolam 1984, this volume).

On account of the notable lack of any regional differentiation obvious on cursory inspection, the striatum has traditionally been described as a homogeneous structure. More recent studies, however, have disclosed a more covert stereometric pattern of compartition subdividing the striatum into an intricate labyrinth of small compartments characterized by peculiarities of histochemistry, afferent connections, and/or cell-packing density (Graybiel & Ragsdale 1983, and 1984, this volume).

Afferent connections of the striatum

The striatum is the main entrance portal for neural inputs to the basal ganglia, and appears to receive its most voluminous afferent connections from the cerebral cortex, intralaminar thalamic nuclei, and substantia nigra. Additional afferents, however, arise from the amygdala and from the dorsal raphe nucleus of the midbrain.

The corticostriate connection. Fibre-degeneration studies have shown that virtually all regions of the neocortex project to the striatum in a topographic pattern that by and large preserves the topology of the cortical mantle (Kemp & Powell 1970). The mosaic of the projection is, however, not sharply defined, and consequently few if any parts of the striatum are projected upon by a single cortical area. The projection from the sensorimotor cortex, most widely distributed, may be the only corticostriatal projection that is bilateral (Carman et al 1965, Künzle 1975).

The scheme of corticostriatal topography indicated by earlier studies by fibre-degeneration methods may require some revision. Autoradiographic findings have shown that the precentral cortex of the monkey projects almost exclusively to the putamen (Künzle 1975) rather than to both putamen and caudate nucleus as previously reported, and the prefrontal cortex projects to the entire length of the caudate nucleus rather than exclusively to the caput of the latter (Goldman & Nauta 1977). Future tracer-transport studies may reveal additional inaccuracies in current schemes, but are unlikely to invalidate the notion that all cortical regions project to the striatum in a reasonably well-defined topographic pattern, and that all parts of the striatum receive fibres from one or more districts of the cortical mantle.

Nucleus accumbens. Unlike the rest of the striatum, this anterior-ventromedial region leaning against the septum, and extending ventrally as the small-celled core of the olfactory tubercle, receives its main cortical input not from the neocortex but from the hippocampus (Raisman et al 1966, Kelley & Domesick 1982). Since it was found to be projected upon also by the amygdala (DeOlmos & Ingram 1972), the nucleus accumbens, together with the parvicellular core of the olfactory tubercle which likewise receives projections from the amygdala, became regarded as the 'limbic striatum'. More recent findings, however, suggest that the striatal region receiving afferents from within the circuitry of the limbic system extends well outside the nucleus accumbens and olfactory tubercle. Of these afferents, those originating from the amygdala appear to have the widest distribution: the amygdalostriatal projection in the rat involves the entire striatum except for an anterior, dorsolateral sector that receives the main projection from the motor cortex (Kelley et al 1982). ('Anterior' here denotes the large part of the striatum that extends rostral to the crossing of the anterior commissure [Fig. 1A] and is customarily focused on in studies of the striatum, to the exclusion of the narrower posterior striatal half [Fig. 1B] extending caudal to the anterior commissure alongside the globus pallidus.) Other afferents to the striatum which originate from limbic structures (hippocampus; fronto-cingulate cortex) or from brainstem structures embedded in the circuitry of the limbic system (ventral tegmental area, dorsal raphe nucleus) are distributed within the same region (c.f. Fig. 1); like the amygdalostriatal projection, they

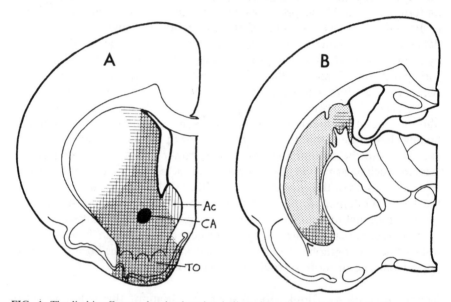

FIG. 1. The limbic-afferented striatal region indicated by various line and stipple patterns in frontal sections involving a rostral (A) and more caudal (B) level of the striatum. Vertical lines, the region innervated by the ventral tegmental area; horizontal lines, the projection from the prefrontal cortex; stippling, the projection from the amygdala. Not shown are the striatal afferents from the hippocampal formation (confined to the nucleus-accumbens region medial to the anterior commissure), cingulate cortex (a narrow vertical strip along the lateral wall of the lateral ventricle), and dorsal raphe nucleus (throughout the striatum but densest in the limbic-afferented area). Abbreviations in this and other illustrations: Ac, nucleus accumbens; Am, amygdaloid complex: A8 and A10, dopamine cell groups A8 and A10 of Dahlström & Fuxe; AVT, ventral tegmental area; CA, anterior commissure; CM, centrum medianum; C–P, caudato-putamen; GPe, globus pallidus, external segment; GPi, globus pallidus, internal segment; GPv, ventral pallidum; HL, lateral habenular nucleus: LC, locus ceruleus; MC, motor cortex; NC, caudate nucleus; ncF, nucleus of Forel's field H; NR, red nucleus; NRd, dorsal raphe nucleus; NRm, median raphe nucleus; OT, olfactory tubercle; PT, pyramidal tract; Put, putamen; SC, superior colliculus; SNc, substantia nigra, pars compacta; SNr, substantia nigra, pars reticulata; Sth, subthalamic nucleus; TO, olfactory tubercle; Tpd, nucleus tegmenti pedunculopontinus, pars diffusa; TPC or Tpc, nucleus tegmenti pedunculopontinus, pars compacta; VA, nucleus ventralis anterior; VL, nucleus ventralis lateralis; VM, ventromedial thalamic nucleus; ZI, zona incerta; 1, ansa lenticularis, ventral division; 2, ansa lenticularis, dorsal division, or fasciculus lenticularis; 3, ansa lenticularis, middle division. (Reproduced from Kelley et al 1982 by permission of Pergamon Press Ltd.)

largely avoid the anterodorsolateral striatal sector. The anterior half of the striatum thus appears divisible into (a) a ventromedial region projected upon by a variety of limbic and limbic system-associated structures ('limbic striatum' of Kelley et al 1982) and (b) a dorsolateral sector receiving only very sparse limbic afferents but massively innervated by the motor cortex ('non-limbic striatum').

Thalamostriate connections. The only thalamostriate projections known thus far originate from the intralaminar or non-specific thalamic cell groups, in particular the parafascicular nucleus and centrum medianum (the so-called CM–PF complex). That CM–PF projects to the putamen was observed first by Vogt & Vogt (1941) in the human, and confirmed later in experimental animals. The topographic pattern of the thalamostriate projection has not been elaborated in detail, but it appears that the more rostral intralaminar nuclei project to more rostral parts of the striatum; the nucleus accumbens, for example, receives its thalamic afferents from the parataenial nucleus (Swanson & Cowan 1975).

The nigrostriatal projection. The existence of a massive nigrostriatal projection has been suspected since the beginning of the century from observations of rapid cell-atrophy in the substantia nigra following extensive destruction of the striatum, but attempts to demonstrate the connection directly by fibre-degeneration methods succeeded only after the introduction of the Fink–Heimer method in 1966. Several years earlier, however, Andén et al (1964) by the monoamine-histofluorescence method had provided the first direct evidence of a nigrostriatal fibre system originating from the dopamine neurons of the substantia nigra. Hökfelt & Ungerstedt (1969) reported evidence that nigrostriatal fibres end in axon terminals containing small granular vesicles; other electron-microscope studies showed that these terminals synapse with dendritic spines (Kemp 1968a) and cell bodies of spiny neurons, and with dendrites of spidery neurons (Fox et al 1971/72).

Andén et al (1966) and Ungerstedt (1971) subdivided the nigrostriatal dopamine system into (1) a nigrostriatal system from the dopamine neurons of the pars compacta of the substantia nigra (cell group A9 of Dahlström & Fuxe 1964) to the larger part of the striatum, and (2) a mesolimbic system (Ungerstedt 1971) from dopamine cell group A10 to the nucleus accumbens and olfactory tubercle. Cell group A10 in the rat forms a large, wedge-shaped protrusion from the medial half of the pars compacta of the nigra into the ventral tegmental area of Tsai. This dopamine cell group lies embedded in the mesencephalic trajectory of the medial forebrain bundle (Nauta et al 1978), and thus seems almost certain to receive afferents descending from the preoptic region, hypothalamus, substantia innominata and amygdala.

In later autoradiographic studies (Fallon & Moore 1978, Beckstead et al 1979), the heaviest projection from A10 was found to involve not only the nucleus accumbens and olfactory tubercle but also the ventral quarter of the entire length of the striatum (a region of which the accumbens forms an anterior part), and to extend dorsally from this basal region over nearly the entire medial half of the overlying caudato-putamen (cf. Fig. 1: vertical lines), where it overlaps with the nigrostriatal projection from the medial half of the pars compacta. No converse overlap, however, appears to occur in the nucleus accumbens and olfactory tubercle: in contrast to the more posterior extent of the basal striatal region, this anteroventral district receives nigral afferents almost exclusively from A10 (Nauta et al 1978, Beckstead et al 1979). The nucleus accumbens and olfactory tubercle thus seem distinguished from the rest of the striatum by their poverty in A9 afferents as much as by their wealth in afferents from A10.

From fibre-degeneration studies in the monkey, Carpenter & Peter (1972) concluded that the caudal two-thirds of the nigra projects almost exclusively to the putamen and, hence, that the caudate nucleus is likely to be projected upon by the anterior third of the nigra. In the rat no evidence of such a gradient was found; rather it appears that in that species each nigral locus projects to the entire length of the striatum (Beckstead et al 1979).

Efferent connections of the striatum

The efferent connections of the striatum are established exclusively by delicate, thinly myelinated axons passing medially from their origin in slender fascicles, the 'pencil bundles' of Wilson (1914), that converge on the globus pallidus much like the spokes of a wheel (Fig. 2). These bundles pass through the outer and inner segments of the globus pallidus and continue in part beyond the pallidum to the substantia nigra.

The striatopallidal projection. Throughout their passage through the pallidum the striatofugal fibres maintain an orderly radial arrangement, and since they terminate close to the radial bundles in which they enter the pallidum, the striatopallidal projection must be organized in a fairly orderly radial pattern (Nauta & Mehler 1966). In accord with this general arrangement, the nucleus accumbens as the most anteroventral striatal region projects to the most anteroventral part of the external pallidum (Nauta et al 1978), in particular to the latter's subcommissural part, the ventral pallidum of Heimer & Wilson (1975) (GPv in Fig. 2). Striatofugal fibres terminate massively in both the external and internal pallidal segments but it is difficult to determine whether the same or different sets of fibres innervate the two segments (in Fig. 2 both possibilities have been indicated). A clue to this question is provided by some recent immunohistochemical findings: the striatal projection to the external pallidal segment includes numerous enkephalin-positive but (except for the so-called ventral pallidum, see below) only a few substance P-positive fibres. In the striatal projection to the internal pallidal segment and substantia nigra a reverse ratio prevails (c.f. Haber & Nauta 1983), suggesting that the two

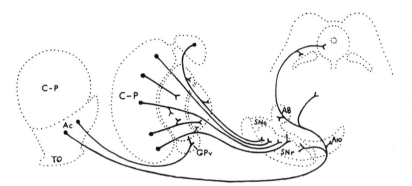

FIG. 2. Diagrammatic representation of striatofugal connections as discussed in the text. Abbreviations: see legend to Fig. 1. (Reproduced from Nauta & Domesick 1979 by permission of Pergamon Press Ltd.)

pallidal segments are innervated in part at least by separate contingents of striatal fibres. As to the mode of termination of the striatopallidal connection, collaterals of fibres composing the radial fascicles combine to form dense sleeve-like plexuses around the dendrites of globus pallidus neurons, on which they establish a nearly uninterrupted lining of *de passage* synaptic contacts (Fox et al 1974).

The striatonigral connection. Most of the remaining distribution of striatofugal fibres establishes the striatonigral connection. From an analysis of the fibre-calibre spectrum in the radial striatofugal fascicles Fox et al (1975) concluded that most if not all striatonigral fibres, instead of composing an independent striatofugal system, are the attenuated and largely unmyelinated end-stretches of striatal efferents that have given off collaterals to the globus pallidus at more proximal points in their course. The observation that the mode of termination of these fibres in the nigra is similar to that of striatofugal fibres in the pallidum (Schwyn & Fox 1974) seems to support this concept. Electrophysiological findings by Yoshida et al (1974) likewise suggest that the pallidum and nigra are innervated in parallel by the same striatofugal fibres. However, persuasive as this combined anatomical-electrophysiological evidence is, the immunohistochemical data mentioned in the preceding paragraph would seem to complicate the issue. The apparent sparseness of enkephalin-positive striatonigral fibres, contrasting with the abundance of substance P-positive striatonigral fibres, suggests that, at least with respect to these two categories of striatal efferents, the substantia nigra could share a large number of striatofugal fibres with the internal, but hardly with the external pallidal segment, the latter receiving only very sparse substance P-positive striatal efferents. It thus appears that the conclusion of Fox et al (1975) may require some qualification bearing on the variety of chemically specified striatofugal fibres.

Striatofugal fibres extending caudally beyond the nigral complex to the mesencephalic tegmentum and central grey substance have been traced only from the accumbens region of the striatum (Nauta et al 1978); the distribution of these longest striatofugal fibres, schematically indicated in Fig. 2, involves to some extent the median raphe nucleus and is also in other respects comparable to the distribution of hypothalamo-mesencephalic fibres.

The topography of the striatonigral projection is remarkable. Autoradiographic findings in the rat (Nauta & Domesick 1979, Domesick 1980) suggest that in that species (a) each striatal locus projects to the entire length of the nigra; (b) the projection preserves the mediolateral coordinate of striatal topology, but (c) it inverts the dorsoventral coordinate: the most dorsal striatal regions project in greatest density to the most ventral zone of the pars reticulata, middle-depth regions of the striatum to middle-depth strata of the pars reticulata; only the most ventral striatal zone including the nucleus accumbens and olfactory tubercle project directly into the pars compacta and the immediately subjacent zone of the pars reticulata. These various striatonigral relationships are indicated schematically in Fig. 2.

The striatonigral connection is emphasized in particular as a 'return loop' reciprocating the dopaminergic nigrostriatal projection. It is, however, certain that fibres of this massive projection synapse not only on dendrites of dopamine neurons, but also on non-dopaminergic neurons of the pars reticulata that project to the ventromedial thalamic nucleus (Somogyi et al 1979). It thus appears that the striatonigral projection includes the first link of a striatofugal 'through-line' leading sequentially over the pars reticulata and thalamus to the cerebral cortex (see below). Although not yet reported, it would by analogy seem likely that the projection also involves pars reticulata neurons projecting to the superior colliculus and pedunculopontine nucleus (see below).

Despite the foregoing considerations, subsequent findings by Somogyi et al (1981b) clearly indicate that the striatonigral projection, to some extent at least, synapses directly upon nigrostriatal neurons. The question arises how strict this reciprocity in the nigro-striato-nigral circuit is. Does each striatal locus project back exclusively to those nigrostriatal neurons by which it is innervated? No generally valid answer to this question can be given at

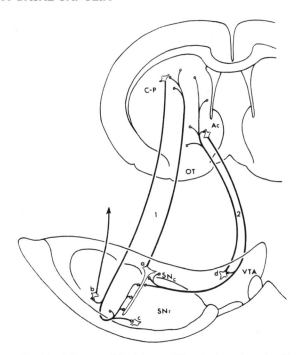

FIG. 3. Diagram of feed-back loops and feed-forward lines in the striatonigral interconnection as proposed in the text. Note: 1) closed loop (1) between caudato-putamen (C-P) and dopamine cells of the substantia nigra's pars compacta (SNc) (neuron a): the striatonigral limb of this loop is here assumed to contact dopamine-cell dendrites protruding into pars reticulata (SNr), as well as non-dopaminergic pars reticulata neurons (b and c) which may be interneurons in loop 1 or may project as feed-forward lines to thalamus, superior colliculus, or midbrain tegmentum. 2) A closed loop (2) also exists between the nucleus accumbens (Ac) and dopamine cell group A10 in the ventral tegmental area (VTA) (neuron d) but the striatonigral limb of this loop ends in greatest volume in pars compacta and upper stratum of pars reticula, contacting dopamine cells innervating the caudato-putamen (neuron a). The apparent arrangements suggest an 'open-loop' relationship between Ac and C-P in which the output of Ac can affect the dopamine innervation of C-P without itself being controlled by the output of C-P. Abbreviations: see legend to Fig. 1. (Reproduced from Domesick 1981 by permission of Raven Press, New York.)

present, mainly because the great majority of striatonigral fibres terminate in the pars reticulata and therefore can directly affect the nigrostriatal neurons of the pars compacta only by synapsing on the ventrally protruding dendrites of the latter; the pattern in which this dendritic plexus overlaps the plexus of striatonigral fibres is unknown. For the nucleus accumbens, however, the evidence suggests a negative answer (Nauta et al 1978, Domesick 1981, Somogyi et al 1981b): although this anterior, ventromedial striatal region receives its nigral innervation almost exclusively from the ventral tegmental area, its reciprocating striatonigral projection only sparsely involves the ventral tegmental area and is distributed in greatest volume to the medial half of the pars compacta (Fig. 3). It thus appears that the nucleus accumbens, at least, is not interconnected with the nigral complex in point-for-point reciprocity, and that impulses from the nucleus accumbens could be thought to affect the nigral innervation, and hence the functional state, of the entire medial half of the striatum. There is at present no evidence that the medial striatal half maintains a similar but converse open-loop connection with the ventral tegmental area.

Globus pallidus

In all mammalian species the globus pallidus (or pallidum) is subdivided into a lateral or external (GPe) and a medial or internal (GPi) segment. In primates both these segments lie lateral to the internal capsule, demarcated from each other by the internal medullary lamina. In many non-primate mammals, however, the internal segment lies largely embedded within the internal capsule and cerebral peduncle and is therefore called entopeduncular nucleus.

The neurons of both pallidal segments are large fusiform or multipolar cells with exceptionally long, smooth dendrites enmeshed by the striatopallidal-fibre plexuses mentioned earlier.

Afferent connections of the pallidum

Only two connections afferent to the pallidum are known: the striatopallidal and a subthalamopallidal projection. Numerous nigral dopamine fibres course through the pallidum in passing to the striatum but no evidence of actual termination of such fibres in the pallidum appears to have been reported.

The striatopallidal connection, apparently the more massive of the afferents of the pallidum, has already been reviewed (p 8).

A subthalamopallidal projection was recognized first in the monkey by Carpenter & Strominger (1967) who traced degenerating fibres from lesions in the subthalamic nucleus mainly to the internal pallidal segment. In a later autoradiographic study in the monkey H. J. W. Nauta & Cole (1978), however, showed that the projection is distributed almost equally to both segments, and terminates densely in band- or sheet-like zones oriented parallel to the internal medullary lamina. The projection to the external pallidal segment can be interpreted in part at least as a reciprocation of the pallidosubthalamic connection.

Efferent connections of the pallidum

Nearly all of the known efferent connections of the globus pallidus or pallidum are established by the ansa lenticularis. Monakow (cf. Nauta & Mehler 1966) subdivided this massive pallidofugal system into (1) a ventral and (2) a dorsal division, both originating from the internal pallidal segment, and (3) a middle division originating from the external segment (the three divisions are indicated by these same numbers in Fig. 4).

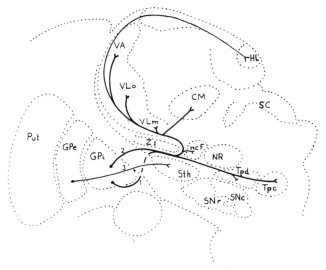

FIG. 4. Diagram of pallidofugal projections as described in the text. Abbreviations: see legend to Fig. 1. (Modified from Nauta & Mehler 1966.)

The pallidosubthalamic connection. Pallidal efferents to the subthalamic nucleus originate from the external pallidal segment (Fig. 4) (cf. Nauta & Mehler 1966); it is still uncertain whether the internal pallidal segment contributes to the connection.

Pallidothalamic projections originate from both pallidal segments. That from the external segment is of only modest size but is of considerable interest because it appears distributed to the thalamic reticular nucleus (H. J. W. Nauta 1979) which is known to project to most or all remaining thalamic nuclei (Scheibel & Scheibel 1966). The much more massive projection from the internal segment is distributed over the so-called thalamic fasciculus to three thalamic districts: (1) the nuclei ventralis lateralis and ventralis anterior (VA–VL complex), (2) the centrum medianum, and (3) the lateral habenular nucleus (Fig. 4). Since the VA–VL complex projects to the precentral cortex, it seems certain that the globus pallidus by its thalamic conduction channel can affect the neural mechanisms of the motor cortex.

The pallidal projection to the centrum medianum may be composed, in part at least, by collaterals branching off from the pallidal efferents destined

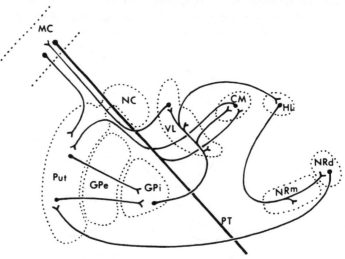

FIG. 5. Diagram emphasizing reciprocities in the connections of the internal pallidal segment (GPi). Note three transthalamic 'return-loops' to the striatum: one involving the centrum medianum (CM), a second one sequentially the VA–VL complex (VL) and the motor cortex (MC), a third one the lateral habenular nucleus (HL) and the dorsal raphe nucleus (NRd). Abbreviations: see legend to Fig. 1. (Reproduced from Nauta & Domesick 1979 by permission of Pergamon Press Ltd.)

for the VA–VL complex. Since the centrum medianum projects predominantly to the putamen, this pallidothalamic connection suggests a transthalamic circuit: globus pallidus → centrum medianum → putamen → globus pallidus (ç.f. Fig. 5, in which this and other reciprocities in the pallidal connections are schematically indicated).

A pallidohabenular projection was first suggested by Wilson (1914), but could be verified only recently in the cat by autoradiographic experiments (H. J. W. Nauta 1974). The connection is established by fibres from the internal pallidal segment that terminate exclusively in the lateral half of the lateral habenular nucleus.

The pallidohabenular connection is noteworthy for two reasons, First, the lateral habenular nucleus projects directly to the median and dorsal raphe nuclei (Herkenham & Nauta 1979) and is probably a principal origin of afferents to the mesencephalic serotonin cell groups. Since the dorsal raphe nucleus is the main source of the serotonin innervation of the striatum, the pallidohabenular connection could form part of a neural circuit: pallidum → lateral habenular nucleus → dorsal raphe nucleus → striatum → pallidum (Fig. 5). Second, as well as pallidohabenular fibres the lateral habenular nucleus receives fibres from the lateral preoptico-hypothalamic region and thus appears to be one of the few structures in which pallidal and limbic system efferents are known to converge (a further example of pallido-limbic confluence is mentioned below under ventral pallidum).

The pallidomesencephalic projection. Projections from the internal pallidal segment to the mesencephalon are established by the relatively small group of ansa lenticularis fibres that continue their caudal course through the subthalamic region into the midbrain tegmentum (Fig. 4). In the monkey, most of these long pallidofugal fibres terminate densely in a relatively large-celled region of the pontomesencephalic tegmentum, the pedunculopontine nucleus (Nauta & Mehler 1966). No pallidofugal fibres have been traced caudally beyond this level.

The ventral pallidum

This recently discovered component of the pallidum merits separate description. The term (Heimer & Wilson 1975) denotes a large rostroventral extension of the external pallidal segment into the basal forebrain region beneath the anterior commissure. Morphologically indistinguishable from the main body of the external segment or dorsal pallidum with which it is continuous behind the commissure, the ventral pallidum differs from the latter histochemically: in addition to the dense plexus of enkephalin-positive striatopallidal fibres which it shares with the dorsal pallidum, it is selectively pervaded by an almost equally dense plexus of substance P-positive striato-pallidal fibres which sharply delimit it from the dorsal pallidum (Haber & Nauta 1983). Studies by the autoradiographic fibre-labelling method—which does not discriminate between striatofugal fibres of different peptide content—have shown that the entire ventral pallidum as defined by its substance P-positive fibre plexus receives its striatal innervation exclusively from the ventromedial, limbic-innervated sector of the anterior striatum (Nauta et al 1978 and unpublished studies). A further reason for regarding the ventral pallidum as the limbic subdivision of the external pallidal segment could be found in its efferent relationships: recent anterograde and retrograde-labelling studies in the rat have demonstrated that the ventral pallidum projects not only to the subthalamic nucleus and substantia nigra—as does the dorsal pallidum—but also to various limbic structures, namely the mediodorsal thalamic nucleus, amygdala, lateral habenular nucleus and ventral tegmental area (Haber et al 1982). Only by retrograde cell-labelling with wheatgerm agglutinin-conjugated horseradish peroxidase has it been possible to demonstrate an additional ventral–pallidal projection to medial regions of the cerebral cortex, but it must be noted that this projection as well as the one to the amygdala appears to originate almost exclusively from acetylcholinesterase-positive cells (Grove et al 1983) that lie scattered throughout the ventral pallidum and that may be outlying cells of the basal nucleus rather than true pallidal neurons. It is at present unknown whether these neurons are links in a striatofugal conduction line, or components of other, possibly limbic, circuits.

Subthalamic nucleus

This nucleus appears foremost as a satellite to the pallidum: it receives a massive projection from the external pallidal segment and in turn projects massively to both pallidal segments. It also, however, receives afferents from the motor cortex (Hartmann-von Monakow et al 1978), and gives rise to considerable projections to the substantia nigra pars reticulata (H. J. W. Nauta & Cole 1978, Ricardo 1980) and the VA–VL complex of the thalamus, as well as a sparse projection to the pedunculopontine nucleus (H. J. W. Nauta & Cole 1978, Edley & Graybiel 1983). Thus far, only the two last-mentioned, lesser efferents are known to connect the subthalamic nucleus to structures outside the domain of the basal ganglia, and it is noteworthy that both converge with stronger projections from the internal pallidal segment.

Substantia nigra

The ultrastructural similarity of pallidum and substantia nigra has been emphasized by many investigators (cf. Fox et al 1974). However, unlike the cells of the globus pallidus, nigral neurons can be classified into a dopaminergic and a non-dopaminergic group. Dopamine neurons are the predominant elements of the pars compacta or dopamine cell group A9 of the nigra, as well as of the outlying nigral cell group A10, whereas non-dopamine neurons prevail in the pars reticulata (Dahlström & Fuxe 1964). This cytochemical dualism has no evident corollary in size or shape of the neurons: nearly all the cells of both pars compacta and pars reticulata conform to the same basic types in both Nissl and Golgi material. There is, however, experimental evidence that the dopamine neurons are distinguished by their much more abundant Nissl substance, and therefore darker appearance in Nissl preparations (Domesick et al 1983).

Afferent connections of the nigral complex

The most massive of the afferent connections, the striatonigral projection, was discussed above. Additional basal ganglia inputs to the nigra originate from the external pallidal segment (Hattori et al 1975, Grofová 1975) and subthalamic nucleus (H. J. W. Nauta & Cole 1978, Ricardo 1980). The only known afferent connection from the cerebral cortex was traced by Beckstead (1979) from several prefrontal areas to the pars reticulata. Afferents to the nigra from structures within the circuitry of the limbic system have been traced from the lateral preoptico-hypothalamic region (Swanson 1976); according to a later study (Nauta & Domesick 1978) this projection may involve dopamine cell group A10 (and thus indirectly the limbic sector of the striatum) more than A9. Although not anatomically demonstrated thus far, further afferents to the nigral complex may originate from more caudal points within the limbic circuitry. Cell group A10 in particular is likely to receive input from mesencephalic and bulbar cell groups projecting rostrally over the medial forebrain bundle. Such cell groups include—but may not be restricted to—the mesencephalic raphe nuclei as well as the locus ceruleus and more caudally placed norepinephrine (noradrenaline) cell groups.

Efferent nigral connections

Some of the nigrofugal projections described in the following account are illustrated schematically in Fig. 6. At present, the following can be considered well documented.

The nigrostriatal projection, the most massive of the efferent nigral connections, was discussed earlier in this account. It originates mainly from the pars compacta (dopamine cell group A9) of the nigra as well as from the outlying nigral dopamine cell group A10, but also from more dispersed pars reticulata cells which are likely to be the dopamine cells scattered throughout the pars reticulata.

Nigral projections to subcortical structures implicated in the circuitry of the limbic system were first identified in histofluorescence studies by Ungerstedt (1971) in the form of dopamine fibres from A10 in the ventral tegmental area (AVT) to the central nucleus of the amygdala and the bed nucleus of the stria terminalis. Later autoradiographic findings (Fallon & Moore 1978, Beckstead et al 1979) indicated that this projection also involves the anterior amygdaloid area, the nucleus of the diagonal band, and the septal region.

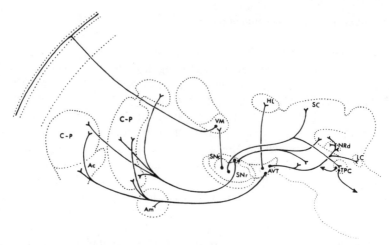

FIG. 6. Projections from the nigral complex. Besides the nigrostriatal projection, note projections from the ventral tegmental area (AVT) to habenula, mesencephalic tegmentum, dorsal raphe nucleus and locus ceruleus, and from pars reticulata to ventromedial thalamic nucleus, superior colliculus and pedunculopontine nucleus. The dopaminergic mesocortical system is not included in this diagram. Abbreviations: see legend to Fig. 1. (Reproduced from Nauta & Domesick 1979 by permission of Pergamon Press Ltd.)

Nigral projections to limbic cortex. Dopamine terminals in the anteromedial cortex of the rat were first reported by Thierry et al (1973) and were shortly thereafter attributed to a direct nigrocortical connection (Fuxe et al 1974) sometimes separately referred to as the mesocortical system. Beckstead (1976) reported evidence that this projection in the rat originates mainly from the ventral tegmental area and is distributed to the cortical area that also receives largely overlapping projections from the anteromedial and mediodorsal thalamic nuclei. Additional nigrocortical projections to the entorhinal area likewise appear to originate largely from dopamine cells in the ventral tegmental area (Fallon & Moore 1978, Beckstead et al 1979).

Nigrothalamic projections. A substantial nigrothalamic projection originating from the pars reticulata is distributed mainly to the ventromedial thalamic nucleus and in lesser volume to the paralamellar zone of the mediodorsal nucleus (Carpenter et al 1976). By this non-dopaminergic connection the nigra could exert a widespread effect on the neocortex, for the ventromedial thalamic nucleus in the rat projects profusely to the plexiform layer of a cortical region larger than the anterior half of the cortical convexity (Herkenham 1976).

Fewer nigrothalamic fibres have been traced autoradiographically from the ventral tegmental area to the lateral habenular nucleus. In the light of biochemical findings this projection seems likely to originate in part or entirely from the dopamine cell group A10 (cf. Beckstead et al 1979).

The nigrotectal projection. In the cat and rat a non-dopaminergic nigrotectal projection has been demonstrated that originates from the pars reticulata and is distributed to the middle grey layer of the superior colliculus in a regular band-pattern suggestive of a highly selective mode of termination (Graybiel 1978, Beckstead et al 1979).

Descending nigral projections. A non-dopaminergic projection has been traced from the pars reticulata to the pedunculopontine nucleus, where it must converge with efferents from the internal pallidal segment and subthalamic nucleus (cf. Beckstead et al 1979). The projection is closely associated with the nigrotectal pathway and may be composed in part of collaterals of the latter's fibres.

A further descending projection of the nigral complex, probably dopaminergic in part at least, originates in the ventral tegmental area and is distributed to the locus ceruleus as well as to a paramedian zone of the midbrain that includes the dorsal raphe nucleus and surrounding regions of the central grey substance, the anterior part of the median raphe nucleus, and an extensive medial region of the tegmentum. Further nigral projections spread caudally from the pars compacta over more lateral tegmental regions (cf. Beckstead et al 1979).

Conclusions

(1) The neural pathways conveying the output from the basal ganglia to the bulbospinal motor apparatus are still incompletely known. Since the most massive exit channel leads from the internal pallidal segment by way of the ansa lenticularis and thalamus to the precentral cortex, an action of the basal ganglia via the descending projections of the latter remains a plausible notion. There exists a smaller, cortex-independent ansal subdivision but its recipient, the pedunculopontine nucleus of the midbrain, projects mainly rostrally and only sparsely to the hindbrain (Edley & Graybiel 1983). Unlike the pallidum, the substantia nigra is rarely emphasized as a potential exit gate from the basal ganglia; yet nigral efferents could articulate with the tectobulbar-tectospinal system and with pathways descending from the midbrain reticular formation and locus ceruleus. Striatal efferents to the midbrain that bypass both pallidum and nigra are known to exist, but their origin appears to be confined to the nucleus accumbens and their distribution appears to coincide largely with that of known exit routes from the limbic system (Nauta et al 1978).

(2) The striatum of the rat can be subdivided into (a) a ventromedial limbic system-afferented quadrant that includes, but extends well beyond, the nucleus accumbens, and (b) a dorsolateral 'non-limbic' quadrant that receives its main cortical input from the motor cortex. The striatopallidal lines originating from these two sectors differ, at least in part: the 'non-limbic' striatum is the main origin of the classical exit line leading over the larger dorsal subdivision of the external pallidal segment to the subthalamic nucleus, whereas the limbic-afferented striatum projects over the subcommissural part of the external pallidal segment (ventral pallidum) not only to the subthalamic nucleus but in addition to various subcortical limbic structures: amygdala, mediodorsal thalamic nucleus, habenula and ventral tegmental area. These latter relationships suggest routes whereby impulses of limbic origin are led back into the circuitry of the limbic system after being channelled successively through the striatum and external pallidal segment.

(3) It seems plausible to suggest the limbic-afferented striatal sector as an interface between the respective neural mechanisms underlying motivational and more strictly motor aspects of movement. This suggestion is reinforced by the evidence that at least one component of the 'limbic striatum', the nucleus accumbens, seems enabled by its striatonigral efferents to modulate not only the source of its own dopamine innervation but also that of a large additional region of the striatum.

REFERENCES

Andén N–E, Carlsson A, Dahlström A, Fuxe K, Hillarp NA, Larsson K 1964 Demonstration and mapping out of nigro-neostriatal dopamine neurons. Life Sci 3:523-530

Andén N–E, Dahlström A, Fuxe K, Larsson K, Olson L, Ungerstedt U 1966 Ascending monoamine neurons to the telencephalon and diencephalon. Acta Physiol Scand 67:313-326

Beckstead RM 1976 Convergent thalamic and mesencephalic projections to the anterior medial cortex in the rat. J Comp Neurol 166:403-416

Beckstead RM 1979 An autoradiographic examination of cortico-cortical and subcortical projections of the mediodorsal-projection (prefrontal) cortex in the rat. J Comp Neurol 184:43-62

Beckstead RM, Domesick VB, Nauta WJH 1979 Efferent connections of the substantia nigra and ventral tegmental area in the rat. Brain Res 175:191-217

Bolam JP Synapses of identified neurons in the neostriatum. This volume, p 30-42

Carman JB, Cowan WM, Powell TPS, Webster KE 1965 A bilateral cortico-striate projection. J Neurol Neurosurg Psychiatry 28:71-77

Carpenter MB, Peter P 1972 Nigrostriatal and nigrothalamic fibres in the rhesus monkey. J Comp Neurol 144:93-116

Carpenter MB, Strominger NL 1967 Efferent fiber projections of the subthalamic nucleus in the rhesus monkey. Am J Anat 121:41-72

Carpenter MB, Nakano K, Kim R 1976 Nigrothalamic projections in the monkey demonstrated by autoradiographic technics. J Comp Neurol 165:401-416

Dahlström A, Fuxe K 1964 Evidence for the existence of monoamine-containing neurons in the central nervous system. Acta Physiol Scand Suppl 232:1-55

DeOlmos JS, Ingram WR 1972 The projection field of the stria terminalis in the rat brain. J Comp Neurol 146:303-334

DiFiglia M, Pasik P, Pasik T 1976 A Golgi study of neuronal types in the neostriatum of monkeys. Brain Res 114:245-256

Domesick VB 1980 Further observations on the anatomy of nucleus accumbens and caudatoputamen in the rat: similarities and contrasts. In: Chronister RB, De France JF (eds) The neurobiology of the nucleus accumbens. Haer Institute for Electrophysiological Research, Brunswick, ME, p 7-39

Domesick VB 1981 The anatomical basis for feedback and feedforward in the striatonigral system. In: Gessa GL, Corsini GU (eds) Apomorphine and other dopaminomimetics: basic pharmacology. Raven Press, New York, vol 2:27-39

Domesick VB, Stinus L, Paskevich PA 1983 The cytology of dopaminergic and non-dopaminergic neurons in the substantia nigra and ventral tegmental area in the rat: a light- and electron-microscopic study. Neuroscience 8:743-765

Edley SM, Graybiel AM 1983 The afferent and efferent connections of the feline nucleus tegmenti pedunculopontinus, pars compacta. J Comp Neurol 217:187-215

Fallon JH, Moore RY 1978 Catecholamine innervation of the basal forebrain. IV: Topography of the dopamine projection to the basal forebrain and neostriatum. J Comp Neurol 180:545-580

Fox CA, Andrade AN, Hillman DE, Schwyn RC 1971 The spiny neurons in the primate striatum: a Golgi and electron microscopic study. J Hirnforsch 13:181-201

Fox CA, Andrade AN, Schwyn RC, Rafols JA 1971/72 The aspiny neurons and the glia in the primate striatum: a Golgi and electron microscopic study. J Hirnforsch 13:341-362

Fox CA, Andrade AN, LuQui IJ, Rafols JA 1974 The primate globus pallidus: a Golgi and electron microscopic study. J Hirnforsch 15:75-93

Fox CA, Rafols JA, Cowan WM 1975 Computer measurements of axis cylinder diameters of radial fibers and 'comb' bundle fibers. J Comp Neurol 159:201-224

Fuxe K, Hökfelt T, Johansson O, Jonsson G, Lidbrink P, Ljungdahl A 1974 The origin of the dopamine nerve terminals in limbic and frontal cortex. Evidence for mesocortical dopamine neurons. Brain Res 82:349-355

Goldman PS, Nauta WJH 1977 An intricately patterned prefronto-caudate projection in the rhesus monkey. J Comp Neurol 171:369-386

Graybiel AM 1978 Organization of the nigrotectal connection: an experimental tracer study in the cat. Brain Res 143:339-348

Graybiel AM 1984 Neurochemically specified subsystems in the basal ganglia. This volume, p 114-144

Graybiel AM, Ragsdale CW 1983 Biochemical anatomy of the striatum. In: Emson PC (ed) Chemical neuroanatomy. Raven Press, New York, p 427-504

Grofová I 1975 The identification of striatal and pallidal neurons projecting to substantia nigra. An experimental study by means of retrograde axonal transport of horseradish peroxidase. Brain Res 91:286-291

Grove EA, Haber SN, Domesick VB, Nauta WJH 1983 Differential projections from AChE-positive and AChE-negative ventral-pallidum cells in the rat. Soc Neurosci Abstr 9:16

Haber SN, Nauta WJH 1983 Ramifications of the globus pallidus in the rat as indicated by patterns of immunohistochemistry. Neuroscience 9:245-260

Haber SN, Groenewegen HJ, Nauta WJH 1982 Efferent connections of the ventral pallidum in the rat. Soc Neurosci Abstr 8:169

Hattori T, Fibiger HC, McGeer PL 1975 Demonstration of a pallidonigral projection innervating dopaminergic neurons. J Comp Neurol 162:487-504

Hartmann-von Monakow K, Akert K, Künzle H 1978 Projections of the precentral motor cortex and other cortical areas of the frontal lobe to the subthalamic nucleus in the monkey. Exp Brain Res 33:395-403

Heimer L, Wilson RD 1975 The subcortical projections of the allocortex: similarities in the neural associations of the hippocampus, the piriform cortex, and the neocortex. In: Santini M (ed) Golgi centennial symposium: perspectives in neurobiology. Raven Press, New York, p 177-193

Herkenham M 1976 The nigro-thalamo-cortical connection mediated by the nucleus ventralis medialis thalami: evidence for a wide cortical distribution in the rat. Anat Rec 184:426

Herkenham M, Nauta WJH 1979 Efferent connections of the habenular nuclei in the rat. J Comp Neurol 187:19-48

Hökfelt T, Ungerstedt U 1969 Electron and fluorescence microscopical studies on the nucleus caudatus putamen of the rat after unilateral lesions of ascending nigro-neostriatal dopamine neurons. Acta Physiol Scand 76:415-426

Kelley AE, Domesick VB 1982 The distribution of the projection from the hippocampal formation to the nucleus accumbens in the rat—an anterograde-horseradish and retrograde-horseradish peroxidase study. Neuroscience 7:2321-2335

Kelley AE, Domesick VB, Nauta WJH 1982 The amygdalostriatal projection in the rat—an anatomical study by anterograde and retrograde tracing methods. Neuroscience 7:615-630

Kemp JM 1968a An electron microscopic study of the terminations of afferent fibres in the caudate nucleus. Brain Res 11:464-467

Kemp JM 1968b Observations on the caudate nucleus of the cat impregnated with the Golgi method. Brain Res 11:467-470

Kemp JM, Powell TPS 1970 The cortico-striate projection in the monkey. Brain 93:525-546

Künzle H 1975 Bilateral projections from precentral motor cortex to the putamen and other parts of the basal ganglia. Brain Res 88:195-210

Leonard CM 1969 The prefrontal cortex of the cat. I: Cortical projection of the mediodorsal nucleus. II: Efferent connections. Brain Res 12:321-343

Namba M von 1957 Cytoarchitektonische Untersuchungen am Striatum. J Hirnforsch 3:24-48

Nauta HJW 1974 Evidence of a pallidohabenular pathway in the cat. J Comp Neurol 156:19-28

Nauta HJW 1979 Projections of the pallidal complex: autoradiographic studies in the cat. Neuroscience 4:1853-1873

Nauta HJW, Cole M 1978 Efferent projections of the subthalamic nucleus: an autoradiographic study in monkey and cat. J Comp Neurol 180:1-16

Nauta WJH, Domesick VB 1978 Crossroads of limbic and striatal circuitry: hypothalamo-nigral connections. In: Livingston KE, Hornykiewicz O (eds) Limbic mechanisms. Plenum Press, New York, p 75-93

Nauta WJH, Domesick VB 1979 The anatomy of the extrapyramidal system. In: Fuxe K, Calne DB (eds) Dopaminergic ergot derivatives and motor function. Pergamon Press, Oxford, p 3-22

Nauta WJH, Mehler WR 1966 Projections of the lentiform nucleus in the monkey. Brain Res 1:3-42

Nauta WJH, Smith GP, Faull RLM, Domesick VB 1978 Efferent connections and nigral afferents of the nucleus accumbens septi in the rat. Neuroscience 3:385-401

Raisman G, Cowan WM, Powell TPS 1966 An experimental analysis of the efferent projection of the hippocampus. Brain 89:83-108

Ricardo JA 1980 Efferent connections of the subthalamic region in the rat. I. The subthalamic nucleus of Luys. Brain Res 202:257-272

Scheibel ME, Scheibel AB 1966 The organization of the nucleus reticularis thalami: a Golgi study. Brain Res 1:43-62

Schwyn RC, Fox CA 1974 The primate substantia nigra: a Golgi and electron microscopic study. J Hirnforsch 15:95-126

Somogyi P, Hodgson AJ, Smith AD 1979 An approach to tracing neuron networks in the cerebral cortex and basal ganglia. Combination of Golgi staining, retrograde transport of horseradish peroxidase and anterograde degeneration of synaptic boutons in the same material. Neuroscience 4:1805-1852

Somogyi P, Bolam JP, Smith AD 1981a Monosynaptic cortical input and local axon collaterals of identified striatonigral neurons. A light and electron microscopic study using the Golgi–peroxidase transport–degeneration procedure. J Comp Neurol 195:567-584

Somogyi P, Bolam JP, Totterdell S, Smith AD 1981b Monosynaptic input from the nucleus accumbens–ventral striatum region to retrogradely labelled nigrostriatal neurones. Brain Res 217:245-263

Swanson LW 1976 An autoradiographic study of the efferent connections of the preoptic region in the rat. J Comp Neurol 167:227-256

Swanson LW, Cowan WM 1975 A note on the connections and development of the nucleus accumbens. Brain Res 92:324-330

Thierry AM, Blanc G, Sobel A, Stinus L, Glowinski J 1973 Dopaminergic terminals in the rat cortex. Science (Wash DC) 180:499-501

Ungerstedt U 1971 Stereotaxic mapping of the monoamine pathways in the rat brain. Acta Physiol Scand Suppl 367:1-48

Vogt C, Vogt O 1920 Zur Lehre der Erkrankungen des striären Systems. J Psychol Neurol (Leipz) 25:627-846

Vogt C, Vogt O 1941 Thalamusstudien I–III. J Psychol Neurol (Leipz) 50:31-154

Wilson SAK 1912 Progressive lenticular degeneration; a familial nervous disease associated with cirrhosis of the liver. Brain 34:295-509

Wilson SAK 1914 An experimental research into the anatomy and physiology of the corpus striatum. Brain 36:427-492

Yoshida M, Rabin A, Anderson M 1974 Monosynaptic inhibition of pallidal neurons by axon collaterals of caudato-nigral fibres. Exp Brain Res 15:333-347

DISCUSSION

Yoshida: In our studies on the inhibitory efferent pathway of the caudate nucleus (Yoshida et al 1971, 1972, Yoshida & Precht 1971) we found that most of the caudatonigral fibres have axon collaterals which terminate in the internal segment of the globus pallidus (or entopeduncular nucleus, ENT). The evidence is as follows. We recorded inhibitory postsynaptic potentials (IPSPs) in ENT during paired stimulation of the substantia nigra and the caudate nucleus. As the interval between the two stimuli became shorter, the test IPSPs became smaller and finally occluded almost completely.

Nauta: Your findings are certainly compatible with the histochemical evidence that both the internal pallidal segment and the substantia nigra contain a dense plexus of substance P-positive striatal efferents that could be sister collaterals from the same striatofugal trunk-fibres. The present histochemical data on the external pallidal segment suggest that only its subcommissural part, the ventral pallidum, could collaterally share a significant number of striatofugal efferents with the substantia nigra. This suggestion is, of course, based on the assumption that all ramifications of a single axon contain the same peptide or peptides. Moreover, we cannot discount the possibility that another histochemical marker, not as yet used for this purpose, would reveal relationships not demonstrable by enkephalin-like or substance P-like immunoreactivity.

Carpenter: I have always regarded the relationship between basal ganglia and the limbic system as somewhat tenuous. Until now the major recognized connection between these systems has been the pathway from the pallidum to the lateral habenular nucleus, which varies greatly in different animals. In the monkey this connection is relatively small but in the rat and the cat it is a much more abundant projection. However, the projection from the basal lateral amygdala to a large area of the striatum reinforces the concept of a basal ganglia/limbic system relationship in a new way. When I looked again at some of our autoradiographic material in the rhesus monkey it showed exactly what Walle Nauta has just demonstrated. Our series was not as complete as his but the distribution was of the same order. So I now believe that the striatum really does consist of limbic and non-limbic subdivisions.

Kitai: With Drs Imai and Steindler, we have double-labelled dorsal raphe neurons by injecting retrograde markers in either the limbic or basal ganglia structures. We found that some dorsal raphe neurons were double-labelled by injections in the locus ceruleus and the hippocampus. This would mean that some single raphe neurons project to both the hippocampus and the locus ceruleus.

Using the same technique we found that single dorsal raphe neurons project to the substantia nigra and to the striatum. There were no raphe neurons projecting to the striatum and the hippocampus. Interestingly, we found that some dorsal raphe neurons project to the striatum and the amygdala and others to the hippocampus and the amygdala. This would then mean that the amygdala may be the structure which forms a bridge between the limbic system and the striatum. We also know from anatomical and electrophysiological observations that the amygdala nucleus has strong inputs to the accumbens nucleus, which some anatomists consider to be an extension of the striatum.

Graybiel: Dr Somogyi, would you mind reviewing the electron microscope evidence for direct contact of terminals from various parts of the striatum onto the dopamine-containing neurons in the pars compacta and ventral tegmental area?

Somogyi: Professor Nauta kindly referrred to our results on the monosynaptic input of nigrostriatal neurons (Somogyi et al 1981). I must admit that we studied cells labelled by retrograde transport of horseradish peroxidase (HRP) and although many of the cells were presumably dopaminergic neurons, we did not identify their transmitter. By injecting HRP into the dorsal striatum and placing a lesion in the ventral striatum we could show both anterogradely labelled boutons and degenerating boutons making synaptic contacts with the same retrogradely labelled nigrostriatal neurons. These results provide evidence for the convergence of monosynaptic input from the dorsal striatum and the ventral striatum–nucleus accumbens region to the same nigrostriatal neurons. This is in agreement with the predictions made by Professor Nauta from light microscope studies. The only morphological evidence that dopaminergic neurons receive monosynaptic striatal input has been obtained by Wassef et al (1981), who showed degenerating boutons in synaptic contact with tyrosine hydroxylase-positive dendrites in the substantia nigra. Whether these neurons projected to the striatum was not studied.

At the same time there is also good evidence that the strionigral projection terminates on non-dopaminergic cells. By placing lesions in the dorsal part of the striatum we found degenerating terminals in monosynaptic contact with nigrothalamic neurons (Somogyi et al 1979), confirming the electrophysiological results of Deniau et al (1976). Furthermore the dendrites in the nigra which receive monosynaptic striatal input are of two types. One type is sparsely contacted by the several types of boutons and in separate immunocytochemical experiments this type has been found to contain tyrosine hydroxylase (Wassef et al 1981, Z.F. Kisvárday, A.D. Smith, P. Somogyi, unpublished observations). The other type of dendrite, first described by Schwyn & Fox (1974), is densely and uniformly ensheathed by synaptic terminals. Some of the later types of dendrites have been identified as originating from nigrothalamic neurons (Somogyi et al 1979). The input from the striatum to the nigra is therefore not only topographically organized but also terminates on several types of neuron at each level.

Kitai: Drs Chang, Kita and I labelled subthalamic neurons intracellularly with HRP and found that their terminals in the substantia nigra are asymmetrical and contain small vesicles. They are quite different from striatal inhibitory GABAergic terminals.

Porter: My comment concerns the output systems from the basal ganglia and the ways in which those parts of the basal ganglia that have motor connections, rather than hippocampal or other limbic connections, may be able to exert their actions on motor systems. Professor Nauta referred to the thalamic nuclei which convey part of that output back to the motor cortex areas as taking the information to the motor cortex. We will have to be very careful about the terminology we use for motor areas of cerebral cortex. The VA–VL complex projects ahead of the cytoarchitectonic area 4 zone which is regarded as the primary motor cortex. Many of us would like to regard the more forward projections as going to a premotor zone and therefore that particular output doesn't go directly to the major source of corticospinal neurons. There must be some other connectivity underlying the influence on corticospinal neurons.

At some time during this meeting we should attempt to see what evidence exists for the other generalization that Professor Nauta raised, namely that the extrapyramidal motor structures might work through a pyramidal or corticospinal system. Information about connectivity within the cortex is accumulating that could provide the substrate for that idea, but very little functional evidence exists on that particular question.

Evarts: Bob Porter's comments bring to mind the matter of differences between different animal species (rats, cats or monkeys) in terms of the relative functional roles of limbic and non-limbic parts of the basal ganglia. Before this meeting I asked Ann Graybiel about species differences in the relative densities of projections from basal ganglia to limbic and non-limbic structures and her views were in agreement with those expressed by Carpenter (1981), who noted that in the primate the basal ganglia projection to the habenula is relatively slight whereas in the rodent it is massive: Herkenham & Nauta (1977) reported that with HRP injections in the lateral part of the lateral habenular nucleus virtually all the cells in many sections of the entopeduncular nucleus were labelled with HRP. Thus, there is a major quantitative difference between primates and rodents in basal ganglia projections to habenula, which in turn provides one of a number of links to limbic structures. As to non-limbic structures, these have certainly come to occupy a relatively larger fraction of the basal ganglia in the primate. We have never been led astray by the rat in terms of pure neurochemistry, but David Marsden has pointed out that to study certain manifestations of basal ganglia disease one *must* use primates. Rats don't get parkinsonism, perhaps because there are certain non-limbic cortical structures and projections that they lack.

Carpenter: Parent & De Bellefeuille (1982) have shown by double-labelling techniques that the pallidal neurons projecting to the rostral–ventral tier thalamic nuclei are the same ones that give descending collaterals to the pedunculopontine nucleus. They also find that the cells projecting to the habenular nuclei are peripallidal. These cells surround the medial pallidal segment and extend into the lateral hypothalamus. Herkenham & Nauta (1977) report that almost all cells of the entopeduncular nucleus are labelled after HRP injections of the lateral habenular nucleus, which suggests that there is quite an important difference between rat and primate.

Graybiel: This seems to vary with the species. The original work on the descending connections of the pedunculopontine (TPc) nucleus by you (Carpenter & Strominger 1967) and by Drs Nauta & Mehler (1966) showed a dense projection in primates down to the TPc nucleus. In the cat we were surprised to see only a sparse projection to TPc from the equivalent nucleus (the entopeduncular nucleus) (Moon Edley & Graybiel 1983). Apparently, the projection is even smaller in the rat. But when we injected tritiated amino acids into the substantia nigra, we labelled a dense projection to TPc. This tends to suggest that the pars reticulata of the substantia nigra and the internal segment of the globus pallidus have somewhat different functions—or different cellular compositions—in primates, cats and rats.

As for how much information is sent downstream from this pedunculontine region, I think Dr Nauta said it perfectly: it was our great hope that the TPc would be a major zone by which the basal ganglia could influence the bulbospinal mechanism. In fact, some years ago we put large injections of tritiated amino acids into the TPc region of the cat and these labelled a dense projection to Magoun's inhibitory area—that is, the part of the magnocellular reticular formation just dorsal to the inferior olivary complex (Graybiel 1977). But when Sandra Moon Edley and I followed this up, it turned out that the smaller the injections were, and therefore the better confined they were to TPc, the smaller the labelled descending pathways. In other words, if there is a projection from the pedunculopontine neurons to the caudal part of the reticular formation, my suspicion is that in the cat its densest zone of termination would be in the medial magnocellular zone at medullary levels, and that the pathway is not massive.

DeLong: The matter of species differences in anatomical organization is very important in relation to functional considerations. It seems likely that the differences between rodents and primates may be substantial. For example, Parent & De Bellefeuille (1982) have found significant differences in pallidal efferents in rats, cats and monkeys. Perhaps the rat entopeduncular nucleus is not the equivalent of the internal segment of the globus pallidus of primates. In the rat and other lower species the pars reticulata of the nigra may be the equivalent of the inner segment of the globus pallidus in the primate. The nigra of the rat may be the main output system for motor-related functions, rather than the entopeduncular nucleus. We needn't try to equate the nigra of the rat and the nigra of the primate. Similarly the caudate in the cat is quite different from the caudate of the primate. The cat caudate receives input from the sensorimotor areas whereas in the primate that all goes to the putamen.

Evarts: Earlier in this discussion I said that the rat had never led us astray, but now I recall that investigators at the NIH were led astray in 1978 when they injected rats with *N*-methyl-4-phenyl-1,2,3,6-tetra-hydropyridine (NMPTP), a substance that produces parkinsonism in primates. They failed to observe a parkinsonian-like disorder in the rats and abandoned the project. Only later did they discover that the effects of NMPTP are quite different in the rat and the primate.

Rolls: If there is some anatomical segregation of inputs into the striatum, to what extent are anatomists willing to suggest that segregation might be maintained throughout the striatum and its outputs to pallidum, and then through the thalamus to reach different areas of the cortex? For example, does the supplementary motor cortex receive inputs primarily from the putamen through the thalamus? Does the lateral premotor cortex receive inputs primarily from the caudate nucleus, which of course will have some visual inputs? Does any part of the striatum come into contact with the prefrontal cortex, perhaps via the ventral pallidum projecting through the mediodorsal nucleus of the thalamus to the prefrontal cortex?

Di Chiara: I wanted to emphasize the issue of the output pathways from basal ganglia through areas which do not necessarily project back to the cerebral cortex. The concept that the basal ganglia are provided with an output which does not feed back to the cerebral cortex was initially suggested, as Professor Nauta showed, by anatomical studies. But many of those who work on primates, and many neurologists, do not seem to take this into consideration. Indeed, the output pathways which bypass the cerebral cortex might be of some importance in humans, not only in normal motor behaviour but also, and even more so, in abnormal motor behaviour such as Huntington's chorea and some symptoms of Parkinson's disease. Which are the pathways conveying such output? Dr Graybiel indicated as a possible candidate for this output the pedunculopontine nucleus. But she said that this nucleus has few projections downstream. It mostly goes back up to the nigra, subthalamus, etc. An area which certainly can provide a downstream projection is the superior colliculus and the adjacent mesencephalic reticular formation. There is a huge projection from the nigra to the deep layers of the superior colliculus, and these in turn project directly to the spinal cord, although only to the cervical segment. But the superior colliculus might also influence spinal mechanisms through the mesencephalic reticular formation, which is connected to the spinal cord by a polysynaptic pathway with relays in the pontine reticular formation.

REFERENCES

Carpenter MB 1981 Anatomy of the corpus striatum and brain stem integrating systems. In: Brookhart JM et al (eds) Motor control. American Physiological Society, Bethesda, MD (Handb Physiol sect 2 The nervous system vol 2) part 2:947-995

Carpenter MB, Strominger NL 1967 Efferent fibers of the subthalamic nucleus in the monkey. A comparison of the efferent projections of the subthalamic nucleus, substantia nigra and globus pallidus. Am J Anat 121:41-72

Deniau JM, Feger J, Le Guyader C 1976 Striatal evoked inhibition of identified nigro-thalamic neurones. Brain Res 104:152-156

Graybiel AM 1977 Direct and indirect preoculomotor pathways of the brainstem: an autoradiographic study of the pontine reticular formation in the cat. J Comp Neurol 175:37-78

Herkenham M, Nauta WJH 1977 Afferent connections of the habenular nuclei in the rat; a horseradish peroxidase study with a note on the fiber-of-passage problem. J Comp Neurol 173:123-145

Moon Edley S, Graybiel AM 1983 The afferent and efferent connections of the feline nucleus tegmenti pedunculopontinus, pars compacta. J Comp Neurol 217:187-215

Nauta WJH, Mehler WR 1966 Projections of the lentiform nucleus in the monkey. Brain Res 1:3-42

Parent A, De Bellefeuille L 1982 Organization of efferent projections from the internal segment of globus pallidus in primate as revealed by fluorescence retrograde labeling method. Brain Res 245:201-213

Schwyn RC, Fox CA 1974 The primate substantia nigra: a Golgi and electron microscopic study. J Hirnforsch 15:59-125

Somogyi P, Hodgson AJ, Smith AD 1979 An approach to tracing neuron networks in the cerebral cortex and basal ganglia. Combination of Golgi-staining retrograde transport of horseradish peroxidase and anterograde degeneration of synaptic boutons in the same material. Neuroscience 4:1805-1852

Somogyi P, Bolam JP, Totterdell S, Smith AD 1981 Monosynaptic input from the nucleus accumbens–ventral striatum region to retrogradely labelled nigrostriatal neurons. Brain Res 217:245-263

Wassef M, Berod A, Sotelo C 1981 Dopaminergic dendrites in the pars reticulata of the rat substantia nigra and their striatal input. Combined immunocytochemical localization of tyrosine hydroxylase and anterograde degeneration. Neuroscience 6:2125-2139

Yoshida M, Precht W 1971 Monosynaptic inhibition of neurons of the substantia nigra by caudate-nigral fibers. Brain Res 32:225-228

Yoshida M, Rabin A, Anderson ME 1971 Two types of monosynaptic inhibition of pallidal neurons produced by stimulation of the diencephalon and substantia nigra. Brain Res 30:235-239

Yoshida M, Rabin A, Anderson ME 1972 Monosynaptic inhibition of pallidal neurons by axon collaterals of caudato-nigral fibers. Exp Brain Res 15:333-347

Synapses of identified neurons in the neostriatum

J. P. BOLAM

University Department of Pharmacology, South Parks Road, Oxford, OX1 3QT, UK

Abstract. Studies of the basal ganglia and particularly the neostriatum have described a complex array of neuron types, synapses and putative transmitters. One approach to the study of such an area is to examine identified neurons and thus establish the neural circuits that underlie function. Striatal neurons have been identified under the light microscope by one or more of the following methods: (1) structure, based on Golgi impregnation or the intracellular injection of horseradish peroxidase (HRP); (2) projection area, by the retrograde transport of HRP or the tracing of HRP-injected or Golgi-impregnated axons; (3) chemistry, by immunocytochemistry, histochemistry or autoradiography, to reveal the presence of a selective uptake system for a putative transmitter. Examination of identified neurons in the electron microscope allows the characterization of their afferent synapses (by immunocytochemistry or anterograde degeneration) and their local synaptic output.

The afferent and efferent synapses of five classes of identified striatal neurons are discussed: (1) those neurons described in Golgi preparations as medium-size and densely spiny; (2) a large type of striatonigral neuron; (3) GABAergic interneurons; (4) cholinergic neurons; (5) somatostatin-immunoreactive neurons. It is concluded that medium-size densely spiny neurons provide the basic framework of the neural circuits of the neostriatum.

1984 Functions of the basal ganglia. Pitman, London (Ciba Foundation symposium 107) p 30-47

The neostriatum is an extremely complex area of the brain; it receives its major inputs from the cortex, substantia nigra, thalamus, dorsal raphe, amygdala and ventral tegmental area. Its major outputs are to the pallidus and substantia nigra. Within the neostriatum, biochemical, pharmacological and immunocytochemical techniques have identified at least 15 putative transmitters (Graybiel & Ragsdale 1983). Morphological studies at the ultrastructural level have identified more than nine distinct types of synapses (Hassler 1978) and studies of Golgi-impregnated material have identified eight or nine different types of neurons (Chang et al 1982, Bolam et al 1981b). A vast number of pharmacological papers have described complex interactions between many of the putative transmitters (see Dray 1979).

One approach to the study of such a complex area of the brain is to establish circuit diagrams based not just on the morphology of neurons but also (1) on their connections with other neurons within the neostriatum and with neurons in distant brain regions, and (2) on the chemical characteristics of the neurons and their afferent synapses. In other words, the aim is to describe circuit diagrams in which morphology, connectivity and chemistry are directly correlated.

The object of this paper is to summarize our knowledge of the morphology, chemistry and connections of five classes of striatal neurons: (1) those neurons described in Golgi preparations as medium-size and densely spiny; (2) a type of large neuron in Golgi preparations that is very similar to a scarce type of striatonigral neuron; (3) GABAergic interneurons; (4) cholinergic interneurons; (5) somatostatin-immunoreactive neurons.

Medium-size densely spiny neurons

Medium-size and densely spiny neurons are the most freqently impregnated class of cells in Golgi preparations of neostriatum, and most striatal neurons probably belong to this class (Bolam et al 1981a, Chang et al 1982, DiFiglia et al 1976, Dimova et al 1980, Kemp & Powell 1971). The terms 'medium-size' and 'densely spiny' refer to their appearance in Golgi-impregnated material: the perikaryon is of medium size (12–18 µm), with several dendrites emerging from it; these dendrites are initially spine-free but then become densely laden with spines. Not only are they the most common neuron of the striatum but they are also the major projection neuron, sending their axons to the pallidus or the substantia nigra, or both. The fact that they are projection neurons has been demonstrated by two methods. First, the axons of medium-size spiny neurons identified by the intracellular injection of horseradish peroxidase (HRP) have been traced out of the striatum and have been seen to arborize within the globus pallidus (Chang et al 1981). Secondly, by the combination of Golgi impregnation with the retrograde transport of HRP from the substantia nigra, Golgi-stained medium-size spiny neurons have been identified as striatonigral neurons (Somogyi & Smith 1979). In addition to having long projecting axons, these neurons also give off extensive local collaterals within the striatum that form classical symmetrical synapses; thus they can influence the activity of other striatal neurons (Somogyi et al 1981a, Wilson & Groves 1980).

The results of several experimental approaches support the hypothesis that striatal efferent neurons contain, and may use as transmitters, γ-aminobutyric acid (GABA), substance P, enkephalin and related peptides, and taurine (reviews Dray 1979, Graybiel & Ragsdale 1983). Morphological studies using immunocytochemical techniques have demonstrated that neurons ultrastructurally similar to medium-size spiny neurons contain glutamate decarboxylase (EC 4.1.1.15; GAD, the synthetic enzyme for GABA) (Bradley et al 1983, Morelli et al 1983, Oertel & Mugnaini 1983), enkephalin (DiFiglia et al 1982) and substance P (Bolam et al, 1983b). We have also recently shown that after the local injection of [^{3}H]taurine only one class of neuron accumulates the radiolabel; these were neurons that are ultrastructurally similar to medium-size spiny neurons, some of which were also retrogradely labelled with HRP from the substantia nigra (Clarke et al 1984). Thus the possibility exists that some striatonigral neurons also contain taurine and use it as a transmitter. It is unclear whether different subpopulations of medium-size spiny neurons contain different transmitters. However, recent evidence suggests that at least some GAD-immunoreactive neurons also contain enkephalin-like immunoreactivity (Morelli et al 1983).

Input to medium-size densely spiny neurons (see Table 1)

The synaptic input of medium-size spiny neurons can be divided into two broad classes, those boutons which form asymmetrical membrane specializations and those which form symmetrical membrane specializations. The asymmetrical synapses make contact almost exclusively with dendritic spines whereas symmetrical synapses occur on all parts of the neurons, i.e. dendritic spines and shafts, perikarya and axon initial segments. By a combination of Golgi staining and anterograde degeneration or immunocytochemistry three characterized inputs—those from the cortex, terminals immunoreactive for tyrosine hydroxylase (i.e. dopaminergic from the substantia nigra), and GAD-immunoreactive terminals—have been unequivocally shown to form synaptic contacts with medium-size spiny neurons (Somogyi et al 1981a, J. P. Bolam et al, unpublished; T. F. Freund et al, unpublished). Thus, in rats that had received multiple electrolytic lesions of the cerebral cortex, Golgi-impregnated medium-size densely spiny neurons were found to receive synaptic input from axonal boutons undergoing the characteristics of degeneration, i.e. originating from neurons in the cortex. The synaptic specializations of these terminals were typically asymmetrical and made contact with dendritic spines. These same Golgi-impregnated neurons were also retrogradely labelled with HRP from the substantia nigra and were thus identified as striatonigral neurons. In other words, medium-size densely spiny striatonigral neurons are one of the postsynaptic targets of corticostriatal fibres in the striatum (Somogyi et al 1981a). Biochemical and electrophysiological studies suggest that these terminals use glutamate as a transmitter. In similar experiments T. F. Freund et al (unpublished) have found tyrosine hydroxy-

TABLE 1 Characterized inputs to medium-size spiny neurons

Method of identification of spiny neurons	Transmitter or origin of afferent terminals	Type of synapse	Distribution	References[a]
Golgi-impregnated, retrogradely labelled from SN	Cortex	Asymmetrical	Spines	1
Golgi-impregnated, retrogradely labelled from SN	TH-positive (dopaminergic from SN)	Symmetrical	Spines, dendritic shafts	2
Golgi-impregnated	GAD-positive (GABAergic)	Symmetrical	Dendritic shafts, perikarya	3
Ultrastructural	Enkephalin-positive	Symmetrical	Dendritic shafts, perikarya	4,5
Ultrastructural	Substance P-positive	Symmetrical	Dendritic shafts, perikarya	6
Ultrastructural	ChAT-positive (cholinergic)	Mainly symmetrical, some asymmetrical	Spines, perikarya	7
Ultrastructural	5HT-positive (from raphe)	Asymmetrical	Spines	8

SN, substantia nigra; TH, tyrosine hydroxylase; GAD, glutamate decarboxylase; GABA, γ-aminobutyric acid; ChAT, choline acetyltransferase; 5HT, 5-hydroxytryptamine.
[a]References: 1) Somogyi et al 1981a, 2) T. F. Freund et al unpublished observations, 3) J. P. Bolam unpublished observations, 4) DiFiglia et al 1982, 5) Somogyi et al 1982, 6) Bolam et al 1983b, 7) Wainer et al 1983, 8) Pasik et al 1981.

lase-immunoreactive synaptic terminals in synaptic contact with Golgi-impregnated spiny neurons also retrogradely labelled from the substantia nigra; i.e. striatonigral neurons receive direct synaptic input from dopaminergic nigrostriatal neurons. In this case the synaptic specializations were of the symmetrical type, and the contacts were predominantly with the necks of dendritic spines or dendritic shafts. This pattern of innervation of the neurons, i.e. input to spine necks and dendritic shafts, was distinctly different to that observed for GAD-immunoreactive terminals. These terminals also form symmetrical membrane specializations but the synaptic contacts are predominantly with the dendritic shafts and perikarya of Golgi-stained spiny neurons (J. P. Bolam et al, unpublished observations).

For the remaining characterized inputs, ultrastructural features alone have been used to identify the neurons as medium-size densely spiny neurons. These inputs include: (1) substance P (Bolam et al 1983b) and (2) enkephalin (DiFiglia et al 1982, Somogyi et al 1982), both of which form symmetrical synaptic contacts and have a similar distribution to that of GAD-positive terminals; (3) 5-hydroxytryptamine (asymmetrical synapses on dendritic spines) (Pasik et al 1981); (4) somatostatin (symmetrical synapses on dendritic spines) (DiFiglia & Aronin 1982, Takagi et al 1983). Preliminary results obtained using a monoclonal antibody to choline acetyltransferase (EC 2.3.1.6; ChAT, the synthetic enzyme for acetylcholine) have shown immunoreactive terminals in symmetrical synaptic contact with spines and with perikarya that are ultrastructurally similar to medium-size spiny neurons (Wainer et al 1983).

Output of medium-size densely spiny neurons

As mentioned above, the output of spiny neurons is to both local neurons and neurons in distant brain regions. Electron microscope analysis of the local axonal varicosities (within the striatum) of either HRP-filled neurons or Golgi-impregnated neurons retrogradely labelled with HRP from substantia nigra reveals that they form symmetrical synapses with dendritic shafts and

TABLE 2 Input and output of identified aspiny neurons

Neuron	Methods of identification	Input	Output	References[a]
Large aspiny (striatonigral type 2)	1) EM; retrograde labelling from SN 2) Golgi/EM 3) EM	Symmetrical, asymmetrical, some GAD-positive, ENK-positive	SN	1, 2, 3
GABA interneuron	1) [^{3}H]GABA uptake Golgi/EM 2) Golgi/EM 3) GAD-positive 4) AChE and Golgi	Symmetrical, asymmetrical, some GAD-positive	Spines, spiny dendrites, other dendrites	4, 5, 6, 7
Somatostatin-positive	1) Somatostatin immunostaining 2) Golgi/EM 3) AChE/Golgi	Symmetrical, asymmetrical	Spines, dendritic shafts	5, 6, 8, 9
Cholinergic neurons	1) AChE/Golgi 2) ChAT/Golgi 3) Golgi 4) Intracellular HRP	Symmetrical, asymmetrical	Spines, dendrites, medium-spiny perikarya	5, 6, 10, 11

EM, electron microscopy; GAD, glutamate decarboxylase; AChE, acetylcholinesterase; ChAT, choline acetyltransferase; HRP, horseradish peroxidase; ENK, enkephalin; SN, substantia nigra.
[a]References: 1) Bolam et al 1981b, 2) Somogyi et al 1982, 3) J. P. Bolam unpublished observations, 4) Bolam et al 1983a, 5) Takagi et al 1984, 6) Bolam et al 1984a, 7) DiFiglia et al 1980, 8) Takagi et al 1983, 9) DiFiglia & Aronin 1982, 10) Bolam et al 1948b, 11) Wilson et al 1983.

spines (Somogyi et al 1981a, Wilson & Groves 1980). Thus the postsynaptic target neurons that we can identify with most certainty are other neurons that possess spines and so might also be medium-size spiny neurons that presumably project to the substantia nigra or globus pallidus. The synaptic output of medium-size spiny neurons in the globus pallidus has been studied by analysing HRP-injected neurons, the terminals of which have been shown to form symmetrical synaptic contact with dendritic shafts and perikarya of pallidal neurons (Chang et al 1981).

The other approach to studying the distant synaptic connections of striatal efferent neurons is to make lesions in the striatum and then look for degenerating synaptic boutons in contact with identified neurons in target regions. Although we cannot be sure of the cell type or types from which degenerating terminals originate, they are probably medium-size spiny neurons as these constitute the major type of projection neuron. By this approach striatal projection neurons have been shown to make symmetrical synaptic contact with: (1) pallidonigral neurons identified by the retrograde transport of HRP (S. Totterdell et al 1984), (2) nigrothalamic neurons (Somogyi et al 1979), (3) nigrostriatal neurons (Somogyi et al 1981b), and (4) dendrites in the substantia nigra that are immunoreactive for tyrosine hydroxylase (tyrosine monooxygenase; EC 1.14.16.2) and therefore probably dopaminergic (Wassef et al 1981).

Large aspiny striatonigral neurons (see Table 2)

A second type of HRP-labelled striatonigral neuron has been observed in the striatum after injection of HRP into the substantia nigra (Bolam et al 1981b). These neurons are rare and are distinctly different from medium-size spiny neurons in that they are large, have indented nuclei, are ensheathed in synaptic boutons, and have been found so far only in the ventral regions of the striatum. Indirect comparison with Golgi-impregnated neurons would suggest that they have very long (up to 700 µm) aspiny infrequently branching dendrites. They are similar in many ways to typical pallidal and nigral neurons.

The synaptic input to these neurons consists predominantly of boutons forming symmetrical synapses. Asymmetrical synaptic contacts are also present; these have never been observed on perikarya and are rare on proximal dendrites but their numbers increase distally. Two chemically characterized inputs have been identified by using antisera to either enkephalin or glutamate decarboxylase (Somogyi et al 1982, J. P. Bolam et al unpublished). When these antisera are used in striatal tissue, large neurons are 'outlined' by immunoreactive varicosities which in the electron microscope can be seen to form symmetrical synaptic contacts with the perikarya and dendrites.

The output of these neurons, as stated above, is to the substantia nigra. However, we do not know whether they have local axon collaterals, since Golgi-stained examples with impregnated axons have not been observed and we know nothing of the morphology of their axon terminals.

GABAergic interneurons (see Table 2)

Evidence for the existence of interneurons in the neostriatum that use GABA as a transmitter comes from pharmacological and biochemical studies and from the observation that a subpopulation of striatal neurons selectively accumulate locally administered [^{3}H]GABA (Bolam et al 1983a). These neurons have been further characterized by Golgi impregnation and electron microscopy. They are medium-size neurons with many aspiny dendrites and in the electron microscope they can be seen to possess nuclear indentations, a characteristic that distinguishes them from medium-size spiny neurons. These observations have been confirmed using antisera to GAD; this class of neuron is the most intensely and easily stained of striatal neurons (Oertel & Mugnaini 1983, J. P. Bolam et al unpublished). Evidence that they are interneurons is derived from studies of Golgi-impregnated neurons having the same morphological features (DiFiglia et al 1980, Takagi et al 1984); the local axonal arborization is characteristic of local circuit neurons. In addition, these neurons have never been recognized in ultrastructural studies of striatal efferent neurons (for references, see Bolam et al 1981b). Interestingly, this same class of neuron stains lightly for acetylcholinesterase (EC 3.1.1.7) (Bolam et al 1984a).

The synaptic input to the perikarya is sparse, consisting of predominantly symmetrical synapses with only occasional asymmetrical contacts. Proximal dendrites receive a more frequent input of symmetrical synapses whereas the majority of terminals on distal dendrites are asymmetrical (Bolam et al 1984a, DiFiglia et al 1980, Takagi et al 1984).

Only one of these inputs has been chemically characterized and that is in material immunostained with GAD antisera. Thus some of the boutons that form symmetrical synaptic contact with the perikarya and proximal dendrites of GAD-immunoreactive neurons are themselves immunoreactive for GAD (Ribak et al 1979, J. P. Bolam et al unpublished). The origin of these immunoreactive terminals is unknown but they probably originate either from other GABA interneurons or from GABA-containing medium-size spiny neurons.

Information on the output of GABAergic interneurons is limited. Analysis of the axons of Golgi-impregnated neurons has shown that their boutons form symmetrical synaptic contacts with spines and dendrites (DiFiglia et al 1980, Takagi et al 1984). One of these postsynaptic dendrites originated from the same neuron and thus represents an 'autapse' (Takagi et al 1984). Thus GABAergic interneurons themselves and spine-bearing neurons, which probably represent medium-size spiny neurons, are both targets of GABAergic interneurons. Examination of GAD-immunoreactive terminals could provide information on the output of GABA interneurons; however, we cannot be sure of the origin of such terminals since medium-size spiny neurons also contain GAD-immunoreactive material.

Cholinergic neurons (see Table 2)

Cholinergic neurons are probably the largest neurons of the striatum (diameter of soma up to 40 μm); they have long aspiny or smooth dendrites and represent only 1–2% of the total neuronal population. They have been identified in Golgi-impregnated material, in material immunostained to reveal choline acetyltransferase immunoreactivity, and in material in which Golgi impregnation was combined with ChAT immunocytochemistry (Bolam et al 1984b) or acetylcholinesterase histochemistry (Bolam et al 1984a). Biochemical, pharmacological and morphological studies suggest that these cells are interneurons (Dray 1979), although in the cat putamen a small proportion of large acetylcholinesterase-positive neurons project to the cortex (Parent et al 1981).

The input to cholinergic neurons consists of boutons forming both symmetrical and asymmetrical membrane specializations. The perikaryal input is sparse and the asymmetrical input occurs predominantly in the distal regions of the neuron (Bolam et al 1984a,b, Chang & Kitai 1982, Takagi et al 1984, Wilson et al 1983).

By analysis of the axons of Golgi-impregnated neurons (Takagi et al 1984) or HRP-injected neurons (Wilson et al 1983), the local boutons have been shown to form symmetrical synapses predominantly with dendritic spines or shafts. In addition, our preliminary results from studies using monoclonal antibodies to ChAT indicate that at least one target of cholinergic terminals are neurons identified on an ultrastructural basis as medium-size and densely spiny (Wainer et al 1983).

Somatostatin-positive neurons (see Table 2)

Somatostatin-immunoreactive neurons appear in the light microscope as medium-size neurons, often with a polar appearance and infrequently branching long dendrites (DiFiglia & Aronin 1982, Takagi et al 1983). Dendritic spines have not been observed. Neurons with a similar light microscope morphology and ultrastructure have been recognized in Golgi-impregnated material (Takagi et al 1984) and some of these were shown to stain lightly for acetylcholinesterase (Bolam et al 1984a). The Golgi and the immunocytochemical studies have both shown that this class of neuron possesses an axon that arborizes close to the cell body; these neurons are therefore probably interneurons.

The axosomatic synaptic input is sparse, consisting predominantly of large boutons forming symmetrical synaptic contacts; asymmetrical synapses are observed only rarely. The input to proximal dendrites is similar to that of the perikarya but more distal dendrites receive predominantly asymmetrical synapses.

The output of these neurons has been studied in Golgi-impregnated material (Takagi et al 1984) and in sections incubated to reveal somatostatin immunoreactivity (DiFiglia & Aronin 1982, Takagi et al 1983). The axonal varicosities form symmetrical synaptic contact with dendritic shafts and spines; however, the varicosities of somatostatin-immunoreactive axons traced from their parent cell bodies did not form classical synaptic contacts (Takagi et al 1983). The varicosities were seen in close apposition to dendrites and perikarya but did not fulfil all the criteria of synaptic contacts. On the other hand, somatostatin-immunoreactive boutons detected in 'random searches' of the striatum formed classical symmetrical synaptic contacts with dendritic spines or shafts, but it remains to be established that these boutons come from intrinsic somatostatin-containing neurons.

General features of identified striatal neurons

From the data available on the morphology, chemistry and connections of striatal neurons, certain generalizations can be made. The following comments refer only to medium-size spiny neurons, GABA interneurons, cholinergic neurons and somatostatin neurons and do not include the second type of striatonigral neuron, since this type has more features in common with pallidal neurons than other striatal neurons.

(1) All striatal neurons have a fairly sparse input to the perikarya, consisting predominantly of boutons forming symmetrical synapses.

(2) Synaptic input is more frequent on distal parts of the neurons, where the proportion of asymmetrical synapses is higher.

(3) All these neurons subserve local circuit functions and the synapses formed by local boutons are of the symmetrical type.

These points are of significance because the two identified types of bouton

SYNAPSES OF STRIATAL NEURONS

forming asymmetrical contacts are of extrinsic origin, i.e. from the cortex and 5HT-immunoreactive terminals originating presumably in the dorsal raphe. In addition, although the tyrosine hydroxylase-immunoreactive terminals, i.e. dopaminergic terminals from the substantia nigra, form symmetrical synaptic contacts, they do so in distal regions of at least one type of neuron, the medium-size densely spiny neuron. The remaining types of identified symmetrical synaptic contacts are formed by neurons located in the striatum. Thus extrinsic information (from cortex, substantia nigra and raphe) is received and probably processed in distal regions of neurons, while the local input is received by all parts of the neurons and therefore modifies extrinsic information on all parts but predominantly in parts closer to the cell body.

(4) Every transmitter marker or Golgi-stained or HRP-filled local axon that has been examined makes synaptic contact either with neurons identified by Golgi impregnation as medium-size and densely spiny, or with neurons that have similar ultrastructural features to these neurons.

At present, we must conclude that medium-size spiny neurons provide the basic framework of neostriatal circuitry. They receive input from the cortex, substantia nigra and probably the dorsal raphe; they also receive input from other neurons within the neostriatum, including other medium-size spiny neurons (via local axon collaterals), and probably from the interneurons of the striatum (cholinergic, GABAergic and somatostatin-positive). At least one of their target neurons within the striatum is other spiny neurons; a spiny neuron network is thus established. Outside the striatum they make contact with neurons in the globus pallidus, some of which project to the substantia nigra, and within the nigra they contact neurons that project back to the striatum or to the thalamus.

The results of these studies on striatal circuitry leave many questions unanswered, probably the most important of which are: (1) Are all spiny neurons similar with respect to morphology and chemistry? The answer to this question is probably no, since subtle differences in the morphology of spiny neurons are coming to light (Bishop et al 1982, Danner & Pfister 1981) and there are inhomogeneities in the distribution of transmitters and transmitter markers (Graybiel 1984, this volume). (2) Do all spiny neurons receive all the inputs described above? The answer to this question awaits the application of double or even triple staining techniques at the ultrastructural level. (3) What are the origins and chemical characteristics of the inputs and what are the outputs of the interneurons of the striatum?

In this paper I have discussed only a few of the afferents to the striatum, only five of the neuron types and seven of the putative transmitters. Further work is required to incorporate into the neuronal circuits the other neuron types and putative transmitters and to relate these to the functional aspects of the neostriatum and other regions of the basal ganglia.

Acknowledgements

My own work described in this paper was supported by grants from the Wellcome Trust, E. P. Abraham Cephalosporin Trust and the Medical Research Council. I am supported by an MRC Senior Research Fellowship.

REFERENCES

Bishop GA, Chang HT, Kitai ST 1982 Morphological and physiological properties of neostriatal neurons: an intracellular horseradish peroxidase study in the rat. Neuroscience 7:179-191

Bolam JP, Powell JF, Totterdell S, Smith AD 1981a The proportion of neurons in the rat neostriatum that project to the substantia nigra demonstrated using horseradish peroxidase conjugated with wheatgerm agglutinin. Brain Res 220:339-343

Bolam JP, Somogyi P, Totterdell S, Smith AD 1981b A second type of striatonigral neuron: a comparison between retrogradely labelled and Golgi-stained neurons at the light and electron microscopic levels. Neuroscience 6:2141-2157

Bolam JP, Clarke DJ, Smith AD, Somogyi P 1983a A type of aspiny neuron in the rat neostriatum accumulates [^{3}H]γ-aminobutyric acid: combination of Golgi-staining, autoradiography and electron microscopy. J Comp Neurol 213:121-134

Bolam JP, Somogyi P, Takagi H, Fodor I, Smith AD 1983b Localization of substance P-like immunoreactivity in neurons and nerve terminals in the neostriatum of the rat: a correlated light and electron microscopic study. J Neurocytol 12:325-344

Bolam JP, Ingham CA, Smith AD 1984a The section Golgi impregnation procedure. 3. Combination of Golgi-impregnation with enzyme histochemistry and electron microscopy to characterize acetylcholinesterase-containing neurons in the rat neostriatum. Neuroscience, in press

Bolam JP, Wainer BH, Smith AD 1984b Characterization of cholinergic neurons in the rat neostriatum. A combination of choline acetyltransferase immunocytochemistry, Golgi impregnation and electron microscopy. Neuroscience, in press

Bradley RH, Kitai ST, Wu J-Y 1983 Putative neurotransmitters in neostriatal neurons: light and electron microscopic study. Soc Neurosci Abstr 9:658

Chang HT, Kitai ST 1982 Large neostriatal neurons in the rat: an electron microscopic study of gold-toned Golgi-stained cells. Brain Res Bull 8:631-643

Chang HT, Wilson CJ, Kitai ST 1981 Single neostriatal efferent axons in the globus pallidus: a light and electron microscopic study. Science (Wash DC) 213:915-918

Chang HT, Wilson CJ, Kitai ST 1982 A Golgi study of rat neostriatal neurons: light microscopic analysis. J Comp Neurol 208:107-126

Clarke DJ, Smith AD, Bolam JP 1984 Uptake of [^{3}H]taurine into medium-size neurons and into identified striatonigral neurons in the rat neostriatum. Brain Res 289:342-348

Danner H, Pfister C 1981 Spine-haltige Neurone im Caudatus-Putamen-Komplex der Ratte. J Hirnforsch 22:75-84

DiFiglia M, Aronin N 1982 Ultrastructural features of immunoreactive somatostatin neurons in the rat caudate-nucleus. J Neurosci 2:1267-1274

DiFiglia M, Pasik P, Pasik T 1976 A Golgi study of neuronal types in the neostriatum of monkeys. Brain Res 114:245-256

DiFiglia M, Pasik T, Pasik P 1980 Ultrastructure of Golgi-impregnated and gold-toned spiny and aspiny neurons in the monkey neostriatum. J Neurocytol 9:471-492

DiFiglia M, Aronin N, Martin JB 1982 Light and electron microscopic localization of immunoreactive leu-enkephalin in the monkey basal ganglia. J Neurosci 2:303-320

Dimova R, Vuillet J, Seite R 1980 Study of the rat neostriatum using a combined Golgi-electron microscope technique and serial sections. Neuroscience 5:1581-1596

Dray A 1979 The striatum and substantia nigra: a commentary on their relationships. Neuroscience 4:1407-1439

Graybiel AM 1984 Neurochemically specified subsystems in the basal ganglia. This volume, p 114-144

Graybiel AM, Ragsdale CW Jr 1983 Biochemical anatomy of the striatum. In: Emson PC (ed) Chemical neuroanatomy. Raven Press, New York, p 427-504

Hassler R 1978 Striatal control of locomotion, intentional actions and of integrating and perceptive activity. J Neurol Sci 36:187-224

Kemp JM, Powell TPS 1971 The structure of the caudate nucleus of the cat: light and electron microscopy. Philos Trans R Soc Lond B Biol Sci 262:383-401

Morelli M, Del Fiacco M, Wu J-Y, Di Chiara G 1983 Immunohistochemical localization of leu-enkephalin and glutamic acid decarboxylase in the nucleus caudatus of the rat. Soc Neurosci Abstr 9:14

Oertel WH, Mugnaini E 1983 Two classes of GABAergic neurons represent the majority of neostriatal neurons in the rat. Soc Neurosci Abstr 9:14

Parent A, Boucher R, O'Reilly-Fromentin J 1981 Acetylcholinesterase-containing neurons in cat pallidal complex: morphological characteristics and projection towards the neocortex. Brain Res 230:356-361

Pasik P, Pasik T, Pecci Saavedra J, Holstein GR 1981 Light and electron microscopic immunocytochemical localization of serotonin in the basal ganglia of cats and monkeys. Anat Rec 199:194A

Ribak CE, Vaughn JE, Roberts E 1979 The GABA neurons and their axon terminals in rat corpus striatum as demonstrated by GAD immunocytochemistry. J Comp Neurol 187:261-283

Somogyi P, Smith AD 1979 Projection of neostriatal spiny neurons to the substantia nigra. Application of a combined Golgi-staining and horseradish peroxidase transport procedure at both light and electron microscopic levels. Brain Res 178:3-15

Somogyi P, Hodgson AJ, Smith AD 1979 An approach to tracing neuron networks in the cerebral cortex and basal ganglia. Combination of Golgi staining, retrograde transport of horseradish peroxidase and anterograde degeneration of synaptic boutons in the same material. Neuroscience 4:1805-1852

Somogyi P, Bolam JP, Smith AD 1981a Monosynaptic cortical input and local axon collaterals of identified striatonigral neurons. A light and electron microscopic study using the Golgi-peroxidase transport-degeneration procedure. J Comp Neurol 195:567-584

Somogyi P, Bolam JP, Totterdell S, Smith AD 1981b Monosynaptic input from the nucleus accumbens-ventral striatum region to retrogradely labelled nigrostriatal neurones. Brain Res 217:245-263

Somogyi P, Priestley JV, Cuello AC, Smith AD, Takagi H 1982 Synaptic connections of enkephalin-immunoreactive nerve terminals in the neostriatum: a correlated light and electron microscopic study. J Neurocytol 11:779-807

Takagi H, Somogyi P, Somogyi J, Smith AD 1983 Fine structural studies on a type of somatostatin-immunoreactive neuron and its synaptic connections in the rat neostriatum: a correlated light and electron microscopic study. J Comp Neurol 214:1-16

Takagi H, Somogyi P, Smith AD 1984 Aspiny neurons and their local axons in the neostriatum of the rat: a correlated light and electron microscopic study of Golgi-impregnated material. J Neurocytol, in press

Totterdell S, Bolam JP, Smith AD 1984 Characterization of pallidonigral neurons in the rat by a combination of Golgi-impregnation and retrograde transport of horseradish peroxidase: their monosynaptic input from the neostriatum. J Neurocytol, in press

Wainer BH, Bolam JP, Clarke DJ et al 1983 Ultrastructural evidence of cholinergic synapses in different regions of rat brain: cholinergic pathways V. Soc Neurosci Abstr 9:963

Wassef M, Berod A, Sotelo C 1981 Dopaminergic dendrites in the pars reticulata of the rat substantia nigra and their striatal input. Combined immunocytochemical localization of tyrosine hydroxylase and anterograde degeneration. Neuroscience 6:2125-2139

Wilson CJ, Groves PM 1980 Fine structure and synaptic connections of the common spiny neuron of the rat neostriatum: a study employing intracellular injection of horseradish peroxidase. J Comp Neurol 194:599-615

Wilson CJ, Chang HT, Kitai ST 1983 Morphology and synaptic connections of a giant aspiny interneuron in the rat neostriatum. Soc Neurosci Abstr 9:658

DISCUSSION

Yoshida: Striatonigral fibres are inhibitory and GABAergic (Yoshida & Precht 1971, Precht & Yoshida 1971). The fibres run through the cerebral peduncle and enter the substantia nigra from the ventral to dorsal direction. When we stimulated the ventral part of the substantia nigra, we were able to record IPSPs, probably produced antidromically in the caudate neurons monosynaptically. It is possible, therefore, that there are a large number of axon collaterals of the caudatonigral inhibitory fibres which terminate in neurons within the caudate nucleus. The GABAergic terminals which you have shown would not necessarily be the terminals of interneurons within the caudate nucleus.

Although this is a different story, I would just like to add one thing. In addition to the caudatonigral fibres, another striatonigral system consists of putameno-nigral fibres. The latter fibres are also inhibitory and GABAergic (Yoshida et al 1981, Yoshida 1981).

Somogyi: We are doing some immunocytochemical experiments localizing GABAergic neurons, using an antiserum directed against GABA. Based on their dendritic and axonal arborization, I can conclude that there is certainly a class of GABA-containing local interneurons in the caudate-putamen of the cat (P. Somogyi, A. Hodgson, unpublished observation).

Bolam: I am convinced that the GABA interneurons exist and that there are also intrinsic collaterals of medium-size spiny neurons that synapse onto other spiny neurons.

Kitai: What is the exact evidence that GABA interneurons exist? I think there are medium spiny neurons and some large aspiny neurons with an axonal morphology similar to the interneurons, in the sense that these neurons possess profusely ramifying axon collaterals within the striatum.

Bolam: There are two approaches to showing that there are GABA interneurons in the striatum. One is to inject tritiated GABA into the striatum itself. Then we get one class of neurons labelled by the tritiated GABA. These are the medium-size aspiny neurons. Neurons of this class have an axonal plexus which is characteristic of an interneuron. We have never seen this type of neuron in retrograde labelling studies. The second approach is by immunocytochemistry, using antibodies directed against glutamate decarboxylase (GAD) or GABA. The same type of neuron is labelled with GAD antibodies, and Dr Somogyi has shown the same type of neuron using antiserum to GABA.

Graybiel: Do any of the projection neurons show immunoreactivity for GABA?

Somogyi: I cannot demonstrate GABA in the spiny neurons by immunocytochemistry. Maybe they don't store GABA in the perikarya.

Graybiel: We can't see large numbers of GAD-positive neurons in the cat striatum either, even with colchicine pretreatment.

Evarts: Could you explain the methods for GAD and GABA?

Somogyi: We developed an antiserum which recognizes GABA in tissue sections (A. Hodgson et al, unpublished results). It reveals all known local GABAergic interneurons in the cerebellum and cerebral cortex. In the striatum we have been able to stain only one type of small aspiny neuron having a large local axon arborization. Golgi impregnation is not the most conclusive way to decide whether a neuron is efferent or not but so far we have not found a main axon leaving the local arbor.

On the other hand with our antiserum we have been unable to stain projection neurons which are thought to be GABAergic, such as the Purkinje cells in the cerebellum or the medium-size spiny neurons in the striatum. The GABAergic efferent neurons may have a different synthetic and/or storage pathway for GABA. This may explain the lack of staining of their perikarya.

Calne: Which neurons are called spiny type I and spiny type II?

Evarts: Spiny I cells (the predominant striatal cells) are small whereas spiny II cells are large.

DiFiglia: We have used enkephalin immunohistochemistry to mark the spiny neurons in the monkey neostriatum. In addition to synapsing with other spiny cells, enkephalin-positive neurons appear to contact aspiny neurons. Immunoreactive enkephalin axons make numerous contacts with the cell bodies and the proximal, secondary and tertiary dendrites of neurons which have nuclei with indentations. We know from previous combined Golgi–electron microscope studies that this feature of the nucleus is characteristic of medium-size aspiny neurons (DiFiglia et al 1980). In addition, immunoreactive enkephalin axons appear to contact axon initial segments. We don't know which cells these axon initial segments belong to, but this interaction may have some physiological significance.

We have thought for a long time that there are morphologically two types of large neurons in the monkey (DiFiglia et al 1976). The large neuron we refer to as spiny type II has long, thick dendrites which are sparsely spined. In Golgi impregnations these cells exhibit long axons which give rise to numerous collaterals. The large globular neuron called aspiny type II has curly, varicose, spine-free dendrites. It is an interneuron and probably corresponds to the large cholinergic cell described by Paul Bolam in the rat. We examined the two types of large neurons in the monkey in a Golgi–electron microscope study and found differences in their cytological appearance and in the types and distribution of their synaptic inputs (Carey & DiFiglia 1982).

Graybiel: How long are the axons of the globular interneurons? Could they be as long as a millimetre?

DiFiglia: In Golgi impregnations of younger animals we see a short axon. Steve Kitai has evidence that in the rat the axons of large neurons may be myelinated but he has never followed one of these axons out of the nucleus. We think that the aspiny type II in the monkey corresponds to the large interneuron that Steve has identified in the rat.

Paul Bolam referred to possible subclasses of medium spiny neurons. These neurons may differ in their efferent projections. In the monkey, immunoreactive enkephalin axons which enter the pallidum from the neostriatum appear to stop at the level of the lateral segment of the globus pallidus (DiFiglia et al 1982a). There is relatively little enkephalin immunoreactivity in the medial pallidal segment. Dendrites in the lateral segment are highly innervated by immunoreactive enkephalin axons which were found to ensheath pallidal dendrites and to comprise 50% of all synapsing axons.

Golgi impregnations show that the axons of spiny neurons in the putamen enter the lateral segment of the globus pallidus where they immediately give rise to numerous fine collaterals which course along and ensheath pallidal dendrites (Fox & Rafol 1976, DiFiglia et al 1982b). This type of efferent system may correspond to the enkephalin-immunoreactive axons which we have observed in our immunohistochemical studies. The projection of striatal fibres into the substantia nigra which Steve Kitai and co-workers have observed with intracellular HRP injections may represent a different population of spiny efferent neurons.

Finally, the importance of identifying afferent inputs to striatal neurons other than the medium spiny type must be emphasized. The most important of these may be the medium aspiny neurons. Recent cytological studies in the monkey suggest that as many as 30% of medium-size caudate neurons and 20% of those in the putamen could be of the aspiny type (DiFiglia & Graveland 1983).

Kitai: As to whether GABAergic neurons are projection neurons or not, I thought that Peter Somogyi and Paul Bolam had indicated that GABAergic neurons are medium-size spiny neurons.

Somogyi: We didn't say that. We just said that there are local aspiny GABAergic interneurons in the neostriatum.

Kitai: But there are many medium-size spiny neurons that are GABAergic. Using double labelling immunocytochemistry we have identified GABAergic neurons that are projection neurons. That is, we injected wheatgerm agglutinin (WGA) into the globus pallidus or substantia nigra and identified these neurons with retrograde uptake of WGA. We then immunocytochemically identified them with both anti-WGA and anti-GAD and came to the conclusion that striatal projection neurons are GABAergic. When we examined these GABAergic neurons in the electron microscope, we found they had no nuclear indentations and had a large nucleus with a small perikaryon, which is very similar to the ultrastructure already reported for the spiny neurons (Chang & Kitai 1982, Wilson & Groves 1980). I think we have good evidence that spiny medium-sized neurons are projection neurons.

As to the large neurons in the rat, we could never follow the axons of these neurons outside the nucleus. These large cells receive excitatory inputs from the cortex, the thalamus and the substantia nigra, as medium spiny neurons do. According to Dr C.J. Wilson in my department, the pattern of spike firing in these large neurons is similar to what Dr Evarts' group observed in the monkey in their chronic extracellular recordings. The firing frequency of these neurons is around 5–8 Hz and spikes are wider with after-depolarization. These may be the cholinergic interneurons described by Parent and his group (Parent & O'Reilly-Fromentin 1982).

Di Chiara: What proportion of the striatal perikarya are GABA interneurons?

Bolam: In our material it is quite variable. In some areas we have had up to 10 or 15%. In other areas there were very few.

Somogyi: We are quantitating the percentage of cell bodies in the striatum in the cat now. The proportion will probably be less than 5%. However the contribution to a neuronal system cannot be assessed in terms of the numbers of the neurons. We should look at the possible local collaterals of the GABAergic spiny neurons against the local GABAergic axon arborization of the interneurons.

Di Chiara: Dr Bolam discussed the neurochemical identity of the medium-sized spiny neurons. We looked for the possible co-localization of GABA and enkephalin in these neurons and found that there is a population in the rat striatum which reacts positively with both glutamate decarboxylase and Leu-enkephalin antibodies. It is difficult to establish the percentage because we looked at consecutive sections and the criteria we use are very restrictive. We have also found that in the substantia nigra there is a great degree of interaction between GABA and dynorphin-enkephalins, so we think that in general opiate peptides are good candidates as modulators of GABA input in the basal ganglia.

Marsden: Dr Bolam, you focused attention on the spiny neuron as the major output neuron. Do all spiny neurons receive cortical glutamergic input onto their spines and a dopaminergic input from the substantia nigra? If that is so, what is the relative distribution of the cortical inputs within the striatum, in comparison to the relative distribution of nigral inputs? Thirdly, where is that thalamic input going?

Bolam: The first question is the one I asked at the end of my talk. I don't know the relative distributions of the cortical and nigral inputs. We are looking at individual neurons and we need to combine our techniques with the techniques Dr Graybiel is using.

Concerning your final point, from morphology there is very little evidence of what the thalamic input is doing, except for the work of Kemp & Powell (1971). They showed that lesions of the cat thalamus caused a reduction in the number of spines of Golgi-impregnated spiny neurons in the caudate, suggesting a direct input to those neurons from the thalamus.

REFERENCES

Carey J, DiFiglia M 1982 A Golgi-electron microscopic study of two types of large neurons in the monkey neostriatum. Neurosci Abstr 8:168
Chang HT, Kitai ST 1982 Large neostriatal neurons in the rat: an electron microscopic study of gold-toned Golgi-stained cells. Brain Res Bull 8:631-643
DiFiglia M, Graveland GA 1983 Neuronal organization in the monkey neostriatum: a quantitative light microscopic study using semi-thick serial sections. Neurosci Abstr 9:659
DiFiglia M, Pasik P, Pasik T 1976 A Golgi study of neuronal types in the neostriatum of monkeys. Brain Res 114:245-256
DiFiglia M, Pasik T, Pasik P 1980 Ultrastructure of Golgi impregnated and gold-toned spiny and aspiny neurons in the monkey neostriatum. J Neurocytol 9:471-492
DiFiglia M, Aronin N, Martin JB 1982a Light and electron microscopic localization of immunoreactive leu-enkephalin in the monkey basal ganglia. J Neurosci 2:303-320
DiFiglia M, Pasik P, Pasik T 1982b A Golgi and ultrastructural study of the monkey globus pallidus. J Comp Neurol 212:53-75
Fox CA, Rafol JA 1976 The striatal efferents in the globus pallidus. In: Yahr MD (ed) The basal ganglia. Raven Press, New York, p 37-55
Kemp JM, Powell TPS 1971 The termination of fibres from the cerebral cortex and thalamus upon dendritic spines in the caudate nucleus: a study with the Golgi method. Philos Trans R Soc B Biol Sci 262:429-439
Parent A, O'Reilly-Fromentin J 1982 Distribution and morphological characteristics of acetylcholinesterase-containing neurons in the basal forebrain of the cat. Brain Res Bull 8:183-196
Precht W, Yoshida M 1971 Blockage of caudate-evoked inhibition of neurons in the substantia nigra by picrotoxin. Brain Res 32:229-233
Wilson CJ, Groves PM 1980 Fine structure and synaptic connections of the common spiny neuron of the rat neostriatum: a study employing intracellular injection of horseradish peroxidase. J Comp Neurol 194:599-616

Yoshida M 1981 The GABAergic systems and the role of basal ganglia in motor control. Adv Biochem Psychopharmacol 30:37-52

Yoshida M, Precht W 1971 Monosynaptic inhibition of neurons of the substantia nigra by caudate-nigral fibers. Brain Res 32:225-228

Yoshida M, Nakajima N, Niijima K 1981 Effect of stimulation of the putamen on the substantia nigra in the cat. Brain Res 217:169-174

Synaptic organization of the basal ganglia: an electroanatomical approach in the rat

J. M. DENIAU and G. CHEVALIER

Laboratoire de Physiologie des Centres Nerveux, Université Pierre et Marie Curie, 4 Place Jussieu, 75230 Paris Cedex 05, France

Abstract. The physiological processes by which basal ganglia participate in the elaboration of movement are still poorly understood. In particular, lack of information about the synaptic organization of the output pathways of the basal ganglia impedes functional analysis of this system. To bridge this gap electroanatomical studies have been undertaken of one of the major output systems of the basal ganglia: the substantia nigra pars reticulata (SNr). This paper reviews recent results and proposes functional perspectives. The findings reported clearly establish that SNr exerts a tonic inhibitory influence on its collicular and thalamic targets. It is suggested that the striatum may achieve a complex-spatial pattern of facilitation in a large spectrum of structures related to ocular and cephalic motor activity by inhibiting the inhibitory nigrothalamic and nigrocollicular branched neurons.

1984 Functions of the basal ganglia. Pitman, London (Ciba Foundation symposium 107) p 48-63

The involvement of the basal ganglia (i.e. striatum, globus pallidus, subthalamic nucleus and substantia nigra) in the control of movement is dramatically attested by the violent motor disorders associated with dysfunction of these forebrain structures. Patients suffering from basal ganglia disease experience various types of abnormal movements and postural deficits which are characteristic of the particular cellular groups affected. The clinical syndromes include akinesia, tremor, rigidity, chorea, athetosis, ballism and dystonia. In animals, experimental lesions or stimulation of these structures have been shown to produce various types of movement and postural disorders. One of the most spectacular effects is the compulsive head-turning induced in rats, cats and even monkeys by unilateral changes in striatal or nigral activity.

In the past, lack of information about the anatomical connections of the basal ganglia has considerably impeded functional analysis of this system. As pointed out by Graybiel & Ragsdale (1979) in their review of the fibre connections of the basal ganglia, 'it is not yet clear even nearly three quarters of a century after S. A. Kinnier Wilson coined the term extrapyramidal system how the basal ganglia should be fitted into a general framework of the supraspinal motor mechanisms'. Thanks to the development of neuroanatomical tracing techniques, and through the combination of anatomical and electrophysiological methods, a great deal of information has now been collected on the pathways linking the basal ganglia with other parts of the brain. From these studies has arisen the concept of a hierarchical organization of basal ganglia in which the input part of the system is occupied by the striatum and the source of outputs by the internal segment of the globus pallidus (or entopeduncular nucleus in rats and cats) and the pars reticulata of the substantia nigra (SNr). It is within this conceptual framework that we have undertaken electroanatomical and physiological studies of the SNr. By analysing the output pathways of the basal ganglia we hope to learn more about the logic of basal ganglia circuitry and thereby further our understanding of the role of these subcortical structures in the elaboration of motor activity.

The SNr as an output system for expression of basal ganglia functions

The particular anatomical features of basal ganglia nuclei (deep subcortical structures, close apposition with fibre bundles, multiple interconnections with bifurcated axons) have complicated experimental analysis of the output pathways of this system. With the use of modern axonal transport and electrophysiological techniques, however, it has been revealed that most of the output pathways of the basal ganglia arise from two small nuclei of this system: the internal segment of the globus pallidus (GPi) and the pars reticulata of the substantia nigra (SNr). Efferent pathways from GPi have been shown to terminate in various thalamic nuclei (ventral group, centre median), in the tegmentum, and in prerubral fields (see review by Graybiel & Ragsdale 1979). In rats and cats, fibres originating from the SNr innervate thalamic nuclei (ventralis medialis, parafascicular, central lateral, paracentral, lateral dorsal, the medial part of ventral anterior and ventral lateral, and the lateral segment of mediodorsal), the superior colliculus, and parts of the reticular formation (Graybiel & Ragsdale 1979, Hendry et al 1979, Gerfen et al 1982).

The GPi has traditionally been recognized as the major efferent system linking the striatum and the subthalamic nucleus to thalamic and tegmental motor centres, but the idea that the SNr might also be involved in mediating striatal and subthalamic influences to motor centres is recent. SNr was considered to be merely the neuropil part of the nucleus where dopaminergic nigrostiatal neurons lying in the adjacent pars compacta (SNc) receive most of their inputs. In consequence the striatonigral projection was believed to have only a regulatory effect on the activity of dopaminergic nigrostriatal cells. However, one could ask whether there was a link between the striatonigral projections and the non-dopaminergic nigrofungal neurons of the SNr.

Using the antidromic activation method (the electrophysiological equivalent of the anatomical axonal retrograde transport method) we identified the nigrothalamic and nigrotectal neurons and found that these cells receive the characteristic inhibitory influence exerted by the striatonigral GABAergic pathway (Deniau et al 1978a). From this and the behavioural evidence that SNr really mediates some of the motor effects evoked from the striatum (DiChiara et al 1979a) has arisen the concept that, as well as GPi, SNr constitutes an important output system for the expression of basal ganglia functions. This view has been further strengthened by the anatomical and electrophysiological evidence that SNr is also involved in relaying the efferents of the subthalamic nucleus (Kanazawa et al 1976, Deniau et al 1978b).

There are striking similarities between SNr and GPi. These embryologically related structures share the same ultrastructural features and, in the light of the above hodological considerations, appear to act in parallel as relays for striatal and subthalamic efferents. The similarities between SNr and GPi can be further extended if we look at the fine organization of their subthalamic and striatal afferents. Indeed, using the antidromic activation method to identify the subthalamopallidal and subthalamonigral neurons in the rat, we have found that most subthalamic cells project to both SNr and GPi by way of axon collaterals (Deniau et al 1978a). This finding, which is reminiscent of the demonstration that striatonigral neurons give axon collaterals to the GPi (Yoshida et al 1972, Chang et al 1981), points to an exact parallel between SNr and GPi and suggests their close functional cooperation in the organization of basal ganglia outflow.

The SNr as a spatial organizer of basal ganglia influences

In recent years particular attention has been given to defining the precise organization of SNr efferents. The major aim of these studies was to determine whether the various nigral efferents (i.e. nigrothalamic, nigrotectal and nigroreticular pathways) should be considered as separate output lines conveying distinct information from the basal ganglia, or whether single SNr neurons project to various structures by way of axon collaterals.

Today there is good electrophysiological and morphological evidence that most SNr cells can simultaneously influence several targets through their bifurcated axons. In electrophysiological studies the various combinations of axonal branching between thalamic nuclei bilaterally, the superior colliculus bilaterally and the reticular formation have been discovered (Deniau et al 1978a, Anderson & Yoshida 1977, Niijima & Yoshida 1982). Morphological support for such axonal branching has been provided by direct visualization of the axonal process of SNr cells intracellularly labelled with horseradish peroxidase. Grofová et al (1982) described the axon of a nigrothalamic neuron which divided into four branches. Three branches ran dorsocaudally towards the superior colliculus and gave off several collaterals in the mesencephalic reticular substance. Some of these collaterals were seen to turn rostrally towards the thalamus. The fourth branch was directed rostrally to enter the subthalamic nucleus. From the axonal features of SNr cells and the extremely wide dendritic field of these cells it can be inferred that SNr is not simply divided into subnuclei conveying separate signals to the thalamus, superior colliculus or reticular formation; on the contrary the experimental findings support the view that each SNr neuron carries out a complex spatial diffusion of highly compiled information from the basal ganglia.

Synaptic interactions between SNr efferents and premotor integrative circuits

To analyse the physiological processes by which basal ganglia participate in the control of movement, we have done electrophysiological experiments to characterize the synaptic influence exerted by SNr efferents and identify the target cells on which it is exerted.

Are the nigral efferents excitatory or inhibitory?

The synaptic effects produced by electrical stimulation of the substantia nigra have been largely controversial. Excitation, inhibition of various latencies and durations, and mixed excitatory and inhibitory effects have been described which produce considerable confusion on this subject. In fact there are considerable difficulties in achieving selective stimulation of the SNr without coactivating the cortical efferent fibres running in the ventral aspect of the nucleus.

By studying the electrophysiological effects of stimulation of the substantia nigra after lesion of the cortical efferent fibres, comparing the latencies of the substantia nigra-evoked responses with the conduction time of action potentials in the nigra efferent pathways, and using weak stimuli, we reached the conclusion that in the rat SNr exerts a purely inhibitory effect on the neurons of the ventromedial thalamic nucleus and of the superior colliculus (Fig. 1) (Deniau et al 1978c, Chevalier et al 1981a, Chevalier & Deniau 1982).

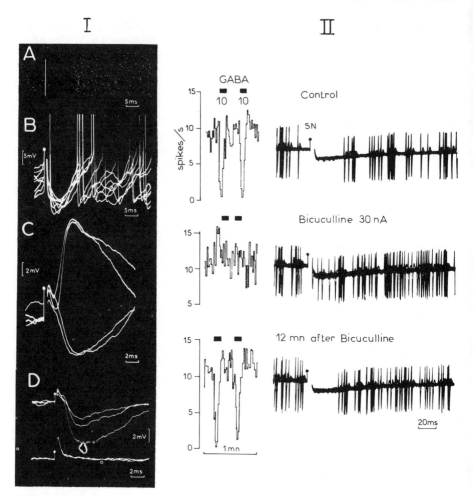

FIG. 1. (I) Typical nigral-evoked monosynaptic inhibitory postsynaptic potential on ventromedial thalamic cells (A, B). Note reversal when a hyperpolarizing current (C) is applied and the fixed latency when stimulation intensity is increased (D).

(II) Typical GABAergic nigral-induced inhibition of a tectal cell. Middle trace: note blockade after microiontophoretic application of the GABA receptor antagonist, bicuculline. SN, substantia nigra.

A similar effect has been described in the nucleus gigantocellularis by Gonzalez-Vegas (1981). This is the same type of inhibitory response as Ueki (1983) recorded on ventromedial thalamic neurons in the cat after stimulation of the substantia nigra. The GABAergic nature of this nigral-induced effect has been assessed by the systemic or local application (Yoshida & Omata

1979, Chevalier et al 1981b) of GABA receptor antagonists (Fig. 1). That nigrotectal and nigrothalamic neurons use GABA or a neurotransmitter has also been shown by a biochemical approach (Di Chiara et al 1979b, Kilpatrick et al 1980).

What does the substantia nigra inhibitory control exerted on tectal and thalamic cells mean?

Identification of the neurons which receive inhibitory nigral influences is just beginning, since only a few SNr efferents have been investigated. However our recent work on nigrothalamic and nigrotectal pathways already offers some insights into the role played by the substantia nigra in the subcortical mechanisms of motor integration.

Nigrocollicular relationship

Unilateral lesions or stimulation of the striatum or substantia nigra produce an asymmetrical posture characterized by compulsive head-turning. Because the superior colliculus is implicated in promoting eye-orienting and head-orienting movements, particular attention has been paid to the nigrotectal pathway as a possible substrate for the head-turning induced by the basal ganglia. Several authors have already attempted to lesion the superior colliculus to evaluate the importance of this structure in the mediation of such motor effects. Their results show clearly that the lateral portion of the superior colliculus, in contrast to the medial one, is involved in mediating the influence of the basal ganglia on head movements (Di Chiara et al 1982, Kilpatrick et al 1982). Since the tectospinal neurons which represent one of the efferent pathways by which the superior colliculus can influence head movements lie in the lateral portion of the superior colliculus (Murray & Coulter 1982, Chevalier et al 1984, Chevalier & Deniau 1984), a nigrotecto-spinal link is possible. To test this hypothesis, we identified the tectospinal neurons in the rat by antidromic activation from the spinal cord and tested their responsiveness to SNr stimulation. Of the 37 tectospinal neurons recorded, 17 exhibited the typical short latency (1.5–2 ms) and short duration (7–15 ms) nigral-induced inhibitory effect (Fig. 2). This inhibition was powerful enough to suppress the spontaneous as well as the synaptically evoked activity of these neurons (Chevalier et al 1984). Interestingly, recent studies have shown that the tectospinal neurons do much more than simply innervate the spinal cord. These cells provide an impressive network of axon collaterals throughout the rhombencephalic reticular core and also innervate the non-specific thalamic nuclei (Grantyn & Grantyn 1982, Chevalier & Deniau 1984). It is noteworthy that most of the areas innervated by these neurons are concerned with eye and head movements. Thus, through the nigrocollicular relationships, it is very likely that basal ganglia can exert a coordinated influence on a wide spectrum of structures devoted to ocular and cephalic motor activity.

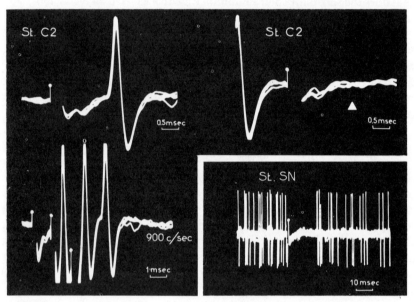

FIG. 2. Nigral-evoked inhibition of an antidromically identified tectospinal neuron. The antidromic invasion is characterized by fixed latency, collision, and high frequency tests. St. C2, stimulation of the cervical cord. St. SN, stimulation of the substantia nigra.

The high level of spontaneous discharge of SNr neurons (40–100/s) has led to the idea that their function is to exert a tonic inhibitory influence on their target cells. As previously mentioned, the nigrotectal neurons themselves are subject to potent inhibitory control from the striatum. When activated, the striatal neurons provoke a phasic arrest of nigrotectal neuron discharges. It was therefore tempting to speculate that, through the striatonigrotectal neuronal circuit, the striatum has a facilitatory effect on tectospinal/tecto-diencephalic (TSD) neurons via a disinhibitory mechanism (Fig. 3). Such a process has been proposed by Hikosaka & Wurtz (1983) to explain why SNr and superior collicular neuronal activities are organized in a mirror fashion when monkeys perform an ocular orienting task. Whereas the neurons of the superior colliculus increase their firing rate before targeting movements, the SNr cells exhibit a phasic arrest of their tonic spontaneous discharge.

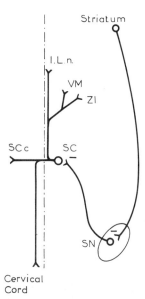

FIG. 3. Schematic representation of the functional relationships between striatum and tectospinal/tectodiencephalic neurons as disclosed by electroanatomical observations. SC, superior colliculus; SCc, contralateral superior colliculus; SN, substantia nigra; VM, ventromedial thalamic nucleus; I.L.n., intralaminar nuclei; ZI, zona incerta.

To test the accuracy of such a disinhibitory mechanism, we examined the effect of a transitory blockade of nigral discharges on the excitability of TSD neurons. SNr and TSD neurons were simultaneously recorded while 50–100 nl of 1 M-GABA solution was injected into the SNr. The arrest of nigral activity was accompanied by vigorous and sustained increase in the firing rate of the TSD cells (Fig. 4). The time courses of those changes in firing rate were

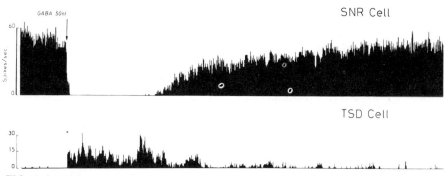

FIG. 4. Disinhibitory effect evoked on a tectospinal/tectodiencephalic (TSD) cell during the GABA-induced arrest of substantia nigra pars reticulata (SNr) activity. Note the close temporal relationship between the changes in SNr and collicular activity.

closely related. Indeed, when the nigral firing rate returned to the control level TSD activity slowed down. Disinhibition of TSD cells by the basal ganglia might provide the spatiotemporal pattern of facilitation in the ocular and cephalic premotor circuits that is required for the initiation of eye-orienting and head-orienting movements.

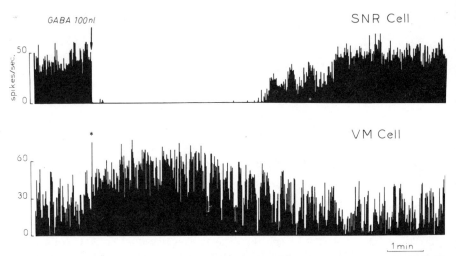

FIG. 5. Disinhibitory effect evoked on a ventromedial (VM) thalamic cell during the GABA-induced arrest of SNr activity.

Nigrothalamic projection

As reported above for the nigrotectal projection, the SNr also exerts a tonic inhibitory influence on thalamic cells. Any decrease in SNr activity, such as can be produced by intranigral injection of GABA, induces an increase in the firing level of ventromedial thalamic neurons (Fig. 5). The activity of these ventromedial cells did not decrease before the SNr cells began to fire again.

Due to lack of precise information on the cortical projection sites of the thalamic neurons receiving this tonic inhibitory nigral influence, it is still very difficult to consider the functional perspectives of the nigrothalamic connection. However, the thalamic nuclei receiving the nigral influence are also innervated by tectal efferents. We therefore propose that, by its action on the nigrotectal and nigrothalamic branched neurons, the striatum may coordinate the disinhibitory influence it exerts at these two levels and facilitate in the diencephalon the transmission of the tectal outflow it generated.

Another promising observation is that all the thalamic nuclei receiving SNr and superior collicular efferents are also recipient zones for cerebellothalamic projections. These three systems might interact at the level of single thalamic neurons. Experimental analyses are now being done to determine whether such interactions exist. We have already demonstrated that cerebellothalamic and nigrothalamic projections converge at the level of single ventromedial cells (Chevalier & Deniau 1982).

The basal ganglia may play a major role, through the nigrothalamic pathway, in the subcortical processes of motor integration by gating the transmission of cerebellothalamic and tectothalamic information at the level of the non-specific thalamic nuclei.

REFERENCES

Anderson M, Yoshida M 1977 Electrophysiological evidence for branching nigral projections to the thalamus and superior colliculus. Brain Res 137:361-364

Chang HT, Wilson CJ, Kitai ST 1981 Single neostriatal efferent axons in the globus pallidus: a light and electron microscopic study. Science (Wash DC) 213:915-918

Chevalier G, Deniau JM 1982 Inhibitory nigral influence on cerebellar evoked responses in the rat ventromedial thalamic nucleus. Exp Brain Res 48:369-376

Chevalier G, Deniau JM 1984 Spatio-temporal organization of a branched tecto-spinal/tecto-diencephalic neuronal system. Neuroscience, in press

Chevalier G, Deniau JM Thierry AM, Feger J 1981a The nigro tectal pathway; an electrophysiological reinvestigation in the rat. Brain Res 213:253-263

Chevalier G, Thierry AM, Shibazaki T, Feger J 1981b Evidence for a GABAergic inhibitory pathway in the rat. Neurosci Lett 21:67-70

Chevalier G, Vacher S, Deniau JM 1984 Inhibitory nigral influence on tecto spinal neurons. A possible implication of basal ganglia in orienting behavior. Exp Brain Res 53:320-326

Deniau JM, Hammond C, Riszk A, Feger J 1978a Electrophysiological properties of identified output neurons of the rat substantia nigra (pars compacta and pars reticulata): evidence for the existence of branched neurons. Exp Brain Res 32:409-422

Deniau JM, Hammond C, Chevalier G, Feger J 1978b Evidence for branched subthalamic nucleus projections to substantia nigra, entopeduncular nucleus and globus pallidus. Neurosci Lett 9:117-121

Deniau JM, Lackner D, Feger J 1978c Effect of substantia nigra stimulation on identified neurons in the VL-VA thalamic complex: comparison between intact and chronically decorticated cats. Brain Res 145:27-35

Di Chiara G, Porceddu M, Gessa G 1979a Substantia nigra as an output station for striatal dopaminergic responses: role of a GABA-mediated inhibition of pars reticulata neurons. Naunyn-Schmiedeberg's Arch Pharmacol 306:153-159

Di Chiara G, Porceddu ML, Morelli M, Mulas ML, Gessa GL 1979b Evidence for a GABAergic projection from the substantia nigra to the ventromedial thalamus and to the superior colliculus of the rat. Brain Res 176:273-284

Di Chiara G, Morelli M, Imperato A, Porceddu ML 1982 A reevaluation of the role of superior colliculus in turning behaviour. Brain Res 237:61-77

Gerfen CR, Staines WA, Arbuthnott GW, Fibiger HC 1982 Crossed connections of the substantia nigra in the rat. J Comp Neurol 207:283-303

Gonzales-Vegas JA 1981 Nigro-reticular pathway in the rat—an intracellular study. Brain Res 207:170-173

Grantyn A, Grantyn R 1982 Axonal patterns and sites of termination of cat superior colliculus neurons projecting in the tecto-bulbo-spinal tract. Exp Brain Res 46:243-256

Graybiel AM, Ragsdale CW 1979 Fiber connections of the basal ganglia. Prog Brain Res 51:239-283

Grofová I, Deniau JM, Kitai ST 1982 Morphology of the substantia nigra pars reticulata projection neurons intracellularly labeled with HRP. J Comp Neurol 208:352-368

Hendry SHC, Jones EG, Graham J 1979 Thalamic relay nuclei for cerebellar and certain related fiber systems in the cat. J Comp Neurol 185:679-714

Hikosaka O, Wurtz RH 1983 Visual and oculomotor functions of monkey substantia nigra pars reticulata. Relation of substantia nigra to superior colliculus. J Neurophysiol 49:1285-1301

Kanazawa I, Marshall GR, Kelly JS 1976 Afferents to the rat substantia nigra studied with horseradish peroxidase, with special reference to fibres from the subthalamic nucleus. Brain Res 115:485-491

Kilpatrick IC, Starr MS, Fletcher A, James TA, MacLeod NK 1980 Evidence for a Gabaergic nigrothalamic pathway in the rat. I. Behavioral and biochemical studies. Exp Brain Res 40:45-54

Kilpatrick IC, Collingridge GL, Starr MS 1982 Evidence for the participation of nigrotectal γ aminobutyrate containing neurones in striatal and nigral derived circling in the rat. Neuroscience 7:207-222

Murray EA, Coulter JD 1982 Organization of tectospinal neurons in the cat and rat superior colliculus. Brain Res 243:201-214

Niijima K, Yoshida M 1982 Electrophysiological evidence for branching nigral projections to pontine reticular formation, superior colliculus and thalamus. Brain Res 239:279-282

Ueki A 1983 The mode of nigrothalamic transmission investigated with intracellular recordings in the cat. Exp Brain Res 1:116-124

Yoshida M, Rabin A, Anderson M 1972 Monosynaptic inhibition of pallidal neurons by axon collaterals of caudato nigral fibers. Exp Brain Res 15:333-347

Yoshida M, Omata S 1979 Blocking by picrotoxin of nigral evoked inhibition of neurons of ventromedial nucleus of the thalamus. Experientia (Basal) 35:794

DISCUSSION

Evarts: Successive inhibitory relays (from striatum to globus pallidus and SNr and then from globus pallidus and SNr to their targets) mean that increased striatal activity, by inhibiting globus pallidus and SNr, can disinhibit and give rise to *increased* activity in thalamic or tectal neurons. What might be the metabolic correlates of pharmacological effects in these successive inhibitory circuits? Ronald Hammer at the National Institutes of Health has studied the effects of reserpine on metabolic activity (as indicated by deoxyglucose uptake) in the basal ganglia of the rat, and has observed that reserpine causes a marked increase in the metabolic activity of the lateral habenula. The lateral habenula receives dense projections from the entopeduncular nucleus, so is analogous to the targets you have been speaking about, Dr Deniau. You showed that SNr controls sets of neurons in the tectum; the entopeduncular nucleus has a corresponding role in controlling cells in the habenula. In the normal rat, the striatum is metabolically quite active and the globus pallidus is much less active; reserpine reduces striatal metabolic activity while increasing metabolism in the globus pallidus. This high striatal metabolism in the normal rat is present in spite of striatal cells having very few impulses (i.e. the striatal cells lack 'spontaneous' activity).

In contrast, the metabolically inactive globus pallidus has virtually continuous impulse activity. It has been known for some time that deoxyglucose uptake is not necessarily related to impulse activity *per se* and, in the particular case of striatum and globus pallidus, metabolism is actually *inversely* related to impulse frequency. But though cells in the striatum are relatively silent, they receive continuous inputs from the cerebral cortex and these cortical inputs act on striatal cells via glutaminergic excitatory synapses. There is thus a great deal of synaptic excitation reaching the striatum from the cortex; this synaptic excitation apparently leads to a great deal of metabolic activity in spite of the fact that it results in very little impulse activity. The effect of reserpine in the rat is to reduce metabolic activity in striatum while increasing metabolic activity in the globus pallidus, and this effect is reversed by L-dopa.

The lateral habenula, which is densely innervated from the entopeduncular nucleus, is normally relatively inactive metabolically. With reserpine it becomes metabolically the most active structure in this region. Unfortunately, one can't be sure whether this increased metabolic activity in the lateral habenula is due to something like you are describing, Dr Deniau. Are the lateral habenular cells more active impulse-wise as a result of reserpine, and is this why we see this increased metabolic activity? Or is the lateral habenula receiving greater inhibitory GABAergic synaptic drives after reserpine than in the normal rat? In your experiments it might be possible to get the answer by using localized GABA injections and then looking at the deoxyglucose uptake in the targets.

Calne: Dr Martin in our group has scanned a number of patients with unilateral Parkinson's disease. I am sure the damage is bilateral in these patients but on the predominantly affected side we see some increased activity in the pallidal region. With the fluorodeoxyglucose (FDG) technique, however, the resolution is only about 8 mm so we can't see changes in many of the structures we are discussing. The frustrating feature of this work is that while we can see asymmetry of FDG in patients with asymmetrical disease, our preliminary studies with normal subjects who are moving the limbs of one side very vigorously show no asymmetrical activity in the pallidal region.

Somogyi: There are branching nigral neurons which project to the thalamus and the striatum. Presumably they are also GABAergic so there should be a nigrostriatal GABAergic pathway. Is there any electrophysiological evidence of this?

Deniau: We have shown in previous electrophysiological and anatomical analyses that nigrothalamic neurons send an ascending axon collateral to the telencephalon. This collateral courses through the striatum. This projection is probably GABAergic and inhibitory. However the exact terminal site and the synaptic effect of this non-dopaminergic nigrotelencephalic projection still remains to be investigated.

Somogyi: Were you able to trace any axon collaterals to thalamic nuclei other than the ventral medial nuclei?

Deniau: We have not looked at other nuclei.

Rizzolatti: Is there any input from substantia nigra to the intermediate layers of the superior colliculus? The eyes are controlled by the intermediate layers of this structure.

Graybiel: In both cat and monkey, axons from the pars reticulata terminate in the intermediate grey layer. The terminal field is broken up into regularly spaced clusters (Graybiel 1975, 1978, Jayaraman et al 1977). Hikosaka & Wurtz (1983) have strong evidence that this pathway has to do with the control of eye movements. Recently, Dr Illing and I have found that the frontal eye fields project to the same set of patches in the colliculus that receive a nigral input (Illing & Graybiel 1983). The overlap seems significant because when we look at the equally patchy inputs to the intermediate grey layer from somatic sensory cortex or from the bulbospinal junction, for example, we find that they project to different regions from those receiving nigral and frontal eyefield input. Dr Deniau's concept that the substantia nigra could coordinate a number of different systems is very interesting in terms of these patterns of overlap and non-overlap.

Di Chiara: One might say that the intermediate and deep layers of the superior colliculus are a site of integration for different sensory modalities— auditory and tactile as well as visual. In addition these layers receive motor impulses from the nigra. This is important because it makes the superior colliculus a site of sensorimotor integration. For this reason I like to think that the functional significance of the nigrocollicular engagement is very different from the pallidothalamic one. The globus pallidus and the substantia nigra both project to the thalamus but they do not both project to the superior colliculus. Only the nigra does. This applies not only to rats but also to primates and should have a distinct functional significance for both species.

Rizzolatti: There are large species differences. I agree with you about rats and cats but I doubt that this is true for monkeys. For example, in the intermediate layers of the monkey superior colliculus we were unable to find polymodal neurons (Gentilucci et al 1984). They may be in the deep layers but it is very difficult to find them.

Marsden: In the area of thalamus to which the axon collateral from the tectospinal tract projects there is a problem with the terminology in the rat compared with the primate. What is the corollary in the primate of VM in the rat?

Nauta: A likely candidate, in my opinion, is the area in the monkey that Olszewski called VAmc, a dorsomedial part of the VA complex close up against the internal medullary lamina.

Marsden: Does this mean there is some degree of overlap of projections of the nigral output system and the entopeduncular (or globus pallidus internal segment) output system at the thalamic level, or are they still segregated?

Deniau: In rats and cats the VM nucleus of the thalamus is known to receive projections from the entopeduncular nucleus. There might be some degree of overlap between nigrothalamic and pallidothalamic projections in this thalamic area.

DeLong: The ventromedial nucleus in rodents has projections to layer 1 of the cortex, I believe. Is that true of the magnocellular part of VA as well, or are we dealing with a very different nucleus in rats and monkeys?

Carpenter: I thought that the magnocellular part of the ventral anterior nucleus of the thalamus had a special projection to the orbital cortex. At one time this was thought to be the pathway that triggered the recruiting response (Carmel 1970).

Evarts: Künzle & Akert (1977) have shown that corticothalamic fibres from the frontal eye field project to thalamic nucleus ventralis anterior pars magnocellularis (VAmc) and to the lateral part of the nucleus medialis dorsalis (MD); given the usual reciprocal relations between thalamus and cerebral cortex it would follow that VAmc and MD project to the frontal eye field. Would you agree with that?

Carpenter: There is a special relationship between the substantia nigra and the frontal eye fields in the primate. The substantia nigra (pars reticulata) projects to the paralaminar part of the dorsomedial nucleus of the thalamus which has reciprocal relations (Künzle & Akert 1977) with the frontal eye field. This field, in addition to projecting back to the paralaminar parts of the dorsal medial nucleus, has projections directly to the superior colliculus. Thus, there are reinforcing and reciprocal connections that establish close relationships between frontal eye fields and the superior colliculus.

Evarts: The lateral part of the substantia nigra pars reticulata (in the monkey, at least) then seems to project to VAmc.

Nauta: Herkenham (1976) in an autoradiographic study in the rat demonstrated a very extensive projection from VM to the upper half of the first cortical layer, densest in the frontal region but extending caudally far enough to invade even the visual cortex. In addition to this expansive unilaminar projection he found evidence of a VM projection to the frontal cortex that resembled in its distribution the more common type of thalamocortical connection involving layers VI, IV and III.

DeLong: There might be subdivisions in the region.

Yoshida: In the thalamus, our physiological studies show that neurons which receive EPSPs from the cerebellothalamic pathway, IPSPs from the entopeduncular (ENT)–thalamic pathway, and IPSPs from the nigrothalamic pathway, are distinctively segregated. ENT-related thalamic neurons are located more medio-ventro-caudally than cerebellum-related thalamic neurons, and nigra-related neurons are located more medio-ventro-caudally than ENT-related neurons. There are not many thalamic neurons which receive convergent responses from these three major afferents. Your observation that many neurons receive the convergence might be explained by the difference between your technique and ours. We used an intracellular recording technique for more than 200 neurons whereas your major units were recorded extracellularly (Uno & Yoshida 1975, Uno et al 1978, Ueki et al 1977, Ueki 1983).

Deniau: In our extracellular and intracellular work on the ventromedial nucleus of the thalamus we have shown that neurons receiving the nigral inhibitory influence can also receive cerebellar input. Generally these neurons are not reliably driven by the cerebellar stimulation. They respond to deep cerebellar nuclei with a very small depolarization. From our experience, the analysis of cerebellar influence on VM cells is more accurate when we use extracellular recording techniques. Since the substantia nigra exerts a tonic inhibitory influence on the VM cells it is possible that such an influence decreases the effectiveness of the cerebellar input in this thalamic area. This might explain the difficulty of discovering the cerebellar input of these neurons.

Marsden: It is very impressive that injection of GABA into the substantia nigra pars reticulata causes the cells in the deep layers of the superior colliculus to start firing enormous bursts. What is the nature of the output signal of those bursting cells in the superior colliculus? Are they neurons that command a movement or are they activating other areas which command movement? If they are command neurons, isn't this the first precedent for a clear-cut neuron system activated directly by basal ganglia output that commands movement?

Calne: What happens chemically when they discharge?

Graybiel: In the monkey, Hikosaka & Wurtz (1983) have evidence that the substantia nigra inhibits saccade-related cells in the intermediate layers. When they inject the GABA antagonist bicuculline locally into the colliculus, the monkey makes saccades to the site in the visual field represented at the site of injection. Local injections of the GABA agonist muscimol reduce the frequency of such saccades and make them hypometric.

REFERENCES

Carmel PW 1970 Efferent projections of the ventral anterior nucleus of the thalamus in the monkey. Am J Anat 128:159-184

Gentilucci M, Bon L, Camarda R, Matelli M, Rizzolatti G 1984 Organizzazione funzionale degli strati intermedi e profondi del collicolo superiore di scimmia. Boll Soc Ital Biol Sper, in press

Graybiel AM 1975 Anatomical organization of retinotectal afferents in the cat: an autoradiographic study. Brain Res 96:1-23

Graybiel AM 1978 Organization of the nigrotectal connection: an experimental tracer study in the cat. Brain Res 143:339-348

Herkenham M 1976 The nigro-thalamo-cortical connection mediated by the nucleus ventralis medialis thalami: evidence for a wide cortical distribution in the rat. Anat Rec 184:246

Hikosaka O, Wurtz RH 1983 Visual and oculomotor functions of monkey substantia nigra pars reticulata. IV. Relation of substantia nigra to superior colliculus. J Neurophysiol 49:1285-1301

Illing RB, Graybiel AM 1983 Organisation and interrelationships of frontal, nigral and bulbospinal afferents to the superior colliculus and periaqueductal gray matter in the cat. Soc Neurosci Abstr 9:1084

Jayaraman A, Batton RR III, Carpenter MB 1977 Nigrotectal projections in the monkey: an autoradiographic study. Brain Res 135:147-152

Künzle H, Akert K 1977 Efferent connections of cortical area 8 (frontal eye field) in *Macaca fascicularis*. A reinvestigation using the autoradiographic technique. J Comp Neurol 173:147-164

Ueki A 1983 The mode of nigro-thalamic transmission investigated with intracellular recording in the cat. Exp Brain Res 49:116-124

Ueki A, Uno M, Anderson ME, Yoshida M 1977 Monosynaptic inhibition of thalamic neurons produced by stimulation of the substantia nigra. Experientia 33:1480-1481

Uno M, Yoshida M 1975 Monosynaptic inhibition of thalamic neurons produced by stimulation of the pallidal nucleus in cats. Brain Res 99:377-380

Uno M, Ozawa N, Yoshida M 1978 The mode of pallido-thalamic transmission investigated with intracellular recording from cat thalamus. Exp Brain Res 33:493-507

Functional organization of the basal ganglia: contributions of single-cell recording studies

M. R. DELONG, A. P. GEORGOPOULOS, M. D. CRUTCHER, S. J. MITCHELL, R. T. RICHARDSON and G. E. ALEXANDER

Departments of Neurology and Neuroscience, The Johns Hopkins University School of Medicine, Baltimore, Maryland 21205, USA

Abstract. Studies of single-cell discharge in the basal ganglia of behaving primates have revealed: (1) characteristic patterns of spontaneous discharge in the striatum, external (Gpe) and internal (Gpi) globus pallidus, pars reticulata and pars compacta of the substantia nigra, and the subthalamic nucleus (STN); (2) phasic changes in neural discharge in relation to movements of specific body parts (e.g. leg, arm, neck, face); (3) short-latency (sensory) neural responses to passive joint rotation; (4) a somatotopic organization of movement-related neurons in Gpe, Gpi, and STN; (5) a clustering of functionally similar neurons in the putamen and globus pallidus; (6) a greater representation of the proximal than of the distal portion of the limb; (7) changes in neural activity in reaction-time tasks, suggesting a greater role of the basal ganglia in the execution than in the initiation of movement in this paradigm; (8) a clear relation of neuronal activity to direction, amplitude (?velocity) of movement, and force; (9) a preferential relation of neural activity to the direction of movement, rather than to the pattern of muscular activity. Some of these findings suggest that the basal ganglia may play a role in the control of movement parameters rather than (or independent of) the pattern of muscular activity. Loss of basal ganglia output related to amplitude may account for the bradykinesia in Parkinson's disease. The presence of somatotopic organization in the putamen and globus pallidus, together with known topographic striopallidal connections, suggests that segregated, parallel cortico-subcortical loops subserve 'motor' and 'complex' functions.

1984 Functions of the basal ganglia. Pitman, London (Ciba Foundation symposium 107) p 64-82

The technique of single-cell recording in behaving animals provides a direct, fine-grained and powerful approach to the understanding of the functions of the basal ganglia. In this review we briefly summarize our recent studies in behaving primates (*Macaca mulatta*) and discuss the relevance of these findings to basal ganglia function and organization.

Neuronal discharge patterns

The frequency and pattern of neuronal discharge are strikingly different in the various nuclei of the primate basal ganglia, as illustrated in Fig. 1. In the striatum most cells exhibit very low discharge rates (DeLong & Strick 1974, Crutcher & DeLong 1984a), whereas most neurons in the globus pallidus and the pars reticulata of the substantia nigra (SNr) discharge tonically at high rates (DeLong 1971, DeLong & Georgopoulos 1979, DeLong et al 1983). The similarities in discharge patterns of neurons in SNr and in the inner segment of the globus pallidus (GPi), together with their close morphological and anatomical similarities (see DeLong & Georgopoulos 1979, 1981), led us to propose that the SNr and GPi represent medial and lateral portions, respectively, of a single structure, which has been subdivided by the internal capsule. In an analogous manner, the similarity of rates and patterns of discharge between cells in the laminae of the globus pallidus ('border' cells) and cells in the nucleus basalis of Meynert (NBM) suggested that 'border' cells might correspond to displaced neurons of the NBM located within the laminae of the globus pallidus (DeLong 1971). The proximity of the NBM and globus pallidus, and the considerable intermingling of their neurons, especially in the rat and cat, potentially complicate the interpretation of lesion, stimulation and single-cell recording studies. It is therefore of the utmost importance to distinguish between these different populations of cells (DeLong 1971, DeLong & Georgopoulos 1981).

The striking differences in the patterns of neuronal activity in the individual nuclei of the basal ganglia not only provide useful and reliable electrophysiological landmarks but may also in themselves ultimately provide important clues about the operations in each structure. As yet, however, only general statements can be made about the nature of these functions. The sustained high-frequency discharge of output neurons of the basal ganglia in GPi and SNr, for example, indicates a continuous tonic influence of these nuclei on their target neurons in the thalamus and brainstem. On this tonic discharge a phasic modulation may be superimposed during movement as a result of phasic inputs from the putamen and subthalamic nucleus.

Neuronal relations to movement and functional organization

In each nucleus of the basal ganglia, except the caudate nucleus and the pars compacta of the substantia nigra (SNc), a clear correlation between cell discharge and active movements of specific body parts, e.g. leg, arm and orofacial structures, has been found (DeLong 1971, DeLong & Georgopoulos 1979, DeLong et al 1983, Crutcher & DeLong 1984a). A further finding of these studies was that cells related to movements of particular body parts were grouped together within each structure in a somatotopic manner. Thus, in the putamen, as shown in Fig. 2, neurons whose activity was related to leg, arm and face movements were located in those regions receiving afferents from the leg, arm and face areas of sensory and motor cortices respectively

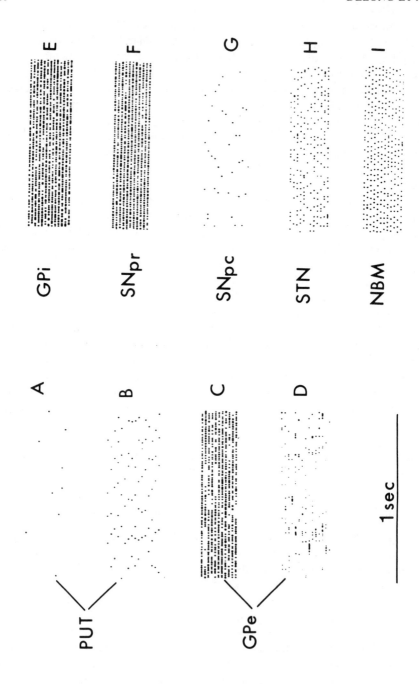

(see Fig. 7 in Künzle 1977). In both GPe and GPi a similar somatotopic organization is present: cells related to leg movements are located dorsally, those related to orofacial movements ventrally, and arm-related cells between them. A somatotopic organization of movement-related cells has also been found in the primate subthalamic nucleus, which closely corresponds to that predicted from the anatomical projections from the motor cortex (Hartmann-von Monakow et al 1978). In the SNr, neurons related to orofacial movements have been observed in the lateral portions of the nucleus. The somatotopic grouping of the neurons within these structures probably results from the orderly topographic projections to the basal ganglia from the motor and sensory cortices (Künzle 1975, 1977) and the topographic projections between the nuclei of the basal ganglia (see DeLong & Georgopoulos 1981 for a review).

In contrast to cells in other nuclei of the primate basal ganglia, only a small proportion of cells in the SNc exhibited significant phasic changes in discharge in relation to active (or passive) movements of the limbs, and even these neurons did not appear to encode specific information about movement parameters (DeLong & Georgopoulos 1979, DeLong et al 1983). This finding suggests that the nigrostriatal dopamine system, rather than conveying specific information about movement, may exert a general tonic modulatory effect on the striatum. In a recent study Schultz et al (1983) observed modulation of a larger proportion of SNc neurons in relation to vigorous proximal (but not distal) limb movements of large amplitude. As in our studies, however, there did not appear to be any specific relation of neuronal discharge to parameters of movement.

FIG. 1. Examples of characteristic neural activity in the different nuclei of the basal ganglia of the primate, *Macaca mulatta*. Neural activity is shown in raster form during the control period of a visuomotor step-tracking task (Georgopoulos et al 1983) in which the animal held a handle within a central window while waiting for the target to move. Each line corresponds to a trial and each dot corresponds to a single spike. In the putamen two characteristic types of activity are found: (A) neurons with very low discharge rates (<1/s) and (B) cells with slow (4–8/s) tonic discharge. The latter are far less frequent than the former. In the external segment of globus pallidus (GPe) two types of activity can be distinguished: one (C), the most common, with recurrent periods of high-frequency discharge separated by periods of silence, and another (D), less frequent, with a low mean frequency of discharge but with intermittent brief high-frequency bursts of spikes. Most cells in the inner pallidum (GPi) (E) show a sustained, high-frequency discharge without pauses. Cells in the primate substantia nigra (SN) discharge at different rates according to their location (DeLong et al 1973). Most neurons in the pars reticulata (SNpr) discharge continuously at high rates (F) with a pattern similar to those of GPi, whereas neurons located in the pars compacta (SNpc) discharge at rates of less than 10/s(G). Most cells in the subthalamic nucleus (STN) (H) discharge tonically, with a tendency to fire in doublets or triplets. Most neurons located below GP in the nucleus basalis of Meynert (NBM) and also within the laminae of GP ('border' cells; DeLong 1971)(I) have a distinctive regular pattern of discharge.

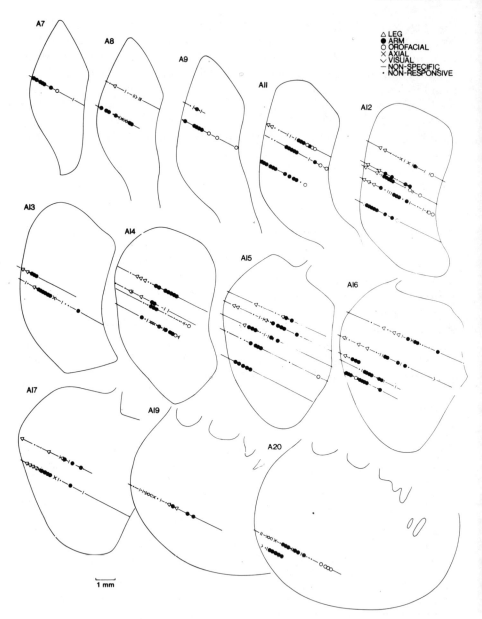

FIG. 2. Locations of neurons related to movements of different body parts in one monkey. Data from both hemispheres are plotted on outline drawings of coronal sections of the left putamen. Each drawing shows the locations of all the neurons studied within 0.5 mm of that anteroposterior level, ranging from anterior 7 (A7) to anterior 20 (A20). For each section, lateral is to the left. (Reproduced, with permission, from *Experimental Brain Research*.)

A striking finding of the studies in the putamen, globus pallidus and subthalamic nucleus has been the considerable anteroposterior extent of the leg, arm and face motor representation, as illustrated for the putamen in Fig. 2. This distribution probably results from the long anteroposterior extent of terminal fields of projections to the putamen from a given region of the sensory and motor cortices (Jones et al 1977, Künzle 1975, 1977). On both anatomical and physiological grounds, therefore, it appears that the sensorimotor representation of leg, arm and face in the basal ganglia consists of separate longitudinal zones related to individual body parts. Input to a given area of the putamen, for example the arm area, appears to come from the different arm areas in somatotopically organized motor, premotor, somatosensory and parietal cortices (see DeLong & Georgopoulos 1981, p 1052-1054, Jones et al 1977).

Within the putamen, clustering of neurons with similar functional properties has been observed (Crutcher & DeLong 1984a, Liles 1979). As illustrated in Fig. 3, discrete clusters of two to five neurons with similar relations to active movements or responses to somatosensory stimulation were typically encountered over a distance of 100–500 µm along a given penetration. Different clusters of leg, arm and orofacial neurons were seen throughout most of the anteroposterior extent of the putamen. Some clusters contained neurons whose activity was related only to active arm movements, while others contained neurons related to both passive and active arm movements. Clusters of neurons with sensory driving were organized by joints, i.e. all or most neurons in a cluster responded best to passive movements of the same joint. Within the 'arm' area of the putamen there appear to be multiple clusters of neurons related to each joint. Additional evidence for this type of clustered, somatotopic organization includes the recent demonstration of discrete foci within the putamen from which motor responses can be evolved by microstimulation (Alexander & DeLong 1983). The somatotopic organization of these low-threshold motor-response zones within the putamen is identical to that revealed by the functional analysis of neuronal sensorimotor response properties.

It remains to be determined whether the clusters of functionally similar neurons in the putamen correspond with (1) the patches of terminal label characteristic of corticostriate and thalamostriate projections (Jones et al 1977, Goldman & Nauta 1977), (2) the cellular islands of the striatum described by Goldman-Rakic (1982) or (3) the patterns of dopamine histofluorescence, acetylcholinesterase activity and enkephalin immunoreactivity (Graybiel et al 1981, Graybiel & Hickey 1982). The findings of the single-cell studies suggest that there may be a functional correlate to the anatomically defined patches. It is possible, therefore, that the clusters of functionally similar neurons may represent the basic functional units of the

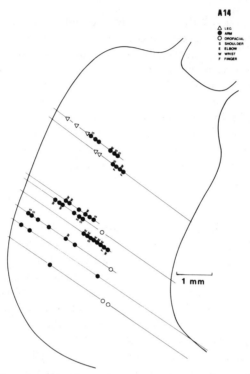

FIG. 3. Locations of neurons related to active movements or passive manipulations of the leg, arm, mouth or face at a single coronal level (anterior 14) from one hemisphere of a different animal from that in Fig. 1. Arm symbols with adjacent letters indicate the locations of neurons responsive to passive manipulations of the shoulder (S), elbow (E), wrist (W), or fingers (F). Arm symbols without adjacent letters show the locations of neurons related to active, but not passive, arm movements. Lateral is to the left. (Reproduced, with permission, from *Experimental Brain Research*.)

striatum, analogous to the functional columns of the neocortex (Crutcher & DeLong 1984a).

Neuronal responses to somatosensory stimulation

A significant proportion of neurons in the putamen, globus pallidus and subthalamic nucleus which are related to active movements of specific body parts also respond to somatosensory stimulation of the same body part (Crutcher & DeLong 1984a, DeLong & Georgopoulos 1979). For the putamen, 41% of cells related to arm movement responded to somatosensory stimuli. Essentially all parts of each limb are represented by putamen neurons with response areas ranging from one to several joints. However, driving from the proximal portions of the arm was more commonly observed than from the distal portions. Of 112 cells responding to somatosensory stimulation, 43%, 31%, 13% and 13% were best related to manipulations of the shoulder, elbow, wrist and hand respectively. The preponderance of somatosensory responses was from deep rather than superficial structures and the majority of responses were to joint rotation. Of the same 112 arm neurons, 92 responded to joint rotation, 10 to muscle palpation, and 4 responded to tapping of tendons or muscles. Only six neurons had cutaneous receptive fields in the glabrous skin of the hand and none responded to light touch of the hairy skin of the arm. The responses of 'arm' neurons to controlled passive displacements of the elbow produced by application of a load during a behavioural task were also studied. Of the 38 neurons which responded to passive manipulations of the elbow or shoulder, 74% responded to load application at latencies between 25 and 50 ms. These short-latency 'sensory' responses typically exhibited highly specific directional and amplitude relations. These findings indicate that the putamen receives somatosensory inputs of a specific and restricted nature, which may be used to control or monitor ongoing movements.

Onset times of neuronal discharge

It has been found in reaction-time tasks that most changes in neuronal discharge in the basal ganglia begin before the onset of movement but after the first electromyographic changes, although changes in discharge in about one-fourth of the cells begin before the first EMG changes (Crutcher & DeLong 1984a, Anderson & Horak 1981, DeLong & Georgopoulos 1981, Georgopoulos et al 1983). Overall, changes in neuronal discharge seem to occur later in the basal ganglia than in the motor cortex. For example, about 50% of the cells in the motor cortex changed activity before the earliest EMG changes (about 80 ms before movement) in reaction-time tasks (Evarts 1974, Thach 1978); this percentage was respectively 24%, 11% and 29% in GPe, GPi and subthalamic nucleus (Georgopoulos et al 1983), and 19% in the putamen (Crutcher & DeLong 1984b). It appears, therefore, that neuronal activation in the motor cortex may precede activation in the movement-related portions of the basal ganglia. This is not unexpected since a major input to these portions of the basal ganglia comes from the motor cortex. It is interesting, in this respect, that microstimulation of the globus pallidus was found to slow an ensuing movement when delivered after the first changes in EMG activity but before movement began (Horak & Anderson 1980). Moreover, lesions (Horak & Anderson 1980) and cooling (Hore & Villis

1980) of globus pallidus resulted in slowing of movements without affecting reaction times. Taken together, these studies are consistent with the view that the basal ganglia in this paradigm are more involved in the execution or facilitation of movement than in the timing of its initiation (Anderson & Horak 1981, Aldridge et al 1980, Hallett & Khoshbin 1980, DeLong et al 1983, Crutcher & DeLong 1984b). It is possible, however, that the small population of neurons that are active early in such reaction-time tasks may play some role in the initiation of movement and that, in other types of paradigms (e.g. self-initiated movements), or in different portions of the basal ganglia (e.g. the caudate), neurons may be involved more specifically in movement initiation.

Relations to movement parameters

In order to evaluate quantitatively an earlier finding (DeLong 1971) of a relationship of neural activity in basal ganglia neurons to the direction of arm movements, and a possible relationship of neural activity to the speed of movement (DeLong & Strick 1974), we examined the relation of neuronal discharge to movement parameters in animals trained to perform a step-tracking task in which the amplitude, speed and direction of movement were varied (DeLong & Georgopoulos 1979, Georgopoulos et al 1983, DeLong et al 1983). We recorded neuronal activity in GPe, GPi, subthalamic nucleus and substantia nigra. Cells were studied only if they were related to arm movements made by the animal outside the behavioural task. Significant neuronal relations to both the direction and amplitude of movement were observed in GPe, GPi and subthalamic nucleus during both the movement time and the initial premovement time (i.e. the 100 ms before movement began), the period during which most of the changes in EMG began to occur. In general, the frequency of cell discharge was a linear function of the movement amplitude. The incidence of significant amplitude effects was highest in the movement time, but the effects were also present in the initial premovement time. The effects of movement amplitude became most apparent when a wide range of amplitudes was used. The relations between cell discharge and peak velocity of movement in the step-tracking task were similar to those described above for the amplitude of movement. This is not surprising since amplitude and peak velocity were highly correlated.

The finding of significant directional effects in the globus pallidus can be explained by the observation that neurons in the putamen (Crutcher & DeLong 1984b) and the subthalamic nucleus (Georgopoulos et at 1983), which both project to globus pallidus, are strongly related to the direction of movement. The directional relations in the putamen and subthalamic nucleus can, in turn, be accounted for by the inputs to these structures from the cerebral cortex, since both precentral and parietal cortical cells show significant relations to the direction of movement (Evarts 1966, Schmidt et al 1975, Georgopoulos et al 1982, Kalaska et al 1983).

To determine whether the activity of neurons in the basal ganglia is related to the direction of movement *per se*, or to the activity of muscles, monkeys (*Macaca mulatta*) were trained in a separate study (Crutcher & DeLong 1984b) to perform a visuomotor tracking task which required elbow flexion/extension movements with assisting and opposing loads. This task dissociated the direction of arm movement from the pattern of muscular activity. Of 120 neurons related to arm movement in the putamen, 50% were related to the direction of arm movement whereas only 13% showed a pattern of activity 'like muscle'. These results indicate that neurons in the putamen are predominantly related to the direction of arm movement rather than to the activity of individual muscles. For the majority of 'directional' neurons these relations to movement did not appear to result from proprioceptive feedback. In an ongoing study in our laboratory similar relations to direction of movement have been found in globus pallidus (Mitchell et al 1983), as shown in Fig. 4. These results, together with those related to amplitude of movement (Georgopoulos et al 1983), indicate that the basal ganglia may be involved in the control (or monitoring) of parameters of movement, rather than muscles.

In addition to the neuronal relations to movement amplitude and direction, several studies (Crutcher & DeLong 1984b, DeLong 1972) have shown a relation to steady and dynamic load. These effects, however, appear to be less strong than those in the motor cortex (see Crutcher & DeLong 1984b for a review).

The neuronal responses to movement can be regarded as the net result of various factors, including the direction and amplitude of movement. It is possible that a change in cell discharge is related to movement direction, and that this change is further modulated according to the amplitude of movement. These relations to the direction and amplitude of movement may be separately controlled, since many neurons showed significant directional effects without amplitude effects, and others showed amplitude-related changes for only one direction of movement. These results may have implications for the broader issue of cerebral control of movement, since the observed neuronal relations in the putamen may largely reflect the nature of the inputs to this structure from the motor, premotor, and somatosensory cortices. It is therefore possible that similar neural relations to parameters of movement, independent of muscular activity, may also be found in these areas of the cortex. In fact, a dissociation between muscular pattern and direction of intended movement was observed by Thach (1978) in the motor cortex. The basal ganglia may, therefore, function as a component of a more distributed system controlling parameters of movement.

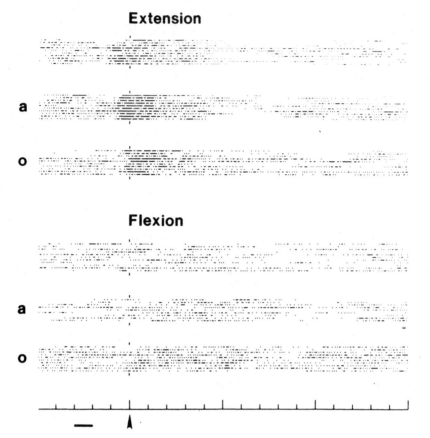

FIG. 4. Raster display of neuronal activity in GPe related to direction of movement during performance of phasic flexion and extension elbow movements under conditions of no load (top block of trials in each category), assisting (a), and opposing (o) loads. The arrow indicates onset of movement as determined from the first change in velocity. For each trial the stimulus to move occurred during the period indicated by the horizontal bar (i.e. 200-300 ms before movement onset). Each division of the time scale is 100 ms.

Of potential clinical relevance is the finding of significant neural relations to the amplitude of movement. This may explain the observation that patients with diseases of the basal ganglia frequently have difficulty in controlling the amplitude of their limb movements. For example, in patients with Parkinson's disease, single-step movements of large amplitude are impaired: these movements fall short of the target (Flowers 1978), which is ultimately reached by a series of small-amplitude movements (Draper & Johns 1964, Flowers 1978). The mechanism of this phenomenon was partially elucidated by Hallett

& Khoshbin (1980), who observed that parkinsonian patients were unable to increase the amplitude of the agonist burst in step-tracking movements. Thus, large-amplitude movements were achieved by several small-amplitude steps. Loss of pallidal output related to amplitude may account for the overall bradykinesia of parkinsonian patients. On the other hand, abnormal modulation of this output may lead the uncontrolled movements of patients with involuntary movements, such as chorea and hemiballismus.

Cortico-basal ganglia relations

The finding that the somatotopic motor representation established in the putamen is subsequently maintained in the globus pallidus suggested that there exist segregated pathways through the basal ganglia for the control of different body parts. These findings led us (DeLong & Georgopoulos 1981) to

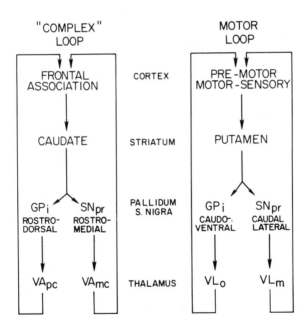

FIG. 5. Schematic depiction of the postulated segregation of pathways from the 'association' (complex loop) and the sensorimotor areas (motor loop) through the basal ganglia and thalamus. VA_{pc}, n. ventralis anterior pars compacta; VA_{mc}, n. ventralis anterior pars magnocellularis; VL_o, n. ventralis pars oralis; VL_m, ventralis lateralis pars medialis. (Reproduced, with permission, from *Experimental Brain Research*.)

re-examine the evidence for the widely held view that the basal ganglia serve as a 'funnel' from association areas to the motor cortex (Evarts &Thach 1969, Kemp & Powell 1970). On reviewing the relevant anatomical studies, we concluded that the available evidence is most consistent with the view that influences from the sensorimotor and premotor cortices are ultimately directed, via pathways through the putamen, pallidum and thalamus, largely on premotor areas (area 6), whereas influences from the association areas are directed, via a separate pathway passing through the caudate, pallidum and thalamus, to the prefrontal cortex. We thus proposed the existence of segregated, parallel cortico-subcortical loops subserving 'motor' and 'complex' functions, as shown schematically in Fig. 5. In this scheme, just as the leg, arm and face representations remain segregated throughout the cerebral cortex, basal ganglia and thalamus, so are the influences from the cortical association areas separately routed through these subcortical nuclei by virtue of non-overlapping, topographically organized projections. There appears to be little or no convergence within the basal ganglia of pathways which originate from the association areas and those from the somatotopically organized sensorimotor areas. To what extent such convergence may occur at thalamic or cortical levels remains uncertain. This view of the functional organization of the basal ganglia provides a framework wherein disturbances, not only in motor but also in more complex behaviour, may result from damage to different portions of basal ganglia nuclei.

REFERENCES

Aldridge JW, Anderson RJ, Murphy JT 1980 Sensory-motor processing in the caudate nucleus and globus pallidus: a single-unit study in behaving primates. Can J Physiol Pharmacol, 58:1192-1201
Alexander, GE, DeLong MR 1983 Motor responses to microstimulation of the putamen in the awake monkey. Soc Neurosci Abstr 9:16
Anderson ME, Horak FB 1981 Changes in firing of pallidal neurons during arm reaching movements in a reaction time task. Soc Neurosci Abstr 7:241
Crutcher MD, DeLong MR 1984a Single cell studies of the primate putamen. I. Functional organization. Exp Brain Res 53:233-243
Crutcher MD, DeLong MR 1984b Single cell studies of the primate putamen. II. Relations to direction of movements and pattern of muscular activity. Exp Brain Res 53:244-258
DeLong MR 1971 Activity of pallidal neurons during movement. J Neurophysiol 34:414-427
DeLong MR 1972 Activity of basal ganglia neurons during movement. Brain Res 40: 127-135
DeLong MR, Georgopoulos AP 1979 Motor functions of the basal ganglia as revealed by studies of single cell activity in the behaving primate. Adv Neurol 24:131-140
DeLong MR, Georgopoulos A 1981 Motor functions of the basal ganglia. In: Brookhart JM et al (eds), The nervous system. American Physiological Society, Bethesda, MD (Handb Physiol) p 1017-1061
DeLong MR, Strick PL 1974 Relation of basal ganglia, cerebellum, and motor cortex units to ramp and ballistic limb movements. Brain Res 71:327-335

DeLong MR, Crutcher MD, Georgopoulos AP 1983 Relations between movement and single cell discharge in the substantia nigra of the behaving monkey. J. Neurosci 3:1599-1606

Draper, IT, Johns RJ 1964 The disordered movement in parkinsonism and the effect of drug treatment. Bull Johns Hopkins Hosp 115:465-480

Evarts EV 1966 Pyramidal tract activity associated with a conditioned hand movement in the monkey. J Neurophysiol 29:1011-1027

Evarts EV 1974 Precentral and postcentral cortical activity in association with visually triggered movement. J Neurophysiol 37:373-381

Evarts EV, Thach WT 1969 Motor mechanisms of the CNS: cerebrocerebellar interrelations. Annu Rev Physiol 31:451-498

Flowers K 1978 Some frequency response characteristics of parkinsonism on pursuit tracking. Brain 101:19-34

Georgopoulos AP, Kalaska JF, Caminiti R, Massey JT 1982 On the relations between the direction of two-dimensional arm movements and cell discharge in primate motor cortex. J Neurosci 2:1527-1537

Georgopoulos AP, DeLong MR, Crutcher MD 1983 Relations between parameters of step-tracking movements and single cell discharge in the globus pallidus and subthalamic nucleus of the behaving monkey. J Neurosci 3:1586-1598

Goldman PS, Nauta WJH 1977 An intricately patterned prefronto-caudate projection in the rhesus monkey. J Comp Neurol 171:369-386

Goldman-Rakic PS 1982 Cytoarchitectonic heterogeneity of the primate neostriatum: subdivision into island and matrix cellular compartments. J Comp Neurol 205:398-413

Graybiel AM, Hickey TL 1982 Chemospecificity of ontogenetic units in the striatum: demonstration by combining ^{3}H-thymidine neuronography and histochemical staining. Proc Natl Acad Sci USA 79:198-202

Graybiel AM, Pickel VM, Joh TH, Reis DJ, Ragsdale CW Jr 1981 Direct demonstration of a correspondence between the dopamine islands and acetylcholinesterase patches in the developing striatum. Proc Natl Acad Sci USA 78:5871-5875

Hallett M, Khoshbin S 1980 A physiological mechanism of bradykinesia. Brain 103:301-314

Hartmann-von Monakow K, Akert K, Künzle H 1978 Projections of the precentral motor cortex and other cortical areas of the frontal lobe to the subthalamic nucleus in the monkey. Exp Brain Res 33:395-403

Horak FB, Anderson ME 1980 Effects of of the globus pallidus on rapid arm movement in monkeys. Soc Neurosci Abstr 6:465

Hore J, Villis T 1980 Arm movement performance during reversible basal ganglia lesions in the monkey. Exp Brain Res 39:217-228

Jones EG, Coulter JD, Burton H, Porter R 1977 Cells of origin and terminal distribution of corticostriatal fibers arising in the sensory-motor cortex of monkeys. J Comp Neurol 173:53-80

Kalaska JF, Caminiti R, Georgopoulos AP 1983 Cortical mechanisms related to the direction of two-dimensional arm movements: relations in parietal area 5 and comparison with motor cortex. Exp Brain Res 51:247-260

Kemp JM, Powell TPS 1970 The corticostriate projection in the monkey. Brain 93:525-546

Künzle H 1975 Bilateral projections from precentral motor cortex to the putamen and other parts of the basal ganglia. An autoradiographic study in Macaca fascicularis. Brain Res 88:195-209

Künzle H 1977 Projections from the primary somatosensory cortex to the basal ganglia and thalamus in the monkey. Exp Brain Res 30:481-492

Liles SL 1979 Topographic organization of neurons related to arm movement in the putamen. Adv Neurol 23:155-162

Mitchell SJ, Richardson RT, Baker FH, DeLong MR 1983 Activity of neurons in the globus pallidus in relation to direction of movement or pattern of muscular activity. Soc Neurosci Abstr 9:951

Schmidt EM, Jost RG, Davis KK 1975 Reexamination of the force relationship of cortical cell discharge patterns with conditioned wrist movements. Brain Res 83:213-223

Schultz W, Ruffieux A, Arbisher P 1983 The activity of pars compacta neurons of the monkey substantia nigra in relation to motor activation. Exp Brain Res 51:377-387

Thach WT 1978 Correlation of neural discharge with pattern and force of muscular activity, joint position and direction of intended next movement in motor cortex and cerebellum. J Neurophysiol 41:654-676

DISCUSSION

Smith: Have you studied motivational behaviour as well as tracking behaviour in your studies of the pars compacta cells?

DeLong: No. We were interested in neuronal activity only in relation to movement. We saw a few cells that varied their activity in relation to the animal's state but we didn't investigate them systematically. W. Schultz has seen changes when a stimulus was presented to an animal sitting quietly: it made more vigorous movements. The role of the basal ganglia in orienting movements might be more interesting than what we saw without orienting.

Rolls: You gave the impression that primarily there are movement-related neurons in the striatum and in the pallidum. Have you seen neurons in the pallidum that you would describe as not related to movement? When we record in the tail of the caudate nucleus, for example, which receives inputs from the inferior temporal visual cortex, we find many visual neurons which habituate rapidly (Caan et al 1984). In the head of the caudate nucleus we find many neurons that respond to environmental cues (Rolls et al 1983, Rolls 1984); again these are not simply movement-related. In the ventral striatum we find neurons that respond to visual stimuli (Rolls et al 1982, Rolls 1984). Are these forms of activity represented in separate parts of the pallidum, and is segregation of these sorts of function maintained through the striatum to reach different cortical areas via the pallidum and thalamus?

DeLong: I agree that many cells in these nuclei respond to external stimuli and are not simply movement-related. Basically the pathway that we are looking at comes through the putamen. The putamen receives its input from the motor, premotor and somatosensory areas. The putamen and those regions of the pallidum which receive putamen projections seemed to us the best place to look for movement-related activity. The caudate nucleus receives primarily association cortical input and that seems to be channelled through more rostral-medial portions of the pallidum and nigra. Thus, the information from association areas seems to be segregated from that of the sensorimotor areas. It seems likely that the movement-related cells in the more caudoventral regions of the inner pallidal segment are, ultimately, influencing premotor areas, particularly the supplemental motor area, whereas activity in the pathway through the caudate and rostrodorsal pallidum is influencing more rostral prefrontal regions. That is the basic split that I wanted to show. We have focused on the regions which by virtue of their known inputs appear to be most closely related to movement.

Marsden: But in other parts of the globus pallidus which do not receive information from the motor putamen have you detected units concerned with behaviour other than movement?

DeLong: We haven't looked yet. We went out of our way to identify cells related to arm movements both within the task and with the arm outside the manipulandum. Our real concern was to study cells that were both task-related and specifically related to arm movement.

Rolls: For neurology one needs as much information from neurophysiology as possible about whether this is a movement-related system. If it is not just this, one might expect to find, in the parkinsonian patient, more complex types of disorder. These might be cognitive disorders or other disorders related to the functions of the prefrontal cortex from which the head of the caudate nucleus receives such major inputs. Our work suggests that, at least as far as the striatum, there is the possibility of segregation according to which cortical area each part of the striatum receives inputs from (Rolls 1983, 1984). So there must be a possibility of different sorts of disorder.

DeLong: I think we are in complete agreement.

Wing: One way to address the question of whether dopamine-innervated pathways subserve cognitive processes other than those directly concerned with movement is to look at cognitive and perceptual function in relation to variations in the remediation of bradykinesia by L-dopa. R.D. Rafal, M.I. Posner and J.A. Walker (unpublished) in Portland, Oregon, have studied patients with end-of-dose akinesia and observed that cognitive functions, including orienting of attention and memory scanning, were unchanged during the parkinsonian off-medication state, despite slowing of movement.

Rizzolatti: When you record from that part of the putamen which receives its main input from motor and premotor areas, do you find visually responsive neurons? It is now well demonstrated (Rizzolatti et al 1981, Godschalk et al 1981, Weinrich & Wise 1982) that many neurons in the cortical premotor areas respond to visual stimuli. What happens in the putamen?

DeLong: We have not seen any activity related to visual inputs in the parts of putamen that we have looked at. But there is a caudal and ventral region in particular that we have not explored, and we haven't looked in the caudate.

Rolls: In the most ventral part of the putamen there are some visual responses, although that may be the part which receives primarily from the temporal lobe cortex (Caan et al 1984).

Calne: Have you any observations on the ipsilateral side, Dr DeLong?

DeLong: In early studies the best correlations in almost every cell were with contralateral arm movements, but about 20% of cells showed changes of a similar pattern with ipsilateral arm movements. If one looked at this more carefully a much greater percentage might be seen to respond to ipsilateral movements. It is likely that if one used a reaching movement, a greater proportion of neurons would respond to both ipsilateral and contralateral movement.

Calne: Have you any observations on pharmacological manipulation, for example with reserpine or 6-hydroxydopamine?

DeLong: At one time we looked at animals treated with reserpine and phenothiazine to see whether the striatum would show a clear change in spontaneous activity when we blocked the dopamine system. We were very disappointed not to find significant changes.

Calne: Did you look at the pallidum in those experiments?

DeLong: There didn't seem to be any obvious changes but we didn't pursue those studies.

Kitai: Earlier Dr Deniau told us that one of the functions of the striatum is to inhibit the globus pallidus. Obviously the pallidal neurons you are recording from, Dr DeLong, may be affected by this inhibition. Yet all the movement-related cells are excitatory cells; that is, the ones which fire. Have you looked at pallidal cells which get inhibited when movement is initiated?

Evarts: The initial change you see in the striatum is an increase in impulse frequency. The paradox is that you don't see a reciprocal change in the globus pallidus.

DeLong: Many cells in both segments of the pallidum show an initial increase in activity that is well sustained. Some cells show a decrease. That has been a consistent finding in these studies. We are trying to reconcile that with what is known about striatal inputs. It is rather difficult to do. The alternative possibility is that it is the pathway to the subthalamic nucleus that is producing the early increase in activity. If that pathway were excitatory that would give us an easy answer.

Yoshida: We have made recordings from single units in the caudate nucleus of free-moving cats. A lamp and a buzzer were used as conditioning stimuli for cats in a shuttle box with two sections. After the stimuli the animal received an electrical shock unless it jumped over the partition into the other chamber. When the cat was just ready to jump, caudate neurons began to increase their firing rate. The increase continued while the animal was jumping. Some neurons also increased their firing rate when the cat started to turn its head.

Even if the cat did not move at all, many neurons increased their firing rate when the lamp and buzzer went on. This might reflect a preparatory state, or attention or motivation. Cerebellar units (interpositus) did not change their firing rate at all when the stimuli were switched on but these units were very sensitive to movement. [Illustrated by a three-minute 16 mm movie.]

Marsden: Dr DeLong, you mentioned that the caudate or putamen units, or both, might be under some degree of peripheral proprioceptive control. Are the basal ganglia an internal system dealing with the brain itself or are they controlled by peripheral information from limbs? There seems to be some controversy between your group and Dr Porter's group on whether globus pallidus neurons are proprioceptively controlled. If they are, is the timing of the discharge in relation to a peripheral passive movement sufficient to result from transmission of proprioceptive information to the cortex and then from cortex back to basal ganglia? In other words is this an indirect effect from the cortex coming back into the basal ganglia?

DeLong: The latencies are consistent with relay through the somatosensory and motor cortex. The changes there occur before the earliest ones in the basal ganglia. There is no evidence of any direct sensory input to the basal ganglia from the periphery, so this has to be processed through the cortex.

Porter: The conditions in which one studies afferent input, particularly if it has to go through a rather long circuit involving the cerebral cortex, as you indicated, will determine in large part the results one obtains from that test. There are plenty of structures in the central nervous system, but most importantly the cerebral cortex itself, that can control inputs through relays from the spinal cord level through the thalamus and at every stage of afferent transmission. It is clear from the 'observations that Bob Iansek (1980) made in my laboratory that, if an animal is made akinetic with repeated doses of reserpine, it is possible to find quite dramatic inputs from manipulation of its limbs when one records from cells in the globus pallidus. After the same animal has recovered from the effects of the drug and is moving, it is very difficult to recognize those inputs, presumably because some different form of processing the afferent information is now possible, such as the controls that the cortex itself exerts on inputs to structures above the foramen magnum (Iansek & Porter 1980).

DeLong: I agree that the conditions of the experiment are important in determining the findings. The other matter is where the neurons responding to somatosensory stimulation are located in the globus pallidus. We have found the majority to be located in the caudoventral portions of each segment. Cells in the more rostromedial portions are much less likely to respond to such stimuli.

REFERENCES

Caan W, Perrett DI, Rolls ET 1984 Responses of striatal neurons in the behaving monkey. 2. Visual processing in the caudal neostriatum. Brain Res 290:53-65

Godschalk M, Lemon RN, Nijs HGT, Kuypers HGJM 1981 Behaviour of neurons in monkey peri-arcuate and precentral cortex before and during visually guided arm and hand movements. Exp Brain Res 44:113-116

Iansek R 1980 The effects of reserpine on motor activity and pallidal discharge in monkeys: implications for the genesis of akinesia. J Physiol (Lond) 301:457-466

Iansek R, Porter R 1980 The monkey globus pallidus: neuronal discharge properties in relation to movement. J Physiol (Lond) 301:439-455

Rizzolatti G, Scandolara C, Matelli M, Gentilucci M 1981 Afferent properties of periarcuate neurons in macaque monkeys. Behav Brain Res 2:147-163

Rolls ET 1983 The initiation of movements. Exp Brain Res Suppl 7:97-113

Rolls ET 1984 Activity of neurons in different regions of the striatum of the monkey. In: McKenzie J, Wilcox L (eds) The basal ganglia: structure and function. Plenum, New York

Rolls ET, Ashton J, Williams G et al 1982 Neuronal activity in the ventral striatum of the behaving monkey. Soc Neurosci Abstr 8:169

Rolls ET, Thorpe SJ, Maddison SP 1983 Responses of striatal neurons in the behaving monkey. 1: Head of the caudate nucleus. Behav Brain Res 7:179-210

Weinrich M, Wise SP 1982 The premotor cortex of the monkey. J Neurosci 2:1329-1345

Basal ganglia outputs and motor control

EDWARD V. EVARTS and STEVEN P. WISE

Laboratory of Neurophysiology, National Institute of Mental Health, Bethesda, Maryland 20205, USA

Abstract. Several lines of evidence suggest that the role of the basal ganglia in motor control is of a higher order than control of movements *per se*. First the striatum receives inputs from cortical areas that subserve mnemonic and other cognitive processes. Furthermore, the supplementary motor area (a zone that receives outputs from the globus pallidus via thalamus) exhibits changes in neuronal discharge and metabolic activity during movement planning as well as during movement. It is possible that this activity reflects its pallidal inputs. In addition, cells in another part of the basal ganglia, the pars reticulata of the substantia nigra, exhibit activity that reflects mnemonic as well as oculomotor and visual processes. Finally, there are striatal neurons that respond to stimuli when these stimuli trigger movement but not when responses to the stimuli are extinguished. Taken collectively, these observations are consistent with the view that the basal ganglia may provide an interface between motor centres and cortical areas for higher brain function.

1984 Functions of the basal ganglia. Pitman, London (Ciba Foundation symposium 107) p. 83-102

The role of the basal ganglia in motor control must be mediated by the structures that receive basal ganglia output, and this paper will consider some of these structures and their functions and summarize some recent observations concerning state or set-related sensory responses in the putamen, globus pallidus, and substantia nigra pars reticulata. First, we will contrast pallido-thalamocortical and cerebello-thalamocortical projections. It was once thought that these projections overlapped within the thalamus, but it is now known that they have largely separate thalamic and cortical targets. Next we will consider the physiological activity of the supplementary motor area (SMA) of the cerebral cortex, an area that receives a major input from globus pallidus via the thalamus. We will then discuss aspects of basal ganglia and cortical activity related to eye movements, some of which is contingent on the cognitive processes that accompany eye movements, and finally we will describe set-related responses in putamen neurons. In concluding, we will argue—based on the observations that both SMA and basal ganglia have physiological activity that is contingent on the cognitive state of the mover as well as on the movement *per se*—that the basal ganglia may provide a pathway whereby cortical regions (e.g. prefrontal and temporal cortex) involved in certain higher brain functions gain access to motor-control mechanisms (cf. DeLong et al 1984, this volume).

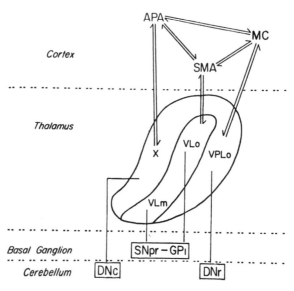

FIG. 1. Summary of anatomical relationships between cerebellar and basal ganglia efferents and primary and non-primary motor cortical areas. This diagram illustrates: (1) a pathway from caudal portions of the deep cerebellar nuclei (DNc), to nucleus X and thence the arcuate premotor area (APA); (2) pathways from the pars reticulata of the substantia nigra (SNpr) and the internal segment of the globus pallidus (GPi), to VLm and VLo and the supplementary motor area (SMA); (3) a pathway from rostral portions of the deep cerebellar nuclei (DNr), to VPLo and the primary motor cortex (MC); and (4) the reciprocal connections between the MC, APA and SMA. See text for details and thalamic abbreviations. (From Schell & Strick 1984.)

Basal ganglia outputs

Figure 1 shows the pathways from the globus pallidus internal segment (GPi) and the substantia nigra pars reticulata (SNr) to cortex via thalamus. The scheme is based on studies of Schell & Strick (1984) who used retrograde transport methods to trace the connections from the thalamus to the cortex, as well as on earlier studies of other investigators who have identified the thalamic terminations of basal ganglia and cerebellar outputs. Two participants in this conference (Professors Nauta and Carpenter) and E.G. Jones and his colleagues have provided much of our knowledge about where cerebellar and basal ganglia fibres terminate in the primate thalamus (see Tracey et al 1980, Carpenter 1981, and Nauta & Domesick, this volume).

The retrograde tracer injections that led to the schema of Fig. 1 were located in two non-primary motor areas and the arcuate premotor area as well as in the primary motor area. Retrograde transport from these three cortical areas showed that they receive inputs from separate thalamic sites. Inputs to SMA originate from the nucleus ventralis lateralis pars oralis (VLo), from the nucleus ventralis lateralis pars medialis (VLm), and from part of the nucleus medialis dorsalis (MD). GPi had been shown to project to VLo and portions of MD and VLm, while SNr had been shown to project to other parts of MD and VLm. Thus, both GPi and SNr have routes to SMA via the thalamus. The arcuate premotor area, in contrast, is projected upon by a thalamic zone called 'nucleus X' that in turn receives inputs from the caudal parts of the deep cerebellar nuclei. Tracer injections into the primary motor area labelled a different thalamic area, the nucleus ventralis lateralis posterolateralis pars oralis (VPLo), a nucleus that receives inputs from the rostral parts of the deep cerebellar nuclei. Thalamic nuclei ventralis anterior pars magnocellularis (VAmc) and ventralis anterior pars parvocellularis (VApc) also receive projections from basal ganglia: VApc receives inputs from GPi and VAmc receives them from SNr, but tracer injections into SMA fail to label VAmc or VApc. Künzle & Akert (1977) have shown that corticothalamic fibres from the frontal eye field end in VAmc and the lateral part of MD, and, assuming that the relationship is reciprocal, there is probably a route from SNr to the frontal eye field through both of these nuclei. In addition to its projections to the thalamus, SNr has a dense projection to the superior colliculus. Thus, SNr has access to both cortical and subcortical structures that control eye movements: the frontal eye field via the thalamus and the superior colliculus via direct projections.

In this paper the primary focus will be on behaviourally-related neuronal activity in SNr and in the cortical areas that receive inputs from GPi and SNr. But brief mention will be made of basal ganglia outputs to the limbic system and of certain species differences in the relative magnitudes of the limbic and non-limbic projections from the basal ganglia. Much of our knowledge about the basal ganglia is based on experiments in rodents and cats. Rodents have been especially important for pharmacological and neurochemical studies, and it was in mice that Carlsson et al (1957) discovered that L-dopa counteracted the catalepsy caused by reserpine. This discovery has since been confirmed in primates. Given the success of this extrapolation one hesitates to call attention to species differences, but certain of these will be noted here because they may be relevant to species differences in motor disturbances caused by basal ganglia dysfunction. In rats, basal ganglia outputs arise from the entopeduncular nucleus (the rodent homologue of GPi) and from the

SNr. The projection from the entopeduncular nucleus to the lateral part of the habenula is quite dense, involving virtually every neuron in certain parts of the entopeduncular nucleus (Herkenham & Nauta 1977), and the thalamic projections appear to be derived from a somewhat smaller proportion of neurons in the entopeduncular nucleus. For monkeys, in contrast, it is the projection to the thalamus that is predominant (Carpenter 1981). It is not that monkeys have lost connections from the basal ganglia to the habenula and the limbic system, but rather that monkeys have a much enlarged pallidothalamic projection. The predominance of the pallido-thalamocortical route for basal ganglia output in the monkey as compared to the predominance of output to limbic structures in the rat may help to explain some of the differences in the motor effects of basal ganglia disturbances in rodents as compared to primates. Thus, Marsden (1975, p 3) has noted that experimental parkinsonism can be produced in primates, but not in other species, and that the same is true for dyskinesias. He concluded that 'Although it is possible to study many aspects of basal ganglia function in laboratory rodents, much of the research into experimental parkinsonism and dyskinesias must be undertaken in primates, for only those animals develop the typical clinical phenomena seen in man.'

An example of species differences is seen in the motor effects of reserpine. When rodents that have become akinetic or 'cataleptic' due to reserpine are aroused sufficiently they are capable of moving about without tremor, and often at virtually normal speed. The movements of the reserpinized monkey, in contrast, exhibit marked tremor and rigidity. Tremor and rigidity may depend on the basal ganglia projection to thalamus and thence to cortex, as suggested by the fact that interruption of this projection by thalamic lesions in patients with Parkinson's disease may cause a striking reduction in tremor and rigidity (Van Buren et al 1973). In contrast, arousal and the motivation to move may depend on circuits involving the limbic system—in which the pallidohabenular projection may be involved.

Basal ganglia and the SMA

The new information about the input from the basal ganglia to the SMA leads one to consider a number of classical observations in a new light. It is of special interest, in the context of these observations, that deficits caused by SMA ablations can closely resemble those observed in parkinsonism. The resemblance is close enough for it to have been said that frontal lobe damage is sometimes incorrectly diagnosed as parkinsonism (Botez 1974). It has long been known that basal ganglia dysfunction leads to a marked akinesia that can, to a certain extent, be compensated for when visual information can be used to guide certain movements. The work of Flowers (1978) demonstrates rigorously what clinicians have known for years: the presence of visual targets for movement, for example feet painted on the floor, can improve the performance of a motor act. In monkeys, Hore et al (1977) have shown that inactivation of GPi by cooling produces similar effects. In parallel, it has been found that SMA damage seriously reduces internally generated movements. Limb movements and speech are rarely produced spontaneously, at least for a month or so after the SMA lesion (Laplane et al 1977, Masdeu et al 1978, Damasio & Van Hoesen 1980, Goldberg et al 1981). This deficit begins as global akinesia, but eventually resolves to contralateral akinesia (Damasio & Van Hoesen 1980). The patients can repeat words they hear, thus demonstrating that sensory guidance of some sort can compensate for the disturbed mechanisms of spontaneously generated speech (Goldberg et al 1981). Presumably this could be true for limb movements as well. Similar observations have also been made in non-human primates: SMA ablations reduced the frequency of spontaneously generated vocalizations in squirrel monkeys (Kirzinger & Jürgens 1982). Thus, both the basal ganglia and SMA damage appear to cause deficits in spontaneously emitted motor acts, and both are therefore implicated in the programming and control of such movements.

This conclusion gains further support from the studies of Roland and his colleagues (1980a, b) on regional cerebral blood flow. They have shown that the SMA is highly activated when human subjects rehearse and prepare a complex sequence of movements of the digits, to be performed without sensory guidance. Evidence for SMA having a role in the preparation for movement is also provided by electrophysiological studies that show single-neuron activity (Tanji et al 1980) or slow-wave (readiness) potentials (Deecke & Kornhuber 1978), specifically during periods of motor preparation. In the study of Tanji et al (1980) the monkey was given an instruction stimulus that provided information about a push or pull movement to be performed after the receipt of a kinaesthetic triggering cue. Many SMA neurons show sustained activity during this instructed delay period and this has been interpreted as reflecting the motor set of the animal. The aspect of motor set in which SMA is involved has not been investigated thoroughly, but much speculation has centred around the idea that one of its roles is to uncouple the precentral motor cortex from a stability-promoting negative feedback (Wiesendanger et al 1973, Tanji et al 1980), thus allowing motor commands to be executed in an open loop manner.

There is also evidence for a related functional specialization of the SMA, a role in the control of complex movements. By complex, what is meant is a sequence of individual movements that must be coordinated in a coherent scheme. The cerebral blood flow study already mentioned contributes to this view. The SMA is not greatly activated when subjects execute a simple repetitive movement of the digits or when they perform a sustained isometric contraction of the digits. By contrast, when subjects perform a sequence of movements, SMA is strongly activated (Roland et al 1980a, b, Orgogozo & Larsen 1979). Similarly, ablation of the SMA causes severe and apparently permanent disruption of complex motor acts, especially for coordinated movements of limbs on both sides of the body in humans (Laplane et al 1977) and in monkeys (Brinkman 1981). Thus, both clinical and laboratory observations support the idea that SMA acts with the basal ganglia as part of an integrated system involved in the preparation for complex movements and that these structures are essential for movements initiated and executed without exteroceptive guidance, i.e. movements that are self-initiated, self-guided and spontaneous.

Basal ganglia and eye movements

Other basal ganglia outputs project to centres for the control of eye movements. One pathway originates from SNr and terminates in the superior colliculus, and a second output can be traced from SNr to the frontal eye field via the thalamus. The basal ganglia receive dense projections from the prefrontal and temporal cortex, areas related to behaviours involving visual and spatial memory. It was thus of great interest when Hikosaka & Wurtz (1983) showed that SNr cells identified as projecting to the superior colliculus have memory-contingent relations to saccadic eye movements. This finding was obtained in an experiment in which the monkey was required to remember the location of a stimulus that was presented briefly while the animal was fixating on a different stimulus; a later saccade was rewarded if it was made to the position of the no-longer-present stimulus. Some SNr cells showed a minimal response if the monkey made a saccade to the stimulus while it was still present or if the monkey continued to fixate, but these same cells could show vigorous responses that were temporally correlated with saccades made to the point where a remembered visual stimulus had been. None of these cells showed a change in activity in relation to spontaneous saccades in darkness. The onset of memory-contingent saccade reponses preceded the saccade onset by up to 280 ms (most frequently by 70–240ms).

In discussing these observations, Hikosaka & Wurtz (1983) noted that basal ganglia functions have been categorized as sensory, motor or cognitive, and they proposed that each of these three functional aspects finds expression in the activity of individual neurons in SNr, which is presumably a final stage of processing in the basal ganglia. It was their view that in single substantia nigra cells, these functions are gated in different ways so that the sensory or motor activities of the cells are specialized for the different contexts in which behaviour occurs. They went on to point out that 'The memory-contingent visual response is a decrease in discharge rate following the onset of a visual stimulus that is evident only when the monkey must make a saccade to the location of the stimulus after it is no longer present. The basic assumption necessary to explain this visual response . . . is the presence of a neural correlate of spatial memory, which would signal the previous location of the visual stimulus' (p 1280).

Among the possible sources of inputs that might underlie memory-contingent responses, Hikosaka and Wurtz suggested that the best currently known candidate for the source of the neural correlate of spatial memory is the prefrontal cortex. If the prefrontal cortex contains these memory elements, the signals related to spatial memory are very likely to be transmitted to the substantia nigra via the caudate nucleus.

Hikosaka and Wurtz pointed out that a role of the basal ganglia specifically in oculomotor control is also evident in Parkinson's disease, noting that an increased latency for saccadic eye movements may be comparable to bradykinesia. In addition, they called attention to the longer time it took patients with Parkinson's disease to make saccades back and forth between targets that were left on all the time, and to the difficulty that patients with Huntington's disease have in making saccades to a constantly present visual target in response to verbal instructions. It was noted that in both of these cases the saccade must be initiated by something other than the onset of a visual target, and that such movements have some similarity to saccades aimed at remembered targets.

As mentioned above, SNr, in addition to its projections to the superior colliculus, projects to thalamic nuclei that in turn project to the frontal eye field, a region that returns strong projections to both the superior colliculus and the caudate nucleus (see Künzle & Akert 1977 for references). The frontal eye field is thus intimately connected with basal ganglia and the activity of its neurons, like activity of SNr neurons, is related to cognitive and attentional factors rather than to eye movements *per se*. Indeed, eye movements in the dark are not preceded by discharge of neurons of the frontal eye field in monkeys, and Goldberg & Bushnell (1981) found that, even with the room lights on, it was impossible to correlate presaccadic discharge of neurons of the frontal eye field with spontaneous saccades made in a familiar environment. However, when these saccades were part of conditioned, visually guided behaviour, such correlations were quite common. Thus, Goldberg & Bushnell (1981) found that the activity of neurons of the frontal eye field before a saccade, like the pause of SNr neurons before a saccade, was contingent on cognitive factors and was not simply linked to the eye movement *per se*.

State or set-related responses in putamen neurons

Microelectrode recordings in monkeys (DeLong et al 1984 this volume) have shown that putamen neurons related to voluntary movement are silent in the absence of movement and, conversely, that tonically active putamen neurons appear unrelated to movement. Recent work (Kimura et al 1983, 1984) has confirmed that tonically active putamen neurons are unrelated to body movements *per se,* but has shown that under certain circumstances such neurons may exhibit highly reliable responses to external events. An auditory stimulus (the click of a solenoid valve) elicited short-latency responses in this type of putamen neuron when the stimulus was a cue for juice delivery and its consumption, but such a stimulus failed to elicit responses when juice delivery had repeatedly failed to follow the sound, with the result that the sound no longer triggered movements associated with consuming the juice. Responses of tonically active putamen neurons were observed in three behavioural conditions: (1) *self-paced movement,* in which a series of elbow movements resulted in a solenoid click and a juice reward; (2) *free-reward,* in which click and juice occurred at regular intervals (every 6s) while the arm position was fixed; and (3) *no-reward,* which was similar to the free-reward condition except that the tube conveying the juice was occluded so that the solenoid click was no longer followed by juice. In the no-reward condition licking was extinguished after the first few unrewarded solenoid clicks.

Figure 2 shows responses of a tonically active putamen neuron in the three behavioural conditions. A sequence of 40 consecutive rewards during self-paced movement was followed by 40 consecutive free-rewards. There was no apparent difference between solenoid-evoked activity in self-paced movement and in free reward, showing that presence or absence of arm movements before the solenoid click made virtually no difference in the neuronal reponse. Then, without any alteration to the 6-s intervals between solenoid clicks, the sequence of 40 no-reward clicks was started. The monkeys reacted to the first few unrewarded clicks after the long sequence of 80 rewarded clicks with the motor responses (head torque and sublingual EMG) that would have been appropriate for consuming the juice, but these motor responses were quickly extinguished as the series of 40 consecutive unrewarded solenoid clicks proceeded. (It should be noted that the monkeys had had much experience with these three sequences, and had learnt that an unrewarded click signalled many more to come.) Tonically active putamen neurons that had been responsive to the click lost their responses within a few repetitions of the unrewarded solenoid click, showing that the characteristic responses of these cells to the solenoid click depended on the state or set of the animal towards consumption of the reward. Though dependent on this state or set of the monkey, the responses in the tonically active neurons were

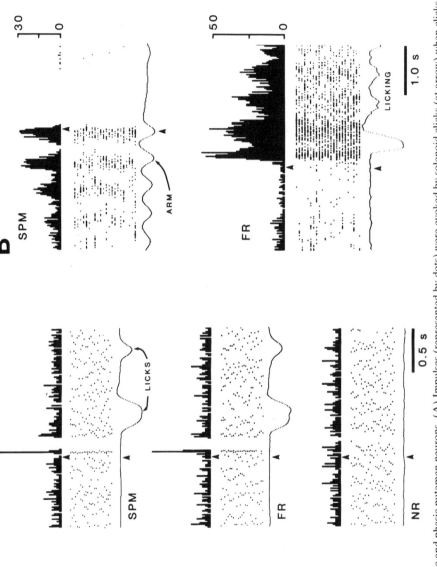

FIG. 2. Tonic and phasic putamen neurons. (A) Impulses (represented by dots) were evoked by solenoid clicks (at arrow) when clicks triggered licking during self-paced movement (SPM) and free-reward (FR), but not during no-reward (NR), when motor responses were extinguished. (B) Phasic discharges of putamen neurons with arm movements (above) and licking (below). (From Kimura et al 1984.)

not related to licking movements *per se*, i.e. the cell did not change its discharge rate in relation to movements not preceded by the auditory stimulus. As shown in Fig. 2A, these neurons responded to the solenoid click with single impulses well in advance of the first in the sequences of licking movements and showed no apparent relation to the subsequent successive licks. By contrast, Fig. 2B shows typical movement-related neurons that had bursts of discharge with each of the series of self-paced arm movements or licking movements.

Tonically discharging putamen cells with set-dependent responses were observed throughout the entire putamen, and unlike the movement-related neurons they appeared to be independent of any somatotopic organization. It may therefore be inferred that the single impulses evoked in tonically active cells about 60 ms after the solenoid click resulted in a synchronous event within a large part of the putamen. Fig. 3A illustrates the responses evoked in eight tonic cells in a single penetration in putamen, and shows the clear synchrony of these responses.

Neuronal activity was also recorded in the globus pallidus in the same three click-reward contingencies, and state-dependent or set-dependent responses were observed in a number of globus pallidus neurons. The globus pallidus cell illustrated in Fig. 3B exhibited a pause of activity in response to rewarded clicks, but this pause disappeared soon after the no-reward sequence started.

Looking at the single impulses evoked in tonically active putamen cells, and contrasting these single impulses with the intense movement-related activity of the phasic putamen cells in Fig. 2B, one might be inclined to minimize the functional significance of the tonically active cells. However, the dopaminergic nigrostriatal neurons are also tonically active and lack phasic relations to movement, but their importance is well established. By analogy, the tonically active putamen cells might also have great functional significance in spite of the small number of stimulus-evoked impulses. It is clear that a high priority in further work on the putamen will be the neurochemical and morphological identification of the tonically active putamen neurons and a further analysis of the state-dependent or set-dependent responses.

Conclusions

This paper has considered behaviour-related physiology of striatum, globus pallidus, SNr and two cortical areas (SMA and the frontal eye field) that receive basal ganglia outputs via the thalamus. In all these regions changes of activity precede movement in certain behavioural situations but not in others. As we have seen, a given neuron in the frontal eye field may discharge in advance of a goal-directed saccadic eye movement and yet fail to discharge in

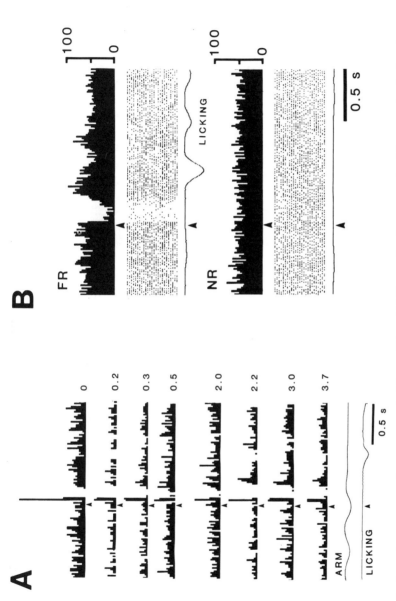

FIG. 3. Set-dependent responses in putamen and globus pallidus. (A) Event-related responses in eight successive tonic putamen neurons (recorded in single microelectrode penetration) when click (at arrow) was followed by juice in free-reward (FR). Numbers at right of histograms are distances (in millimetres) between successive neurons in the penetration. (B) Click-evoked pause of globus pallidus discharge occurred during FR before licking but was absent during no-reward (NR) when movement was extinguished. (From Kimura et al 1984.)

relation to a spontaneously emitted saccade (Goldberg & Bushnell 1981). SNr neurons can turn off before eye movements to a remembered target, but not before those to a visible target (Hikosaka & Wurtz 1983). And SMA neurons show activity modulations before movements triggered by one sensory modality but not before virtually identical movements triggered by another modality (Tanji & Kurata 1982). Similarly, Rolls et al (1983) have shown that many neurons in the head of the caudate nucleus and anterior part of the putamen are activated by a cue when the monkey is set to move in response to the cue, but that the same neuron is unresponsive when the cue is no longer the trigger for movement. Tanji & Kurata (1982) have made the same sort of observation in SMA and we have summarized similar findings for the putamen and globus pallidus.

Granted that the cognitive state of the animal is an important determinant of activity in the basal ganglia and the cortical targets of basal ganglia outputs, what can we say about the routes whereby the outputs of the basal ganglia control motor behaviour, and what part might these pathways play in motor dysfunction. We have discussed four of the basal ganglia output targets: SMA, frontal eye field, superior colliculus and habenula. There are others as well, but for the present we will concentrate on the motor outputs of the basal ganglia–SMA system. In SMA, there are major output pathways to the primary motor area, to the arcuate premotor area, and directly to subcortical structures including the spinal cord. In a sense, the pallidal input to the SMA opens the possibility of a trisynaptic pallidospinal pathway, perhaps directly to motor neurons. These connections suggest an important motor role for the basal ganglia output to the SMA. We do not yet know the relative functional roles of the different output pathways of the basal ganglia, and one cannot expect SMA lesions to duplicate lesions of the basal ganglia, but it seems likely that at least part of the motor symptomatology of patients with basal ganglia disorders may be understood in relation to basal ganglia outputs to SMA.

REFERENCES

Botez MI 1974 Frontal lobe tumors. In: Vinken PJ, Bruyn GW (eds) Handbook of clinical neurology. Elsevier North-Holland, Amsterdam, vol 17:234-280

Brinkman J 1981 Lesions in supplementary motor area interface with a monkey's performance of a bimanual coordination task. Neurosci Lett 27:267-270

Carlsson A, Lindqvist M, Magnusson T 1957 3,4-Dihydroxyphenylalanine and 5-hydroxtryptophan as reserpine antagonists. Nature (Lond) 160:1200

Carpenter MB 1981 Anatomy of the corpus striatum and brain stem integrating systems. In: Brooks VB (eds) Motor control, part 2. American Physiological Society, Bethesda/Williams & Wilkins, Baltimore (Handb Physiol sect 1 The nervous system vol 2) p 947-995

Damasio AR, Van Hoesen GW 1980 Structure and function of the supplementary motor cortex. Neurology 30:359

Deecke L, Kornhuber HH 1978 An electrical sign of participation of the mesial 'supplementary' motor cortex in human voluntary finger movement. Brain Res 159:473-476

DeLong MR, Georgopoulos AP, Crutcher MD, Mitchell SJ, Richardson RT, Alexander GE 1984 Functional organization of the basal ganglia: contribution of single-cell recording studies. This volume, p 64-78

Flowers K 1978 Lack of prediction in the motor behaviour of parkinsonism. Brain 101:35-52

Goldberg ME, Bushnell MC 1981 Behavioural enhancement of visual responses in monkey cerebral cortex. II. Modulation in frontal eye fields specifically related to saccades. J Neurophysiol 46:773-787

Goldberg G, Mayer NH, Toglia JU 1981 Medial frontal cortex infarction and the alien hand sign. Arch Neurol 38:683-686

Herkenham M, Nauta WJH 1977 Afferent connections of the habenular nuclei in the rat. A horseradish peroxidase study, with a note on the fiber-of-passage problem. J Comp Neurol 173:123-145

Hikosaka O, Wurtz RH 1983 Visual and oculomotor functions of monkey substantia nigra pars reticulata. III. Memory-contingent visual and saccade responses. J Neurophysiol 49:1268-1284

Hore J, Meyer-Lohmann J, Brooks VB 1977 Basal ganglia cooling disables learned arm movements of monkeys in the absence of visual guidance. Science (Wash DC) 195:584-586

Kimura M, Rajkowski J, Evarts EV 1983 Event-related responses in spontaneously active putamen neurons. Soc Neurosci Abstr 9:873

Kimura M, Rajowski J, Evarts EV 1984 Tonically discharging putamen neurons exhibit set-dependent responses. Submitted.

Kirzinger A, Jürgens U 1982 Cortical lesion effects and vocalization in the squirrel monkey. Brain Res 233:299-315

Künzle H, Akert K 1977 Efferent connections of cortical area 8 (frontal eye field) in *Macaca fascicularis*. A reinvestigation using the autoradiographic technique. J Comp Neurol 173:147-164

Laplane D, Talairach J, Meininger V, Bancaud J, Orgogozo JM 1977 Clinical consequences of corticectomies involving the supplementary motor area in man. J Neurol Sci 34:301-314

Marsden CD 1975 Introduction. Adv Neurol 10:3-4

Masdeu JC, Schoene WC, Funkenstein H 1978 Aphasia following infarction of the left supplementary motor area. Neurology 28:1220-1223

Nauta WJH, Domesick VB 1984 Afferent and efferent relationships of the basal ganglia. This volume, p 3-23

Orgogozo JM, Larsen B 1979 Activation of the supplementary motor area during voluntary movement in man suggests it works as a supramotor area. Science (Wash DC) 206:847-850

Roland PE, Larsen B, Lassen NA, Skinhoj E 1980a Supplementary motor area and other cortical areas in organization of voluntary movements in man. J Neurophysiol 43:118-136

Roland PE, Larsen B, Lassen NA, Skinhoj E 1980b Different cortical areas in man in organization of voluntary movements in extrapersonal space. J Neurophysiol 43:137-150

Rolls ET, Thorpe SJ, Maddison SP 1983 Responses of striatal neurons in the behaving monkey. 1. Head of the caudate nucleus. Behav Brain Res 7:179-210

Schell GR, Strick PL 1984 The origin of thalamic inputs to the arcuate premotor and supplementary motor areas. J Neurosci 4:539-560

Tanji J, Kurata K 1982 Comparison of movement-related activity in two cortical motor areas of primates. J Neurophysiol 48:633-653

Tanji J, Taniguchi K, Saga T 1980 Supplementary motor area: neural response to motor instructions. J Neurophysiol 43:60-68

Tracey DJ, Asanuma C, Jones EG, Porter R 1980 Thalamic relay to motor cortex; afferent pathways from brain stem, cerebellum, and spinal cord in monkeys. J Neurophysiol 44:532-554

Van Buren JM, Li C, Shapiro DY, Henderson WG, Sadowsky DA 1973 A qualitative and quantitative evaluation of parkinsonians three to six years following thalamotomy. Confin Neurol 35:202-235

Wiesendanger M, Séguin JJ, Künzle H 1973 The supplementary motor area—a control system for posture? In: Stein RB et al (eds) Control of posture and locomotion. Plenum, New York, p 331-346

DISCUSSION

Yoshida: I saw a similar increase in neuronal firing of the caudate nucleus in the cat in the preparatory state before it jumped in the experiments I mentioned earlier (p 80). It is interesting to learn that this happens in the substantia nigra too. Do you think what happens there is a reflection of what is happening in the caudate nucleus?

Evarts: Yes. I think decreases in globus pallidus impulse frequencies reflect increases of activity in the caudate and putamen.

Rolls: In experiments essentially the same as you described, with an environmental stimulus that is behaviourally significant, we see very large responses in caudate neurons (Rolls et al 1983). That input may be coming through the caudate to the putamen. Perhaps there is some interaction throughout the striatum whereby a signal in the caudate can eventually be reflected to a much smaller extent in the putamen, where it may have some function. Do you know the function of the responses you describe?

Evarts: Wurtz pointed out that his observations at the single-unit level fit quite well with observations on deficits after prefrontal lesions in monkeys. The best candidate for a source of spatial mnemonic information to basal ganglia would be the prefrontal projection to the caudate, from which signals could reach the lateral part of the substantia nigra pars reticulata.

Wing: It is tempting to relate Hikosaka and Wurtz's observations of different firing patterns in the pars reticulata (SNr) with pre-cued and spontaneous saccades to your own work on parkinsonian reaction time (Evarts et al 1981). You compared reaction times in simple tasks with a pre-cued response and in tasks where a choice had to be made and found. Simple reaction time was slowed relatively more than choice reaction time. In my paper here I suggest that the parkinsonian deficit is in activating or triggering previously prepared movement. However, the abnormality associated with Parkinson's disease is in the pars (SNc) compacta rather than in the SNr.

Evarts: In the causative sense that deficit would, of course, depend on a loss of cells in the SNc but ultimately this cell loss would be expressed via the SNr pathway that Hikosaka and Wurtz have studied. Hikosaka & Wurtz (1983) discussed the eye movement deficits in patients with parkinsonism and pointed out that internally generated saccadic eye movements are difficult for these patients, possibly because they must hold the instruction (e.g. 'move to the left') and then, on the basis of this verbal input and the mnemonic stores that it conjures up, make the specified eye movement. In patients with basal ganglia disease there are selective deficits in performance of this type of eye movement compared to an eye movement that is triggered by a visual stimulus.

Calne: Clinically that correlates with the difficulty that parkinsonian patients often have in initiating a movement, such as walking, although they can continue once the movement has started.

Rizzolatti: We ablated the postarcuate cortex in the monkey (*Macaca irus*) and found degeneration mostly in nucleus X of Olszewski. In another experiment we ablated the frontal eye field and saw degeneration in the medial dorsal nucleus rather than in the anterior ventralis (VA) (Rizzolatti et al 1983). What is the evidence that VA projects to the frontal eye field?

Evarts: I was referring to the work of Künzle & Akert (1977) on *Macaca fascicularis* in which it was shown that corticothalamic fibres from the frontal eye field project to the thalamic nucleus ventralis anterior pars magnocellularis (VAmc); one may assume that there is a reciprocal projection from VAmc to frontal eye field.

Carpenter: There was a degeneration study and a later autoradiographic study which related the paralaminar parts of the dorsal medial nucleus of the thalamus to the frontal eye fields (Scollo-Lavizzari & Akert 1963, Künzle & Akert 1977). I don't think the magnocellular part of VA is related to the frontal eye field.

Graybiel: Would you or Mahlon DeLong comment further on the relation between your results and those of Bob Wurtz?

DeLong: I don't see them as different. I think different parts of the basal ganglia have different functions. In primates there are parts concerned with the execution of movement, whether it is self-initiated or externally triggered. That is the part that we have been looking at. We have been trying to discover what contribution the basal ganglia make to movement *per se* rather than find their role in initiating or controlling movement in a particular context. At this point it seems that the basal ganglia play a role in controlling the amplitude of movement but don't play a major role in the initiation of externally triggered movements or in the selection of the pattern of muscular activity. This is generally consistent with the clinical observations in patients with Parkinson's disease.

Evarts: In conditions requiring a high degree of attention and memory, Hikosaka & Wurtz (1983) observed that saccades are preceded by pauses of

SNr discharge. In studies of movements triggered by auditory stimuli, my colleagues and I found that globus pallidus neurons showed pauses in activity at the same latencies seen by Hikosaka and Wurtz for SNr responses to a visual stimulus.

It may be useful to distinguish between two factors to which the activity of the inner segment of the globus pallidus (GPi) and that of SNr may be related; one factor is the nature of the triggering event or memory that initiates movement. The properties of this initiating event—its timing, its mnemonic features and so forth—seem to be highly significant for the short-latency pauses of basal ganglia output from SNr and GPi. Movements which occur in circumstances in which memory, attention and learning are less critical may be associated predominantly with longer latency increases of discharge of the sort that DeLong and his colleagues have described. The experimental paradigm that Wurtz has developed to study memory-contingent neuronal activity requires an immense amount of attention and 'mental effort' by the monkeys. The behavioural task that DeLong has used is easier for the monkeys, and the reaction times that are called for are not as short as those called for in the visual-fixation paradigms used by Wurtz. The presence or absence of pauses in SNr or GPi may depend on the nature of the behavioural task and the attentional state of the monkey at the time when movement is initiated.

Marsden: There is still a paradox there. There is evidence from different laboratories that there are units throughout the striatum, and indeed in the pallidum, which attend to cues for movement or attend to periods of preparation for movement. How is that information being used? The paradox is that these units don't seem to 'start' the movement.

Evarts: The pauses that Hikosaka and Wurtz have seen in SNr occurred well in advance of the memory-contingent saccades. In our own work, many globus pallidus neurons showed pauses with latencies of 60–75 ms after a click that triggered a movement. When movement is triggered by an attention-getting external event or by an internal mnemonic process the activity precedes movement in SNr and in globus pallidus.

DeLong: These studies of timing relations are much more easily done for the oculomotor system than for the limbs. There is a very precise onset of movement in the eye, and one is not trying to correlate the activity of a lot of different joints and muscles.

Rolls: Part of the resolution of David Marsden's question may come from considering what the monkey has to do and whether the animal has to use its cortex. If the monkey has to use its frontal eye fields or perhaps other parts of the prefrontal cortex to do Bob Wurtz's memory task, it may be natural for the striatum to be involved, because it is receiving from the cortex. If it is a simpler eye-movement task without a memory component the frontal eye fields may not be necessary and the cortex may not be used. In Mahlon DeLong's task the cortex probably is important because there is a visually triggered movement.

Evarts: In Hikosaka & Wurtz's study (1983), saccade-related SNr neurons that failed to show memory-contingent relations discharged quite late in relation to saccades. These neurons would be similar to the ones that DeLong has described as being related to arm movements but that discharge relatively late in relation to the time at which the muscles contract.

Calne: You showed a pathway through SNr and nucleus ventralis lateralis pars medialis (VLm) to the supplementary motor area, Dr Evarts. In parkinsonian patients there was a time when lesions were deliberately made in VL. It was claimed that if the lesion was precisely on target, very little motor deficit was left. How do you interpret that? Could the surviving patients act as a model and provide any clues to support your views?

Evarts: It was mainly the positive symptoms of tremor and rigidity that were eliminated by this lesion.

Calne: It was claimed (Burns et al 1983) that the rest of motor function was intact but I don't really believe that is true. Even when there was a clinically satisfactory result on one side our patients seldom chose to have surgery on the other side.

Wing: Perret (1968) has tested motor performance in parkinsonian patients who had had surgical lesions in nucleus ventralis lateralis of the thalamus for relief of tremor and rigidity. Postoperatively he found improvements in the speed of aiming movements of the contralateral arm, but there is a suggestion in the data that ipsilateral performance is depressed.

DeLong: The changes are not very spectacular.

Porter: In previous meetings on the basal ganglia we have tended to talk about the relationship between cerebral cortex and basal ganglia in terms of ordered input from zones of the cerebral cortex into the neostriatum. You have drawn our attention to a particular relationship of pallidal output through the thalamus to a circumscribed region of the cerebral cortex. This region may be exclusively the medial part of area 6 which is referred to as the supplementary motor area (SMA). Other people may have some views about how rigidly that is to be circumscribed. But SMA also has inputs from the parietal cortex, the sensory cortex and the cingulate cortex, as well as from further forward in the frontal lobe. The diagram Dr Evarts showed referred to the way in which sensory areas of the cerebral cortex are interrelated (Jones & Porter 1980, Fig. 12); it was not designed to show the motor relationships. There is quite a different picture if one looks at it from the point of view of motor output.

Recordings made in SMA, and to some extent in the postarcuate cortex, are quite clear in indicating that discharges of individual neurons in those cortical areas are related to movement, whether made by the ipsilateral or the contralateral limb. As we get further away from the spinal motor neuron, these bilateral associations are more evident. They may be very important in the processing systems concerned with setting the posture of the animal for a movement at the distal part of the limb, or indeed for aspects of control of that distal movement. Diagrams drawn to look at the relationships of afferent systems, for example, are not necessarily the most appropriate ones to use when one is looking at outputs from those systems.

DeLong: The basal ganglia shouldn't be viewed in isolation from the thalamus and the cortex. We should be thinking about systems involving the basal ganglia rather than about structures such as the caudate, globus pallidus, thalamus, etc. The nucleus ventralis lateralis pars oralis (VLo) and SMA ought to be as important in our thinking as components of the system as are the striatum and the pallidum. The basal ganglia are only part of these systems. As far as I know, VLo has only two extrinsic inputs—one from the pallidum, the other from the cortex. From the studies of Schell & Strick (1984) it now appears that VLo projects to the SMA. Whatever the basal ganglia do, they are doing it in some sort of cooperation with the entire cortex and thalamus. We believe there exist in a general manner two systems, one passing through the putamen and the other through the caudate. The pathways through putamen take origin from the premotor, motor and somatosensory cortices. This pathway continues through the caudoventral pallidum to parts of the thalamus (VLo) that influence SMA. Side by side is a pathway through the caudate which takes origin from most of the association areas. This pathway continues through the more rostrodorsal regions of the pallidum and nigra to portions of the thalamus (VA) which influence more rostral regions of the frontal lobe. The idea of segregated pathways was based on the finding that somatotopically organized areas seem to be present in the putamen, pallidum and thalamus. This maintenance of somatotopy suggested a segregation of information about the control of individual body parts. If that principle of anatomical organization is a general one, information from association areas should remain segregated from the 'motor' representation. This view raised important questions. Is there really complete segregation? Does the pathway passing through the caudate to the premotor area even converge on the more movement-related pathway through the putamen? Future studies should try to resolve these questions.

Arbuthnott: If two kinds of processing are going on, one closely related to movement and another that we might call anticipatory or motivational, which kind is more damaged by the loss of dopamine?

Evarts: We can't say yet. My assumption is that the very earliest processes in the striatum are greatly impaired. One of the problems in this work is to maintain the motivational state in an animal afflicted with experimental parkinsonism. The motivational state in monkeys is so vastly altered after they have been treated with reserpine or *N*-methyl-4-phenyl-1,2,3,6- tetrahydropyridine that it becomes difficult to assess the meaning of one's observations.

Calne: There are some clinical clues. When patients who have difficulty in walking, for example, are given cues such as bands on the floor that they can step over, they get into a self-sustaining pattern of movement. The initial organization of the movement, the formulation of the start, is otherwise extremely difficult.

Evarts: That is a striking correlate of the sort of thing we are talking about. There is a class of neurons that will fire just once for the *initiation* of a sequence of movements, but fail to fire with each individual movement in the sequence.

Calne: If the walking pattern is disrupted by a turn or a narrow area to go through, the problem reappears.

Marsden: Against that, you and others have found that the reaction time in patients with Parkinson's disease varies widely. It is not the best marker of what goes wrong in Parkinson's disease. Getting a movement started is frequently done in the normal length of time, although the time taken to complete the movement is prolonged.

Calne: Once they start walking, many patients move at a normal rate but in a different setting their speed of movement may be impaired.

DeLong: We always have the problem of being clear whether we are talking about the basal ganglia or about Parkinson's disease. Studies (lesion and stimulation) in primates indicate that the major impairment is in the execution rather than the initiation of movements; the latter often seems to be normal or sometimes even a bit earlier than before. The deficits in Parkinson's disease may be due to derangements outside the basal ganglia. For example, loss of dopamine projections to the cortex could cause some of the impairments, as could loss of noradrenergic pathways from the locus ceruleus. There is some very nice recent anatomical and pharmacological work on locomotion by Garcia-Rill. He has found a projection from the substantia nigra in the cat to the locomotor region in the mesencephalon. He has manipulated that by applying GABA agonists and antagonists. This is the very substrate I think we are looking for in trying to understand the role of the basal ganglia in locomotion and in Parkinson's disease. The descending projection from the basal ganglia to this locomotor region would be important in switching on the spinal centres.

Marsden: We are all prisoners of our paradigms. The experimental work in monkeys shows that the major deficit is in the execution rather than the initiation of movement. But one has to relate that to the single-cell studies which are mostly concerned with the initiation of simple movements, not complex sequences of movements. I am not sure that we know what is happening to unit activity in the basal ganglia when an animal goes through a sequence of different movements which comprise a whole unit of motor behaviour.

REFERENCES

Burns RS, Chiueh CC, Markey SP, Ebert MH, Jacobowitz DM, Kopin IJ 1983 A primate model of parkinsonism: selective destruction of dopaminergic neurons in the pars compacta of the substantia nigra by N-methyl-4-phenyl-1,2,3,6-tetrahydropyridine. Proc Natl Acad Sci USA 80:4546-4550

Evarts EV, Teravainen H, Calne DB 1981 Reaction time in Parkinson's disease. Brain 104:167-188

Hikosaka O, Wurtz RH 1983 Visual and oculomotor functions of monkey substantia nigra pars reticula. III. Memory-contingent visual and saccade responses. J Neurophysiol 49:1268-1284

Jones EG, Porter R 1980 What is area 3a? Brain Res Rev 2:1-43

Künzle H, Akert K 1977 Efferent connections of cortical area 8 (frontal eye field) in *Macaca fascicularis*. A reinvestigation using the autoradiographic technique. J Comp Neurol 173:147-164

Perret E 1968 Simple motor performance of patients with Parkinson's disease before and after a surgical lesion in the thalamus. J Neurol Neurosurg Psychiatry 31:284-290

Rizzolatti G, Matelli M, Pavesi G 1983 Deficits in attention and movement following the removal of postarcuate (area 6) and prearcuate (area 8) cortex in macaque monkeys. Brain 106:655-673

Rolls ET, Thorpe SJ, Maddison SP 1983 Responses of striatal neurons in the behaving monkey. 1: Head of the caudate nucleus. Behav Brain Res 7:179-210

Schell GR, Strick PL 1984 The origin of thalamic inputs to the arcuate premotor and supplementary motor areas. J Neurosci 4:539-560

Scollo-Lavizzari G, Akert K 1963 Cortical area 8 and its thalamic projection in *Macaca mulatta*. J Comp Neurol 121:259-270

General discussion 1

Basal ganglia links for movement, mood and memory

Porter: We have been talking about connections in and out of the basal ganglia and between the cerebral cortex and the basal ganglia. We have been asking how this system influences one of the outputs of the total nervous system, in this case movement. We could put similar arguments about some of the more limbic functions of the basal ganglia. We ought to consider the loops Mahlon DeLong described, which for the motor system influence what I shall call the premotor areas upstream of the motor cortex. These loops funnel through the basal ganglia back through the thalamus, in particular the more anterior parts of the ventral lateral thalamus. They go back to premotor cortex, not to area 4, which for the moment I shall regard as the primary source of corticospinal influences. We have also heard that there can be spin-offs from this loop to the reticular formation, to the tectospinal tract or whatever, that could influence the set on which movement could be performed. These spin-offs could indeed influence movement at a lower level than the cortical level.

The question, however, is: how does the operation of this cortico-striato-pallido-thalamo-cortical motor loop influence the more direct outputs to the spinal cord from the motor cortex? One way to answer that question is to inject horseradish peroxidase (HRP) into parts of the motor cortex and look at the contributions that other cortical regions make to a small zone of the motor cortex. Dr Soumya Ghosh, working in my laboratory, made a very small injection into what I call the finger representation of area 4. The size of that injection can be measured to some degree by counting the number of cells in the nucleus ventralis lateralis posterolateralis pars ovalis (VPLo) and the related parts of VLc and VLo that project into area 4. The projection into area 4 comes mainly from VPLo, as Ed Evarts showed. About 70% of the neurons that project from the thalamus into area 4 come from VPLo but additionally there is a small scatter of input from VLo and VLc. When we cut serial sections through the thalamus and counted all the cells, we found that 95% of the thalamic input to a small region of cerebral cortex came from the part of VPLo adjacent to VPLc and its neighbour nuclei. There is a zone or band of labelled cells through there that extends dorsally into the caudal part of VLc, and it extends in a minimal way into VLo. Those three nuclei therefore together make up 95% of the projections from the thalamus, and about 70% of them are from VPLo.

But we mustn't regard the boundaries that people draw around nuclei or cytoarchitectonic regions as being absolutely rigid and presenting only one sort of connection with the cerebral cortex. There are zones of labelled cells that have taken up HRP at their terminations within area 4 and transported it back to their cell bodies. These cell bodies are not evenly distributed over the cerebral cortex but exist in clumps and bands in various parts of the cerebral cortex. Dr Ghosh counted all the corticocortical cells and assigned them to particular areas of cerebral cortex. This allows one to come to some conclusions about the quantitative significance of these corticocortical connections. When we injected HRP into the hand and arm area of area 4, the distribution was rather different from the distribution after an injection limited to the finger area, and there was now a very heavy input from the supplementary motor area (SMA). After a small injection into the finger zone of area 4 we saw a heavy input from the premotor cortex in area 6, a major input from SMA and a sizeable projection from area 3a. These inputs may have functional significance because we know that the premotor zone of area 6 has a lot to do with the set for movements and with visuomotor coordination of the fingers. A lot of evidence for that has come from Ed Evarts' laboratory, from Wise and his colleagues. Maybe for the fingers it is important to have a heavy input from area 6. The input from the sensory cortex is mostly from area 2, with some from area 5 and even area 7. There is a projection from the second somatic sensory area, but the inputs from areas 3b and 1 are very small. (See Fig. 1.)

The input to the whole of the arm area is dominated by SMA, perhaps because this area has to concern itself with the whole context of movements that have to be made, not just with the visual control of the fingers. Area 6 still accounts for a big input in absolute numbers of cells, making the proportional input of area 3a look insignificant. But again there are many cells there. The sensory cortex doesn't really change its proportional input at all.

If we ask the question—what is the relationship between SMA, with all its basal ganglia connections, and one of the possible sources of motor output to the spinal cord through area 4?—we get rather different answers according to whether we are talking about the finger representation or the whole arm representation.

Carpenter: Was this all occurring on the same side?

Porter: Yes, those corticocortical connections were all ipsilateral.

Carpenter: Could you comment on the interrelation between the two sides?

Porter: Plenty of people have shown that the projections from the cortex into the striatum are bilateral and symmetrical. The absolute numbers may not be exactly the same but they are similar.

Carpenter: Does SMA project bilaterally into the motor area?

Porter: It projects to its own SMA on the other side. So there is point-to-point cross-talk across the callosum. But its projection into the motor cortex is

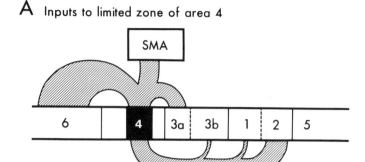

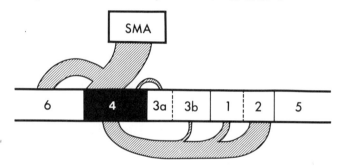

FIG. 1 (*Porter*). Diagram summarizing the essential features of the corticocortical connections with area 4.

more dominant on its own side than it is contralaterally. It was very clear in these studies that there are particular parts of motor area 4 that are influenced from SMA, there are particular zones of area 2 that project into area 4, and there are particular parts of lateral area 6 that do so. It is not a diffuse situation.

Divac: Since basal ganglia seem to project mainly to the SMA, what is known about the functional aspects of this cortical area?

Porter: There are cells in SMA that discharge impulses in relation to natural movements of the limb (Brinkman & Porter 1979). There is a range of activities in SMA which represents the range of activities in movement. That is, some cells in SMA discharge in relation to distal movements of the limb, others discharge in association with proximal movements of the limb, and there are yet others that have a complicated relationship to limb movement that is difficult to characterize. Our recording experiments show that there is somatotopy within SMA. That is, in a general sense, the units representing or related to movements of the arm are situated in a particular part of SMA and do not occupy its whole longitudinal extent. About 90% of units in SMA discharge in relation to that movement, whether it is by the ipsilateral or the contralateral limb. That situation is not grossly affected by sectioning of the corpus callosum. There are still many units in SMA that continue to discharge in relation to the ipsilateral or the contralateral limb even when the corpus callosum is sectioned.

We have been unable to classify sensory responses for units in SMA in the same way as for corticospinal units in area 4, where they are easily recognized. I would interpret that to mean that there has been a lot of processing of this sensory information in other zones before it is reflected on to SMA. That is one reason why disturbances of movement or natural manipulations of the limb are not very good stimuli for driving those cells in SMA. But Tanji has shown that, when animals are required to attend to a particular sensory stimulus and then respond to it, the same cells that I have been describing in terms of their movement responses can show contingent discharges in their attention to an auditory stimulus but not a visual stimulus, a visual stimulus but not a tactile one, and so forth (J. Tanji, 29 Congr Int Union Physiol Sci, Sydney, Aug 1983). In other words it can clearly be shown that those cells are capable of showing changes in their discharge if there is a situation that requires them to make an appropriate response, later on, to a particular stimulus.

If unilateral lesions are made in SMA in monkeys (*Macaca fasciculata*), there is a brief period during which both limbs perform clumsily. That is an evanescent effect which disappears after some training. However, in bimanual tasks, after unilateral lesions of SMA, the two hands fail to do the tasks appropriately by sharing the load. Animals with such lesions demonstrate a disability in such bimanual tasks for as long as a year. They always have trouble in knowing what they are supposed to do with their fingers. They try to do the same thing with both hands (Brinkman 1981).

Graybiel: Can you add a little to your remarks about the timimg relations between the firing of these units and the movements that are made?

Porter: The difficulties in assessing the time relations of neuronal discharges to movement performance get much worse the further away from a motor neuron one is recording. In SMA the recording may be made many synapses away from the motor neuron. Some corticospinal neurons, however, are monosynaptic onto motor neurons and the timing is easily assessed. Further away it becomes extremely difficult and it would be foolish of me to comment on whether, on average, cells in SMA discharge before their connected cells in area 4. In the same sort of diagram as Mahlon DeLong showed for globus pallidus we can find some cells in SMA that are clearly related to a particular motor task and that discharge at least 100–200 ms or even longer ahead of the EMG in the muscles that are associated with that task. But that doesn't mean they are controlling that task.

Evarts: On the subject of timing, we might note that if Wurtz's monkeys had made eye movements analogous to the movements that DeLong's monkeys performed, Wurtz wouldn't have seen such clear pauses of SNr discharge before movement: Hikosaka & Wurtz (1983) showed that the time of suppression of activity in SNr was different in three classes of SNr neurons. In 'memory-contingent' neurons the pause began much earlier than in neurons in which the saccade was visually triggered. We can put this together with Rolls' observations and see that for simple reciprocating movements in which attention and memory are not highly involved, one is going to look at a subclass of globus pallidus or reticulata neurons whose changes will not be early in relation to movement onset.

Yoshida: In the experiments I described earlier (p 80), caudate neurons increased their firing rate when the conditioning stimuli (lamp and buzzer) started, even when the cats did not move at all. There was a tendency for the incremental ratio (increase of firing during conditioning stimuli/firing rate before the stimuli started) to depend on the period of the attentive state (period from onset of conditioning stimuli to the time when cats started to move for jumping). When the period was shorter, the incremental ratio was bigger. So I agree that the caudate neurons must be very sensitive to the attentional state.

The limbic striatum and pallidum

Rolls: We have been concentrating on the motor striatum and motor pallidum. Perhaps we should now discuss the limbic striatum and the limbic pallidum. The nucleus accumbens receives inputs from structures such as the amygdala and the hippocampus (Heimer et al 1982). To try to determine what form of processing takes place there, we have recorded from a population of more than 400 neurons in the nucleus accumbens and olfactory tubercle (i.e. the ventral striatum) in the behaving monkey (Rolls et al 1982, Rolls 1984). We can activate neurons in the nucleus accumbens, but with much more difficulty than in other parts of the striatum such as the head of the caudate nucleus and putamen. Visual stimuli activate neurons of the ventral striatum and so do stimuli with emotional significance, i.e. those associated with reward or punishment. Those two classes of event are different from the stimuli which activate neurons in the putamen or the head of the caudate nucleus (see Rolls 1984). Given these neurophysiological findings, we might suggest that the ventral striatum is a system for relaying information from limbic structures through to certain outputs.

One particular ouput which has been mentioned already is the midbrain locomotor area. It has been suggested that there are connections from the ventral pallidum down towards the midbrain locomotor area (see Yim &

Mogenson 1983). It is known that damage to the nucleus accumbens can influence locomotor activity. Fink & Smith (1980) showed that when the animal is in a novel environment, locomotor activity is higher than normal, but if there has been a 6-hydroxydopamine lesion of the ventral striatum the novelty no longer enhances the locomotion. This may be a way of translating information which has been processed, in this case with respect to memory, through limbic structures so that it influences one sort of output, in this case locomotion.

However, that would not be the only output from this system. Anatomically there is also a connection from ventral striatum to ventral pallidum, and from ventral pallidum the anatomists tell us there is a connection up to the medial dorsal nucleus of the thalamus and then up to the prefrontal cortex (Heimer et al 1982). So limbic structures may influence behaviour through that route as well.

Kitai: Dr Chang and I have been recording from the accumbens neurons and staining those neurons intracellularly. The extrinsic inputs we have stimulated so far, such as the amygdala and VTA, produce monosynaptic excitation. The stained neurons are spiny neurons which are quite similar to the striatal neurons which receive extrinsic inputs. The membrane characteristics, the resistance and the firing patterns are also quite similar.

Nauta: If we subdivide the striatum into limbic and non-limbic compartments and attribute motor function to one and motivational function to the other, we run into the problem that ultimately the two will have to come together. At the moment it is not too difficult to draw circuits for either of these compartments without having much idea of where the meshing takes place. After all, the only proof of motivation is movement.

Marsden: There don't necessarily have to be any meeting points between the two systems. The basal ganglia, in their widest sense, might be doing something that is important for motor behaviour, but which is also equally appropriate to intellectual performance. Some of us believe that there is a meeting point for basal ganglia output in the final expression of movement. Others believe that there are separate basal ganglia systems for movement and thought.

Divac: My bias is to interpret the role of the striatum in terms of its cortical input. Looked at in this way, the neostriatum seems to consist of at least three major subdivisions: one is the part which receives its major input from the association cortex, the other receives it from the somatic sensory and motor cortex, and the third from the allocortex. (The division of the 'neuroendocrine striatum' proposed by Fallon et al [1983] should be further documented.) Broadly speaking, the role of the neostriatum may be to process the cortical input for use by the final common pathway. If so, the portion related to the association areas may 'translate cognition into action' (Divac 1977) whereas the part innervated mainly by the allocortex may translate 'motivation into action'

(Mogenson et al 1980). The part related to the motor and somatosensory cortex is the only one which contains the somatotopically arranged neurons discharging in synchrony with ongoing movements, as shown by DeLong and his collaborators. This part is not innervated by the amygdala and was found in several experiments to behave quite differently from the other parts of the neostriatum (Ingvar et al 1983, Kelly & McCulloch 1981, Palacios et al 1982). Paradoxically, perhaps, I think that the behavioural role of this part is least understood. Turning to the amygdala–neostriatal projections, it may be worth recalling Mishkin's concept (Jones & Mishkin 1972) that amygdala evaluate significance of the perceived stimuli. Thus, perhaps the amygdala 'decide' what goes through the 'non-motor' part of the neostriatum.

Nauta: Dr Marsden's remarks are addressed to an important problem, and prompt me to recapitulate our anatomical findings in the rat. Three successive points need to be made. First, the evidence that most if not all afferents of limbic origin converge on a ventromedial sector of the striatum, here for convenience called 'limbic striatum'. It must be emphasized that this striatal sector receives inputs not only from hippocampus, amygdala and various stations within their subcortical circuits, but also from limbic system-associated neocortical regions such as the frontal anterior cingulate and insular cortex; it follows that the term 'limbic striatum' cannot be translated into 'non-cortical striatum'. As regards other neocortical inputs to the striatum, although the corticostriatal projection as a whole is still incompletely charted, at least it seems clear that the projection from the sensorimotor cortex only marginally involves the limbic striatum and largely remains confined to a dorsolateral sector which in turn receives only sparse fibres of limbic origin and therefore is here, somewhat schematically, called 'non-limbic striatum'.

Second: it now seems certain that the striatopallidal projection from the limbic striatum selectively involves the ventral pallidum (as defined by its dense substance P-positive fibre plexus), whereas the non-limbic striatum projects not at all to the ventral but exclusively to the dorsal pallidum. The limbic versus non-limbic dualism of the striatum thus appears to be replicated in the external pallidal segment (the internal pallidal segment or interpeduncular nucleus is here left out of consideration).

Third: the projection from the external pallidal segment to the subthalamic nucleus originates from both the dorsal and the ventral pallidum. With its contribution to this connection, the ventral (or limbic) pallidum appears to establish an output channel into the extrapyramidal motor system that could be viewed as part of the limbic–extrapyramidal interface. In addition, however, the ventral pallidum has efferent connections that are not shared by the dorsal pallidum: unlike the latter it projects to various subcortical structures implicated in the circuitry of the limbic system: the basolateral nucleus of the amygdala, the mediodorsal and lateral habenular nucleus of the thalamus, and the ventral tegmental area as well as more caudal mesencephalic structures in the path of the medial forebrain bundle. These efferents would seem to allow the ventral pallidum to lead striatopallidal output of limbic antecedents back into the circuitry of the limbic rather than the extrapyramidal system. It seems likely that their activation would affect a functional category other than skeletomuscular function *per se*, in particular perhaps mechanisms of affect and motivation.

Finally, and in response to an earlier comment by Dr Marsden, it cannot be stated at present whether the projections to the subthalamic nucleus and those to subcortical limbic structures are invariably co-activated or can each be engaged separately.

Marsden: The *sequencing* of motor action and the *sequencing* of thought could be a uniform operation carried out by basal ganglia.

Calne: The limbic system and the striatal motor system are two subdivisions that we can handle fairly easily conceptually. Parkinson's disease is an important human model of damage to the basal ganglia that allows us to make observations that we can't make on animals. Roughly speaking, in Parkinson's disease the emotional state is intact but there are various categories of motor problem. What is known about regional changes within the motor striatum and the limbic striatum produced by Parkinson's disease? Is there a relative sparing of the 'limbic striatum'?

Hornykiewicz: We have fairly good biochemical information on the nucleus accumbens, which of course is not the whole ventral striatum. The dopamine innervation of the nucleus accumbens seems to suffer in Parkinson's disease about as much as the innervation of the caudate nucleus (Farley et al 1977). In this respect, however, the putamen is always distinctly more affected than the accumbens and caudate nucleus.

We also have some information on the behaviour in Parkinson's disease of the noradrenergic system innervating the nucleus accumbens. In respect to noradrenaline, the ventral striatum distinguishes itself quite clearly from the dorsal striatum by having a markedly denser noradrenergic innervation. In Parkinson's disease, the noradrenaline concentration in nucleus accumbens is also markedly decreased (Farley & Hornykiewicz 1976). Thus, one can say that, as a rule, in Parkinson's disease there is a severe dopamine deficiency in the predominantly motor putamen, and a distinctly less severe dopamine reduction in the caudate nucleus as well as in the limbic nucleus accumbens where noradrenaline is also reduced. These regional differences may be significant as regards the symptomatology of Parkinson's disease, which is primarily motor but later in the course of the illness may involve cognitive and other disturbances.

Smith: Walle Nauta asked what the connection was between 'limbic' striatum and the 'motor' striatum, but in fact he gave one answer in his talk. In the rat there is input from the nucleus accumbens to the substantia nigra in addition to the projection from the accumbens to the ventral pallidum (Nauta et al 1978). The accumbens sends input to the very same cells in the nigra that project up to the dorsal striatum (Somogyi et al 1981), which is probably the 'motor' striatum. This is one way in which limbic input can feed its way through into what we think is a motor area. We also found convergence back from the dorsal striatum onto the same nigrostriatal neurons that receive the accumbens input (Somogyi et al 1981). So there are parallel pathways with interacting loops which swing round through the nigra so that each part of the striatum may know what the others are doing (see Fig. 2).

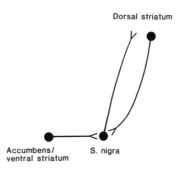

FIG. 2 (*Smith*). Diagram representing the monosynaptic connections that have been revealed by morphological analysis of nigrostriatal neurons in the rat.

Marsden: That is a very non-specific connection, I think, because of the wide distribution of the nigrostriatal dopamine fibres, which are distributed to vast areas of the dorsal striatum.

Somogyi: Perhaps a general influence like that is just what is needed for a limbic influence on motor control.

Smith: The critical question is, what is the identity of the postsynaptic target of that particular neuron, i.e. the nigrostriatal neuron that receives input from the accumbens?

DeLong: Does the projection from the nigra back to the motor area really go to the dorsal lateral quadrant of the striatum, Dr Smith? In Dr Nauta's diagram that area seemed to have little if any input. The dorsal striatum may not be motor striatum.

Smith: I agree. We haven't mapped the area.

Nauta: I was describing the rat brain. The monkey might be different.

Rolls: There are other ways in which different parts of the striatum might interact. The route Dr Smith has just described is one way. The second way might be simply to feed through all the results of what is happening in different parts of the striatum to different cortical areas via the pallidum and thalamus.

The results of what had been processed separately in the basal ganglia might finally be brought together only by corticocortical connections between the different cortical regions influenced by different parts of the striatum.

Another route for interactions arises from the work of Percheron et al (1984) who have found that the dendrites in the globus pallidus and in the substantia nigra are disc-shaped and orthogonal to the plane of the incoming striatal fibres. The discs are very large and only about five are needed to cover, for example, the whole area in which striatal afferents will be coming into the pallidus. This potentially allows some anatomical interaction between different areas of the striatum. We don't know the physiological significance of this yet.

Yoshida: Nishino et al (1984) have made recordings from caudate neurons in the monkey (*Macaca fuscata*). When the monkey looked at food neuronal firing increased greatly, but when the animal was satisfied with the food or when it looked at an inedible object there was no such response from the same neuron. These neurons are distributed very diffusely in the peripheral area of the head of the caudate nucleus.

Marsden: What is the relationship of the substantia innominata to the ventral pallidum?

DeLong: The neurons of the basal nucleus of Meynert (nbM) are to some degree intermingled with those of the ventral pallidum. Moreover, nbM neurons extend up and into the laminae of the pallidum. The constant relationship of the basal nucleus with the pallidum in all species is very striking. It is only in primates that the nbM neurons become more segregated. One certainly wonders if there isn't some cross-talk there, as Dr Nauta has speculated for the ventrotegmental neurons sitting in the path of the median forebrain bundle. I don't know of any evidence that nbM and globus pallidus neurons have anatomical connections or that they share some common input. However, Susan Mitchell in our laboratory has been comparing neurons in the ventral and dorsal pallidum of the monkey with the neurons in the basal forebrain and within the laminae. These are preliminary results but the neurons in the ventral pallidum have properties more like those in the adjacent nbM than those that extend into the laminae, which seem more like adjacent globus pallidus neurons. Some interesting conversations may be going on. If the nbM receives inputs from the basal ganglia, this would provide an unsuspected non-thalamic pathway from basal ganglia to cortex.

REFERENCES

Brinkman C 1981 Lesions in supplementary motor area interfere with a monkey's performance of a bimanual coordination task. Neurosci Lett 27:267-270

Brinkman C, Porter R 1979 Supplementary motor area in the monkey: activity of neurons during performance of a learned motor task. J Neurophysiol 42:681-709

Divac I 1977 Does the neostriatum operate as a functional entity? In: Cools AR et al (eds) Psychobiology of the striatum. Elsevier, Amsterdam, p 21-30

Fallon JH, Loughlin SE, Ribak CE 1983 The islands of Calleja complex of rat basal forebrain. III. Histochemical evidence for a striatopallidal system. J Comp Neurol 218:91-120

Farley IJ, Hornykiewicz O 1976 Noradrenaline in subcortical brain regions of patients with Parkinson's disease and control subjects. In: Birkmayer W, Hornykiewicz O (eds) Advances in Parkinsonism. Editiones Roche, Basel, p 178-185

Farley IJ, Price KS, Hornykiewicz O 1977 Dopamine in the limbic regions of the human brain: normal and abnormal. Adv Biochem Psychopharmacol 16:57-64

Fink JS, Smith GP 1980 Mesolimbic and mesocortical dopaminergic neurons are necessary for normal exploratory behavior in rats. Neurosci Lett 17:61-65

Heimer L, Switzer RD, Van Hoesen GW 1982 Ventral striatum and ventral pallidum. Additional components of the motor system? Trends Neurosci 5:83-87

Hikosaka O, Wurtz RH 1983 Visual and oculomotor functions of monkey substantia nigra pars reticula. III. Memory-contingent visual and saccade responses. J Neurophysiol 49:1268-1284

Ingvar M, Lindvall O, Stenevi U 1983 Apomorphine-induced changes in local cerebral blood flow in normal rats and after lesions of the dopaminergic nigrostriatal bundle. Brain Res 262:259-265

Jones B, Mishkin M 1972 Limbic lesions and the problem of stimulus-reinforcement associations. Exp Neurol 36:362-377

Kelly PAT, McCulloch J 1981 Heterogeneous depression of glucose utilization in the caudate nucleus by GABA agonists. Brain Res 209:458-436

Mogenson GJ, Jones DL, Yim CY 1980 From motivation to action: functional interface between the limbic system and the motor system. Prog Neurobiol 14:69-97

Nauta WJH, Smith GP, Faull RLM, Domesick VB 1978 Efferent connections and nigral afferents of the nucleus accumbens septi in the rat. Neuroscience 3:385-401

Nishino H, Ono T, Sasaki K, Fukuda M, Muramoto K 1984 Caudate unit activity during operant feeding behavior in monkeys and modulation by cooling prefrontal cortex. Behav Brain Res 11(1):21-34

Palacios JM, Kuhar MJ, Rapoport SI, London ED 1982 Effects of γ-aminobutyric acid agonist and antagonist drugs on local cerebral glucose utilization. J Neurosci 2:853-860

Percheron G, Yelnik J, Francois C 1984 The striato-pallido-nigral bundle and 3-dimensional geometry of pallidal and nigral neurons in primates: a convergent system for cortical outputs. In: McKenzie J, Wilcox L (eds) The basal ganglia: structure and function. Plenum, New York

Rolls ET 1984 Responses of neurons in different regions of the striatum of the behaving monkey. In: McKenzie J, Wilcox L (eds) The basal ganglia: structure and function. Plenum, New York

Rolls ET, Ashton J, Williams G et al 1982 Neuronal activity in the ventral striatum of the behaving monkey. Soc Neurosci Abstr 8:169

Somogyi P, Bolam JP, Totterdell S, Smith AD 1981 Monosynaptic input from the nucleus accumbens-ventral striatum region to retrogradely labelled nigrostriatal neurones. Brain Res 217:245-263

Yim CY, Mogenson GJ 1983 Response of ventral pallidal neurons to amygdala stimulation and its modulation by dopamine projections to nucleus accumbens. J Neurophysiol 50:148-161

Neurochemically specified subsystems in the basal ganglia

ANN M. GRAYBIEL

Department of Psychology and Brain Science, Massachusetts Institute of Technology, E25-618, Cambridge, Massachusetts 02139, USA

Abstract. The fibre pathways associated with the basal ganglia include through-conduction lines and side-loops associated with the striatum, pallidum and substantia nigra. Each of these regions is now known to contain subdivisions differing from one another in the neurotransmitter-related compounds they contain. This paper includes an outline of these new findings and a commentary on some of their functional implications.

1984 Functions of the basal ganglia. Pitman, London (Ciba Foundation symposium 107) p 114-149

The division of the supraspinal motor mechanism into 'pyramidal' and 'extrapyramidal' systems was first drawn by S. A. Kinnier Wilson (Wilson 1914, 1929), whose contrast between the manifestations of corticospinal dysfunction and the signs of abnormalities affecting the corpus striatum and midbrain remains a central formulation of clinical neurology. At the time, anatomical findings seemed to support this dichotomy, because the corpus striatum was thought to have independent access to the spinal motor mechanism by way of indirect pathways comparable to the cerebello-rubrospinal connection. By the 1940s, however, the view had emerged that major efferent connections of the cerebellum and corpus striatum travel in the ascending direction and that they converge on the thalamic zone which projects to the motor cortex. This left investigators to grapple with the paradox that the main input to the pyramidal system seemed to come from the extrapyramidal motor system.

There has now been a decided shift away from this view, partly as a consequence of the new and generally more reliable information about extrapyramidal fibre connections which has been gained by the use of axon transport methods. In fact, available evidence suggests that the basal ganglia can engage, through thalamic intermediaries, a considerable part of the frontal cortex, and that a diverse set of subcortical structures comes under the influence of the basal ganglia as well. A corresponding breadth of functional affiliations is now being shown for the pathways leading into the basal ganglia, for these range from the classically recognized corticostriatal connections from sensorimotor and association cortex to channels originating in the hippocampus, hypothalamus and midbrain raphe nuclei.

These new findings are bringing to the known fibre connections of the basal ganglia a richness comparable to the diversity of symptoms observed in disorders of the basal ganglia. They also form a crucial framework for consideration of the neurochemical anatomy of the basal ganglia. In particular, the differentiation now being discovered in the neurotransmitter-related coding in these pathways opens up the possibility of very complex neural interactions and modulations in the basal ganglia, and suggests that the regulation of neurotransmission along different conduction lines may be controlled according to their chemical specification as well as their diverse anatomical affiliations. The potency of dopamine-containing pathways in influencing movement and affect may well reflect this dual complexity.

Anatomical framework

Major facts leading to a modified view of extrapyramidal pathways can be summarized as follows:

(1) The thalamic nuclei receiving direct projections from the corpus striatum (by way of pallidofugal fibres originating in the internal pallidal segment) do not project exclusively to the motor cortex proper (area 4), nor do their cortical projection mainly overlap those of the cerebello-thalamo-cortical pathway. Instead, they appear to lead principally to neocortical regions lying anterior to area 4 (Uno & Yoshida 1975, Kievit & Kuypers 1977, Hendry et al 1979, Tracey et al 1980, DeLong & Georgopoulos 1981, Mehler 1981, DeVito & Anderson 1982, Schell & Strick 1983). Premotor cortex of area 6, including specifically the supplementary motor cortex, is thought to lie in the cortical distribution field of the pallido-thalamo-cortical connection mediated by the ventral thalamic complex (Schell & Strick 1983).

(2) There are efferent pallidal connections that descend at least as far as the parabrachial tegmentum, thus entirely bypassing the direct pallido-thalamo-cortical mechanism (Herkenham & Nauta 1977, 1979, Graybiel & Ragsdale 1979, Moon Edley & Graybiel 1983). These descending pathways include a contingent of fibres terminating in the lateral part of the habenula, a nucleus embedded in the circuitry of the limbic system. Fibres from this habenular subdivision project to the dorsal raphe nucleus, which in turn projects both to

the striatum and to the dopamine-containing cell groups of the midbrain that innervate the striatum and other forebrain regions.

(3) The substantia nigra pars reticulata, long suspected on anatomical grounds to be a caudal continuation of the internal pallidum (Schwyn & Fox 1974), is now known to emit major efferent conduction lines leading to the thalamus, to the intermediate layers of the superior colliculus, and to the midbrain tegmentum (Graybiel 1978, Graybiel & Ragsdale 1979, Carpenter 1981, Beckstead et al 1981). The most intensively studied of these pathways is the nigrotectal connection, which has now been shown to contribute to the control of eye movements, especially those elicited by visual stimulation (Hikosaka & Wurtz 1983a–d).

(4) A remarkable linkage between the basal ganglia and the limbic system is now being documented in studies of ventral subdivisions of the striatum including the ventral caudoputamen, the nucleus accumbens septi and adjoining parts of the olfactory tubercle (Heimer & Wilson 1975, Nauta & Domesick 1978, Kelley et al 1982). These regions (together called the 'ventral striatum' by Heimer & Wilson 1975) receive their major afferent connections from components of the classically defined limbic system (for example, from the subicular cortex of the hippocampal formation and from the amygdala). This ventral striatal district projects massively to a restricted part of the substantia innominata that now is recognized as a specialized extension of the pallidum (the 'ventral pallidum' of Heimer & Wilson 1975). Its efferent connections are mainly directed towards affiliates of the limbic system, including the thalamic mediodorsal nucleus, which innervates prefrontal cortex (Goldschmidt & Heimer 1980, Haber et al 1982, Switzer et al 1982).

(5) Though the striatum is still considered to be the main point of access into the basal ganglia circuitry as a whole, there is growing evidence that this access is by no means exclusive. Alternative points of entrance include prominent pathways leading into the 'loop nuclei' of the corpus striatum: the intralaminar nuclei of the thalamus, the subthalamic nucleus, the dopamine-containing cell groups of the midbrain and the parabrachial pedunculopontine ('TPc') nucleus (see Hartmann-von Monakow et al 1978, Nauta & Domesick 1978, Graybiel & Ragsdale 1979, Carpenter 1981, Moon Edley & Graybiel 1983). For example, there are prominent direct projections from the motor cortex to the intralaminar nucleus centre median and to the subthalamic nucleus; and the pars compacta of the substantia nigra and associated dopamine-containing cell groups are now recognized as major entrance zones for pathways originating in the subcortical limbic continuum including the preoptic area, the hypothalamus and the dorsal raphe nucleus, and probably for descending pathways from prefrontal and premotor regions of the neocortex as well (Swanson 1976, Künzle 1978, Nauta & Domesick 1978, Beckstead 1979).

Families of the extrapyramidal pathways

At first glance, this new information seems mainly to complicate rather than clarify the organization of the extrapyramidal pathways. In fact, however, it is already possible to recognize, at least in outline, distinct families of related lines of conduction leading towards cortical and subcortical regions. Their partial overlap with one another constitutes a systematic stepwise extension of the extrapyramidal mechanism from the immediate realm of skeletomotor activity to levels of neural organization associated with affect and mentation.

The four parts of Fig. 1 were designed to emphasize a progressive shift in the cortical targets of these families from area 4 to the frontal pole, and the corresponding shift in their afferent affiliations from those most closely related to the sensory–motor periphery to those related to the limbic system including the hippocampal formation. The families thus identified include: (a) the cerebello-thalamo-cortical pathways leading principally to area 4 proper and to caudal premotor cortex (Fig. 1A); (b) the pallido-thalamo-cortical pathways leading to premotor fields principally including the so-called supplementary motor cortex in the medial part of area 6 (Fig. 1B); (c) nigro-thalamo-cortical pathways leading to the frontal eye fields (area 8) and to other premotor and possibly to prefrontal areas not yet mapped in the primate (Fig. 1C); and (d) a ventral pallido-thalamo-cortical connection leading to the prefrontal cortex (Fig. 1D). Too little information is available for these groupings to be more than sketched in, an especially vexing point of uncertainty being the possible differences among species in the organization of these pathways. For example, most information about the fibre connections of the ventral striatum and ventral pallidum still comes from work in the rat and cat, but we do not know how much of this information is directly applicable in the primate. For the dorsal striatum, we still do not know enough about the primate to be sure of what differentiation there may be between transthalamic circuits related to the caudate nucleus (which receives the main corticostriatal projection of the prefrontal cortex) and those related to the putamen (which receives the main corticostriatal projection of the motor cortex).

Despite these uncertainties, the general arrangement of these pathways recalls the organizational principles inferred from study of the transthalamic circuits of the posterior association cortex; namely, that the connections are neither strictly hierarchical nor strictly parallel, but formed into highly differentiated lines of conduction that often are interlocked by connections of the next-but-one ('indirectly recursive') form: motor cortex → putamen → thalamus → premotor cortex (Graybiel 1979, Graybiel & Berson 1981).

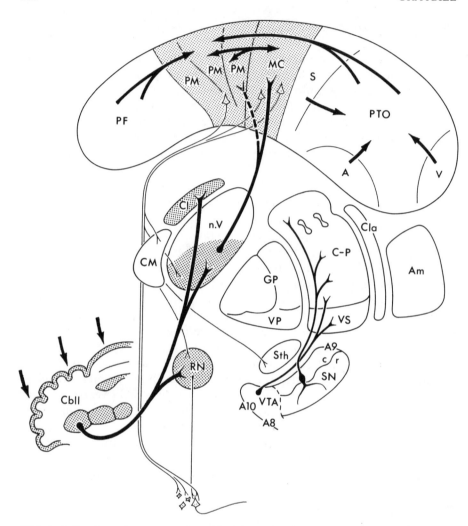

FIG. 1. A–D: schematic representation of four broad families of extrapyramidal pathways.

(A) The cerebello-thalamo-cortical channel leading predominantly to the motor cortex (MC), and probably to the caudal part of the premotor cortex (PM), is shown in heavy lines. Arrows pointing to cerebellum represent its cortical afferents, which are indirect. Cortical association pathways are shown in highly schematic form leading to the motor and premotor fields from the prefrontal cortex (PF) and from the parieto-temporo-occipital association cortex (PTO) which in turn receives inputs from the somatic sensory (S), visual (V) and auditory (A) areas. Many reciprocating pathways have been omitted from the diagram. The corticospinal tract is represented schematically along with its offset connections to the thalamus, to the subthalamic nucleus, and to the red nucleus. Note that the cerebellar pathways lie outside the range of influence of the dopamine-containing efferents of midbrain cell groups A8–A10.

NEUROCHEMICALLY SPECIFIED SUBSYSTEMS IN BASAL GANGLIA

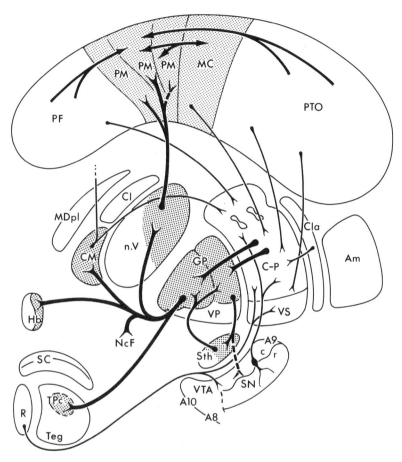

(B) The pallido-thalamo-cortical pathway, in heavy lines, is shown projecting mainly to premotor cortex. This channel originates in the internal pallidal segment, as do efferent connections leading to the habenular complex (Hb), prerubral field (NcF) and midbrain tegmentum (TPc, nucleus tegmenti pedunculopontinus pars compacta or 'pedunculopontine nucleus'). Also indicated in heavy lines are inputs to the pallidum from the striatum (C-P), and the pallido-subthalamo-pallidal loop. The main inputs to the caudoputamen, shown in finer lines, originate in the neocortex, the thalamic centre median–parafascicular complex (CM), the claustrum (Cla), the serotonin-containing dorsal raphe nucleus (R) and the midbrain dopamine-containing nuclei, principally the substantia nigra pars compacta (c, A9). Distinctions between afferents innervating the caudate nucleus and the putamen are not indicated (see text and Graybiel & Ragsdale 1979, 1983). The projection from the ventral tegmental area to the ventral caudoputamen is indicated in (A) and (D).

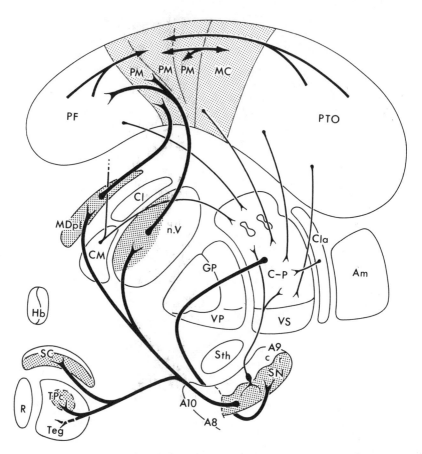

(C) The nigro-thalamo-cortical pathways carried over the medial part of the ventral thalamic complex and the paralamellar part of the mediodorsal nucleus (MDpl) are shown in heavy lines. Also shown in heavy lines are the strionigral pathway and the nigral efferents terminating in the deep layers of the superior colliculus and in the midbrain tegmentum, especially in the TPc nucleus.

Abbreviations for Fig. 1A–D
A, auditory cortex; Am, amygdaloid complex; A8, A9 and A10, cell groups A8, A9, A10 of Dahlström and Fuxe; Cbll, cerebellum; Cl, nucleus centralis lateralis of thalamus; Cla, claustrum; CM, centre median—parafascicular complex of thalamus; C-P, caudoputamen (caudate nucleus and putamen); GP, globus pallidus; Hb, habenular complex; MC, motor cortex; MD, mediodorsal nucleus of the thalamus; MDpl, paralamellar part of the thalamic mediodorsal nucleus; NcF, nucleus campi Foreli; n.V, nucleus ventralis of thalamus; PF,

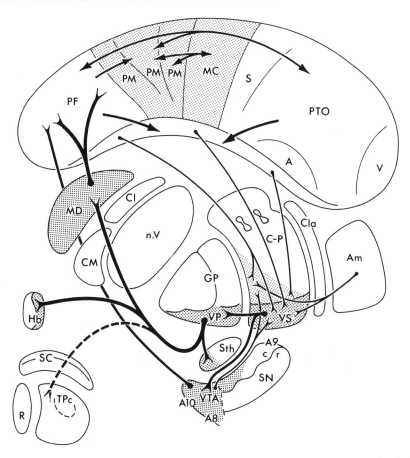

(D) The pathways linking the ventral pallidum (VP) with the prefrontal cortex (and with subcortical sites) are shown in heavy lines. The main inputs to the ventral pallidum originate in the ventral striatum (VS), which in turn receives input from the ventral tegmental area (VTA, A10) and A8 dopamine-containing cell groups of the midbrain, from the amygdala (Am) and from regions of 'prelimbic' and limbic cortex which themselves receive inputs from prefrontal (PF) and posterior (PTO) association cortex. Note the direct corticipetal dopamine system paralleling the indirect strio-pallido-thalamo-cortical pathway that leads through the mediodorsal nucleus.

prefrontal cortex; PM, premotor cortical fields; PTO, parieto-temporo-occipital association cortex; R, raphe nuclei (especially dorsal raphe nucleus); RN, red nucleus; S, somatic sensory cortex; SC, superior colliculus; SNc, substantia nigra pars compacta; SNr, substantia nigra pars reticulata; Sth, subthalamic nucleus; Teg, midbrain tegmentum; TPc, nucleus tegmenti peduncu-lopontinis, or 'pedunculopontine' nucleus; V, visual cortex; VP, ventral pallidum; VS, ventral striatum; VTA, ventral tegmental area (cell group A10).

Relation of the dopamine-containing innervation of the striatum to extrapyramidal families

There are notable parallels between this grouping of the cortical connections of the extrapyramidal motor system into families and the division of the dopamine-containing innervation of the forebrain into nigrostriatal, mesolimbic and mesocortical pathways (Dahlström & Fuxe 1964, Andén et al 1966, Ungerstedt 1971, Thierry et al 1973). First, as shown in Fig. 1A, the cerebellar connections are not within the range of these dopamine-containing pathways. Second, dopamine modulation of the pathways leading to the premotor cortex and to the prefrontal cortex are distinct from one another. It is mainly neurons in the substantia nigra pars compacta (cell group A9) that innervate the striopallidal and strionigral mechanisms, whereas it is mainly neurons in the ventral tegmental area (cell group A10) and in the retrorubral nucleus (cell group A8) that innervate the ventral striopallidal system. Third, the prefrontal family is further distinguished from the others because the prefrontal cortex receives a direct dopamine-containing innervation originating in a subset of A10 neurons. Though this direct mesocortical path seems to reiterate the indirect trans-striatal influence that cell group A10 exerts on the frontal cortex, the cells of origin of the direct and indirect pathways are at least partly non-overlapping (Deniau et al 1980, Fallon 1981, Swanson 1982). The mesocortical system innervating the frontal cortex is further reported to be pharmacologically distinct from the subcortical dopamine-containing innervations because the direct path lacks autoreceptors (Bannon et al 1982). This raises the important general point that distinctly different types of control can be exerted through apparently similar pathways, even when a common neurotransmitter is involved, by virtue of the different pharmacological properties as well as the different intermediate connectivities characterizing the pathways.

It is not yet known whether the premotor pathways originating in the pallidum and substantia nigra pars reticulata are related to distinct dopamine innervations. Other differentiations have been found within the dopamine-containing nigrostriatal projection, however, and these are probably reflected in subdivisions of the pallidal and nigral pathways (see below). For example, in the primate, separate components of the pars compacta lead preferentially to the caudate nucleus and to the putamen (Smith et al 1983). Within the caudate–putamen complex considered as a whole, there is a further differentiation of the dopamine innervation into an early-forming system of 'dopamine islands' and a later-arriving diffusely distributed system (Fuxe et al 1979). The functional significance of this division within the nigrostriatal connection is not yet understood, but there is now evidence, reviewed below (and in Graybiel 1984), that the zones of termination of the dopamine islands form one set of elements in a general mosaic organization of striatal fibre connections.

Neurochemical differentiation within the basal ganglia

The division of the dopamine-containing innervation of the striatum into nigrostriatal and mesolimbic systems has been recognized for nearly 20 years and the mesocortical innervation for a decade. By contrast, internal differentiation in the neurochemical organization of the basal ganglia is a recent discovery, and information about neurochemical subdivisions in the striatum, pallidum and substantia nigra is being added at a rapid rate. The implications of the new findings are potentially very great, for they suggest that individual pathways and groups of pathways making up the circuitry of the basal ganglia may be under differential chemical control and thus open to highly specific intervention in the clinical situation and to finely individuated control under normal physiological conditions.

Early views of the neurochemical organization of the basal ganglia emphasized a relatively simple pattern: that in the striatum there were dopamine-containing inputs affecting, directly or indirectly, cholinergic interneurons and γ-aminobutyric acid (GABA)-containing efferent neurons; and that not only the outputs of the striatum but also the outputs of the two pallidal segments and the pars reticulata of the substantia nigra were carried by GABA-containing fibres. There is now compelling evidence, however, that we must add to this framework a broad spectrum of neurochemical control mechanisms exerted both at the level of the striatum and at the level of the pallidum and the substantia nigra and ventral tegmental area. Most, but not all, of the new findings relate to the differential distribution of neuropeptides in the basal ganglia; and though immunohistochemical methods are still limited, some of these differential distributions can already be linked to the pathway anatomy. There are also suggestions of non-standard forms of neural transmission in the dopamine pathways brought about by release of dopamine from dendrites in the substantia nigra and, apparently, by 'presynaptic' effects on the terminal ramifications of dopamine-containing axons in the striatum.

The pallidum

An important conclusion from recent immunohistochemical work is that the striopallidal pathways innervating the internal and external segments of the globus pallidus are distinct from one another in terms of the neuropeptides they contain (Haber & Elde 1981). The external segment is richly innervated by enkephalin-containing afferents, the internal segment by substance P-containing afferents. This contrasting distribution of enkephalin and substance P has now been demonstrated in the human (Fig. 2). As the main input to the pallidum comes from the striatum, the immediate implication of these findings is that the striatum can exert separate control over the pallido-subthalamo-pallidal loop originating in the external segment and over the pallido-thalamo-cortical and other pallidofugal pathways originating in the internal pallidal segment. Further support for such a differential neurochemical organization comes from the more recent finding that dynorphin is also unevenly distributed within the pallidum, being most heavily concentrated in the internal pallidal segment (Chesselet & Graybiel 1983a,b).

The substance P-rich internal segment can itself be divided into two parts, of which the more medial expresses moderate enkephalin-like immunoreactivity (Haber & Elde 1981, and Fig. 2). This chemical subdivision is of interest because of Carpenter's suggestion that the medial and lateral parts of the internal pallidal segment exert individually distinct controls over the thalamus. Specifically, Carpenter and his colleagues (see Carpenter 1981) have reported that the anza lenticularis originates mainly from the lateral part of the internal segment and innervates especially the rostral part of the pallidorecipient ventral thalamic nucleus, whereas the lenticular fasciculus originates mainly from the medial part of the inner segment and projects to more caudal parts of this pallidorecipient thalamic zone.

The limbic-related ventral pallidum, like the medial part of the internal pallidal segment, contains not only GABA but also enkephalin, dynorphin and substance P (Switzer et al 1982, Haber & Nauta 1983, M.-F. Chesselet & A. M. Graybiel unpublished).

The pars reticulata of the substantia nigra

Marked differences in the distribution of immunoreactivities to these and other neuropeptides are now also being discovered in the substantia nigra pars reticulata (Chesselet & Graybiel 1983a,b). Dynorphin and substance P-like immunoreactivities (like immunoreactivity to glutamic acid decarboxylase) are distributed throughout the pars reticulata, as they are in the internal pallidum. By contrast, there are highly localized distributions of enkephalin-like immunoreactivity in the pars reticulata (shown for the cat in Fig. 3) and of somatostatin-like immunoreactivity in the pars lateralis (see Fig. 3) (Chesselet & Graybiel 1983a,b, Graybiel & Elde 1983). These new findings on the substantia nigra suggest that the pars reticulata–pars lateralis complex of the substantia nigra, no less than the globus pallidus, is divided into districts modulated by different peptide-containing pathways. Though the substance P, dynorphin and the extrinsic GABA are almost certainly derived at least in part from the striatum, the origins of the enkephalin and somatostatin found in the substantia nigra are not certain and may well be non-striatal (Chesselet & Graybiel 1983a,b).

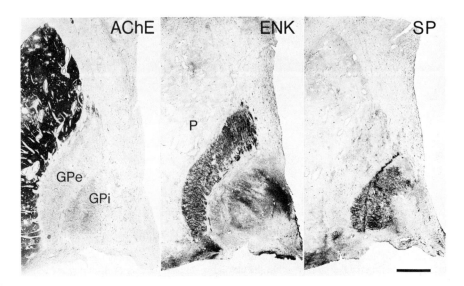

FIG. 2. Transverse sections through the globus pallidus of the adult human brain stained for acetylcholinesterase (AChE), enkephalin-like immunoreactivity (ENK) and substance P-like immunoreactivity (SP). High levels of AChE mark the adjoining putamen (P). Within the globus pallidus, the external segment (GPe) and the internal segment (GPi) are characterized by different peptide immunoreactivities, ENK being dense in GPe and moderate in the medial part of GPi, whereas SP is dense in GPi. Scale bar 3 mm.

Major questions remain to be answered about the findings on both the pallidum and the nigra. It is not known how the peptide-containing pathways are related to those containing GABA; nor is it certain that a full inventory has yet been drawn up for neurotransmitter-specific channels leading to the pallidum and substantia nigra pars reticulata; nor is it clear whether the pathways innervating the pallidum and the pars reticulata are largely collaterals of one another or largely separate or, as suggested elsewhere, a mixture of branched and unbranched axons (see Graybiel & Ragsdale 1983). Finally, though there are hints that some of the peptide-containing pathways reaching the pallidum and the substantia nigra pars reticulata originate outside the striatum (Chesselet & Graybiel 1983a,b), definitive evidence on this matter is not yet available.

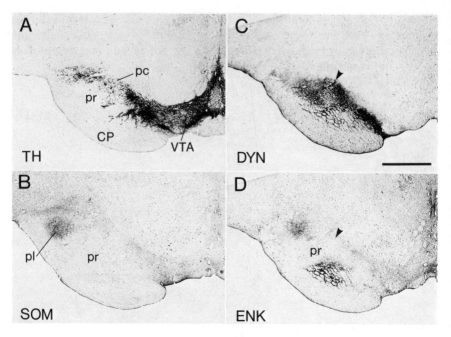

FIG. 3. Heterogeneous distributions of neuropeptides within the substantia nigra, suggesting the existence of several subdivisions within the classically defined borders of this nuclear complex. Sections from adult cat brain. The pattern of tyrosine hydroxylase-like immunoreactivity (TH) is shown in (A) for reference in identifying the pars compacta (pc) and ventral tegmental area (VTA). Strands of high TH extend ventrally into the pars reticulata (pr). (B) shows that somatostatin-like immunoreactivity (SOM) characterizes the pars lateralis of the substantia nigra (pl). (C) and (D): dynorphin-like immunoreactivity (DYN) is present throughout the pars reticulata, but enkephalin-like immunoreactivity (ENK) is mainly confined to a central and ventral part of the pars reticulata. Sections shown in (C) and (D) are serially adjacent to one another, and the arrowheads mark the same blood vessel in the two sections. Note weak ENK above dorsal rim of DYN. CP, cerebral peduncle.

The pars compacta of the substantia nigra

Peptide-mediated control is exerted not only on the pallidum and pars reticulata of the substantia nigra, but also on at least some of the dopamine-containing cells giving rise to the nigrostriatal and mesolimbic tracts. For example, injections of exogenous substance P into cell groups A9 and A10 are reported to have behavioural effects resulting from the interaction of substance P and the dopamine-containing neurons of these nuclei (see Iversen 1982). Cholecystokinin coexists with dopamine in some of the medially situated neurons of this complex, and has been shown to affect the physiological activity of neurons there (Hökfelt et al 1980, Skirboll et al 1981).

There is a growing body of immunohistochemical evidence for the existence of other peptides in parts of the nigral complex that have dopamine-containing neurons. In the cat, careful mapping of the neuropeptides in relation to the disposition of tyrosine hydroxylase (TH)-containing neurons suggests that current definitions of the pars compacta do not reflect adequately the differential peptidergic control of nigral dopamine-containing neurons in different parts of this nigral subdivision (Chesselet & Graybiel 1983a,b). For example, at some levels the lateral zone expressing immunoreactivity to somatostatin does not contain TH-positive neurons, but at other levels it does. It will be of great interest to learn whether these neurochemically distinct districts are related to distinguishable elements in the nigrostriatal system.

The subthalamic nucleus and the nucleus tegmenti pedunculopontinis, pars compacta (TPc)

In sharp contrast to the dorsal and ventral pallidum and the substantia nigra, each of which receives a massive fibre projection from the striatum, the subthalamic nucleus is not distinguished by high concentrations of the peptides enkephalin, dynorphin and substance P. Nor, apparently, do its neurons share with those of the pallido-nigral complex a high content of glutamate decarboxylase (GAD)-like immunoreactivity even in animals pretreated with colchicine (Oertel et al 1982). Nauta & Cuénod (1982) have presented evidence for high affinity uptake and retrograde transport of [^{3}H]GABA to the subthalamic nucleus after injections of this labelled amino acid into both segments of the pallidum, and they suggest on this account that the subthalamopallidal pathway is GABAergic. In earlier work, Yoshida and colleagues suggested that this pathway contains the inhibitory amino acid glycine (Yoshida 1974). The subthalamic nucleus is distinguished by a high content of butyrylcholinesterase (A. M. Graybiel & Ragsdale unpublished), as are the pallidum and the substantia nigra pars reticulata (Chesselet & Graybiel 1983a,b).

The pedunculopontine nucleus (TPc) lies in the dorsolateral part of the tegmental reticular formation and is now recognized as an important 'loop nucleus' of the corpus striatum (Graybiel & Ragsdale 1979, Moon Edley & Graybiel 1983). As shown in Fig. 1, TPc receives fibre projections from the internal pallidal segment and from the substantia nigra pars reticulata. In turn, TPc projects to the pars compacta of the substantia nigra, to the subthalamic nucleus, and to the globus pallidus (especially to the internal pallidal segment, or entopeduncular nucleus). The neurotransmitters used in these pathways are not known, but the TPc region almost certainly overlaps the complex of cholinergic neurons situated in the lateral part of the midbrain tegmentum, and peptide-containing neurons have also been found in this region.

The striatum

Nowhere has neurochemical compartmentalization in the basal ganglia been more vividly demonstrated than in the striatum. Many of the neurotransmitter-related substances that so far have been seen in the dorsal and ventral striatum have been found to express heterogeneity in their local distributions. As a consequence, it is now generally acknowledged that striatal regions no more than a few tenths of a millimetre apart may have sharply different histochemical appearances.

This neurochemical patterning is of increasing interest because of evidence that it reflects a compartmental plan of organization impressed on the afferent and efferent connections of the striatum and on the arrangement of striatal interneurons as well. Unravelling the rules by which the compartments are ordered should therefore help in discovering the way the connections are disposed in relation to one another—a task that up to the present has proved extremely difficult.

In the dorsal striatum, major units of this compartmental organization are the so-called 'striosomes' (striatal bodies). These zones were first identified with conventional acetylcholinesterase histochemistry, when it was found that, in lightly stained sections through the striatum, 300–600 μm-wide pockets of very pale staining were scattered in an otherwise darkly staining neuropil (Graybiel & Ragsdale 1978). Quite similar patterns were found in cat, monkey and human; Fig. 4 illustrates such a cross-section through the adult human striatum.

It is now known that the acetylcholinesterase-poor striosomes correspond to patches or annuli of striatal neuropil having high enkephalin-like immunoreactivity; and that they are distinguished by either distinctly high or distinctly low substance P-like immunoreactivity in the neuropil (Graybiel et al 1981). Fig. 5 illustrates the enkephalin staining pattern for the human. It has further been found that the neuropil of the striosomes is characterized by high GAD-like immunoreactivity (Graybiel et al 1983), by high neurotensin-like immunoreactivity (Goedert et al 1983), and by low somatostatin-like immunoreactivity (Gerfen 1983, M.-F. Chesselet & A. M. Graybiel unpublished). Opiate binding sites in the striatum of adults, and muscarinic binding sites in the immature striatum, have also been shown to have patchy

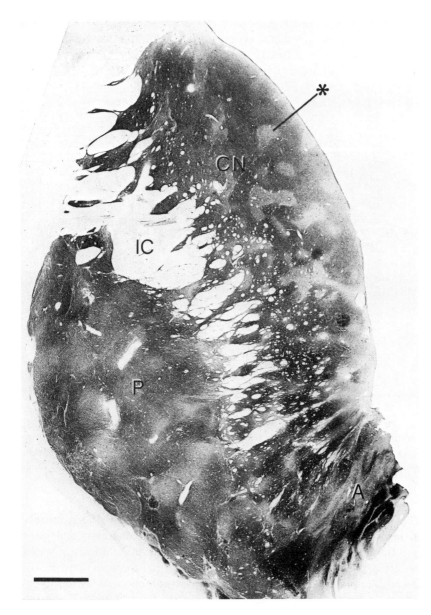

FIG. 4. Pattern of acetylcholinesterase (AChE) staining in a transverse section through the adult human striatum (CN, caudate nucleus, P, putamen, A, nucleus accumbens). Note prominent compartments of low AChE staining (one marked by asterisk). These compartments are called striosomes. IC, internal capsule. Scale bar: 3 mm.

distributions with zones of high binding density matching the striosomes (Herkenham & Pert 1981, Nastuk & Graybiel 1983). Many of these substances have a compartmental arrangement in the nucleus accumbens and olfactory tubercle as well (see Graybiel & Ragsdale 1979, Graybiel et al 1981, Moon Edley & Herkenham 1982, Ragsdale & Graybiel 1983).

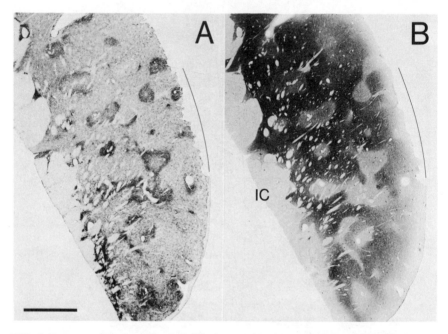

FIG. 5. Demonstration that enkephalin-like immunoreactivity (shown in A) is distributed in a compartmental fashion in the adult human striatum and that the compartments match the AChE-poor striosomes (shown in B). (A) and (B) illustrate serially adjoining sections. Patches of high enkephalin-like immunoreactivity in (A) mainly have ring-and-hollow form. IC, internal capsule. Scale bar: 3 mm.

The striosomal mosaics visualized histochemically have been correlated with local patterning in the distributions of striatal afferent and efferent connections, and there are tantalizing clues that the chemically specified zones are related to a compartmental ordering of these connections. First, the striosomes are now known to correspond to sites innervated by the so-called 'dopamine island' fibres, which comprise a distinct early-arriving contingent of nigrostriatal fibres (Graybiel et al 1983, Graybiel & Ragsdale 1983, Graybiel 1984). Second, the massive corticostriatal projection (thought to be glutamatergic) is divided into fibre systems that avoid the striosomes (as shown in Fig. 6 for certain frontostriatal fibres in the monkey), and other systems that project specifically to the striosomes (see Ragsdale & Graybiel 1981, Graybiel 1983, Donoghue & Herkenham 1983). Third, thalamostriatal fibres from the intralaminar nuclei avoid the striosomes (Graybiel 1982, Herkenham & Pert 1981). Least is known for the efferent organization, but patchy distributions of labelled neurons have been observed in the striatum

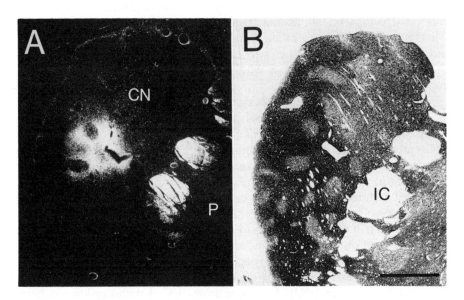

FIG. 6. Evidence that corticostriatal fibres originating in the monkey prefrontal cortex innervate the striatum, so avoiding acetylcholinesterase-poor striosomes. (A): autoradiographically-labelled corticostriatal fibres are shown in white. (B): serially adjoining section stained for AChE. Two prominent striosomes lie in the field of innervation, and these are only very lightly innervated by the labelled fibres. CN, caudate nucleus, P, putamen; IC, internal capsule. Scale bar: 2 mm.

after injections of retrograde tracers into the pallidum or substantia nigra. Patches of weak retrograde labelling and patches of pronounced cell labelling have both been found to lie in spatial register with striosomes (Graybiel et al 1979, Gerfen 1983).

A preliminary but potentially clarifying new finding is that striosomes contain clusters of cell bodies expressing high levels of immunoreactivity to substance P and to dynorphin B (Graybiel & Chesselet 1984). This suggests that some elements of the substance P and dynorphin-containing striopallidal and strionigral projection systems may have preferential origins within the striosomes. By contrast, somatostatin-containing and cholinergic interneurons tend to lie outside these zones.

Commentary

Channels in the extrapyramidal motor system

The high degree of neurochemical differentiation being discovered in the basal ganglia and associated cell groups suggests that many of the major pathways in the extrapyramidal motor system have different chemical signatures. The compartmentalization evident in histochemical studies far exceeds what would have been expected from the pathway anatomy alone. The biochemical anatomy thus has particular value in leading to the prediction that specialized channels exist within broad categories of extrapyramidal conduction.

Neurochemical partitioning is evident already in the striatum, the single largest staging centre of the basal ganglia circuitry. This is all the more remarkable because, as far as is known, the massive corticostriatal projections are not differentiated in this way: they seem to use a common neurotransmitter, glutamate (or aspartate). The implication is that neural processing in the striatum may represent just the opposite of what has been suggested before: rather than being a 'funnelling mechanism' mediating convergence of cortical and thalamic inputs onto the pallidum, the striatum, by its modular nature, may serve to set up channels in the extrapyramidal motor mechanism, much as the complex neural circuitry of the retina establishes channels that are carried through many synapses in the visual system and even into visuomotor regions of the brain (see, e.g., Graybiel & Berson 1981, Schiller 1982). No single trans-striatal pathway has yet been fully specified from its afferent to its efferent side, nor from the striatal output to its extensions in conduction lines leading out from the pallidum and substantia nigra. Given the separate findings so far available, however, it seems likely that within the striatum striosomes represent a means for channelling particular kinds of information from a subset of cortical and subcortical regions to one or more of the target structures of striatal efferents; and that within the globus pallidus and and substantia nigra pars reticulata further channels are established according to the chemical districts now being identified.

A key finding in relation to this proposal is that for each level of subdivision so far identified in the striatum, there are corresponding components of its dopamine-containing innervation. This is true for the broad division of the striatum into dorsal and ventral parts; for the division of the nucleus accumbens into medial hippocampal-recipient and lateral parts; and even for the division of the dorsal striatum into striosomal and non-striosomal compartments (with the striosomes corresponding at least in part to the islandic dopamine system). The newly reported finding that the dopamine-containing innervations of the primate caudate nucleus and putamen arise from different groups of neurons in the pars compacta is probably further evidence of such channelling. The implication is that there may be a spectrum of control exerted by different groups of dopamine-containing neurons on the pathways leading through the striatum.

Extrapyramidal loops

Most inputs to the pallidum originate in the striatum (a major exception being those from the subthalamic nucleus, which are embedded in the pallido-subthalamo-pallidal loop). This restrictiveness of access confers a dominant functional importance on the main input elements—the subdivisions of the striatum—and on the conduction lines established by means of the intrastriatal organizations discussed above. There is increasing evidence, however, that the loop circuits of the basal ganglia also provide points of access to the pallidum and substantia nigra; and there are hints that the neurochemical organization of these loops may be crucial to the function of the 'main' striopallidal and strionigral lines.

First, for the globus pallidus, the pallido-subthalamo-pallidal loop sets up a bypass pathway, quite massive in the primate, by which afferent systems projecting to the subthalamic nucleus could influence the pallidum without first projecting to the striatum. A key aspect of the biochemical anatomy of this bypass is that it would allow escape from modulation by the nigrostriatal dopamine system. Fibre projections to the subthalamic nucleus are limited: aside from inputs from other loop nuclei (and from the pallidum), the principal afferent system arises in the motor cortex. Weaker projections originate in localized parts of the premotor cortex (Hartmann-von Monakow et al 1978). The emphasis on the motor cortex is of special interest, because without the intermediary of the dopamine-modulated striatum, the actions of the motor cortex on the pallidum could have a higher safety factor than those of the trans-striatal path, and could be fast enough to lead in time those of the indirect pathway.

This potential for bypassing the striatum holds also for the second loop system of the globus pallidus, which has the TPc nucleus as its node. In contrast to the subthalamic loop, the pallido-TPc pathway originates in the internal pallidal segment, and the pars reticulata of the substantia nigra also projects to TPc. Like the subthalamic nucleus, TPc receives direct input from the motor cortex. TPc is widely integrated into the circuitry of the basal ganglia, for it projects to the subthalamic nucleus, the thalamic intralaminar nuclei and the pars compacta of the substantia nigra in addition to the pallidum (Moon Edley & Graybiel 1983). The fact that TPc lies in a region containing many cholinergic neurons suggests that some of the cholinergic actions attributed to other pathways in the basal ganglia (and some of the effects of cholinergic drugs on extrapyramidal function) could be mediated through one or more of these TPc loop pathways. In particular, the projection from TPc to the substantia nigra pars compacta may represent an extrastriatal mechanism for affecting the balance between cholinergic and dopaminergic functions in the basal ganglia. These pathways may also, or alternatively, have peptidergic functions, as the TPc region also contains many peptidergic neurons. Only the pedunculopallidal connection has been studied electrophysiologically. It is reported to produce monosynaptic excitatory drive on neurons in both the external and internal segments (Gonya-Magee & Anderson 1983).

The third loop system to be mentioned serves to link the serotonin-containing dorsal raphe nucleus to pallidal and nigral circuitry. Pallidal input to the dorsal raphe nucleus is carried over the pallido-habenulo-raphe pathway, and the dorsal raphe nucleus projects to the substantia nigra (apparently including its pars compacta), to the striatum (especially its ventral parts) and to the pallidum (see Moore 1981, Graybiel & Ragsdale 1983, Steinbusch & Nieuwenhuys 1983). These extrapyramidal connections of the dorsal raphe nucleus are of considerable interest in view of evidence that there are profound changes in neural activity in the dorsal raphe nucleus through the spectrum of sleep states and arousal, and accompanying changes in motor tone (see Trulson & Jacobs 1981). If the raphe loops modulate extrapyramidal function in parallel with these changes in activity, they could be implicated in one of the most remarkable and still unexplained characteristics of choreoathetotic disease: that the abnormal ('involuntary') movements tend to cease during sleep.

Control of the dopamine-containing innervation of the striatum

Because of its known clinical importance, the striato-nigro-striatal circuit is by far the most intensively studied loop pathway of the corpus striatum. There is now electron microscope evidence for at least some direct striatonigral input to the dopamine-containing neurons of the pars compacta, so the 'loop' is at least in part direct (Somogyi et al 1981). On the key question of what other afferent fibres control the loop by projecting to the substantia nigra, the answer so far is that there is remarkably limited access to the pars compacta and to the adjoining ventral tegmental area (A10) and retrorubral nucleus (A8). To take one example, most of the neocortex (save for the prefrontal and premotor cortex: see Künzle 1978 & Beckstead 1979) is thought to emit only modest direct fibre projections, if any, to these dopamine-containing cell groups; in fact, the existence of corticonigral connections has long been a matter of some uncertainty. Most of the known afferents to the pars compacta of the substantia nigra, and to the adjoining cell groups A8 and A10, arise from subcortical structures, especially from basal forebrain, hypothalamic and midbrain regions included in the limbic system as broadly defined. The dopamine-containing pathways leading to the striatum may thus represent a means by which parts of the brain regulating visceroendocrine and affective functions can influence parts concerned with perceptuomotor function.

There is indirect evidence that other mechanisms may open up the range of access routes to the dopamine-containing neurons. Jacques Glowinski's group has emphasized that the release of dopamine within the striatum may be modulated not only by way of pathways directly influencing cell groups A8–A10 but also by other pathways projecting to the striatum and interacting with the dopamine-containing fibres terminating there (see Giorguieff-Chesselet et al 1980, Chesselet 1984). In particular, these investigators propose that the intrastriatal release of dopamine is controlled, in part, by presynaptic actions on the dopamine-containing terminals. This suggests that although afferents projecting to the substantia nigra pars compacta may set the level of firing of nigrostriatal neurons, final control over the effects of the nigrostriatal pathway may nevertheless be exerted at the level of the striatum itself. If this proves to be the case, our views about the functional implications of the standard 'pathway anatomy' of the basal ganglia will have to be revised. Access to the dopaminergic mechanism, in other words, could be broadened to include one or more of the major afferent systems innervating the striatum and acting on the dopamine-containing terminals directly or by way of interneurons within the striatum.

There is a precedent for non-standard control of the dopamine mechanism, at least within the substantia nigra. The principal initial finding was that dopamine is released in the substantia nigra. This led to the suggestion that the release occurs from dopamine-containing dendrites of pars compacta neurons, many of which extend ventrally in prominent strands to infiltrate the pars reticulata (Björklund & Lindvall 1975, Geffen et al 1976, Groves et al 1975, Chéramy et al 1981). Dopamine receptors ('autoreceptors') on dendrites of the pars compacta neurons are thought to mediate self-inhibition of the nigral neurons by the dopamine so released (Geffen et al 1976, Skirboll et al 1979, Roth 1980). The possibility of release of dopamine by collaterals of pars compacta neurons has not been fully excluded (Wassef et al 1981, Grofová et al 1983), but evidence for autoregulation of the nigrostriatal pathway (Aghajanian & Bunney 1973) is scarcely in doubt. Recent findings suggest that this regulation may be differentiated within the spatial domain by virtue of electrotonic coupling of pars compacta neurons (Grace & Bunney 1983a,b,c) and because of non-uniformities in the properties of their dendritic membranes (Llinás et al 1984).

Peptides in the basal ganglia

There is a sharp contrast between the slow and tonic firing of dopamine-containing neurons in the midbrain (and the low rates of spontaneous firing in the dopamine-recipient striatum) and the phasic, high frequency firing of neurons in the pallidum and the pars reticulata of the substantia nigra (DeLong & Georgopoulos 1981). The dopamine-containing innervation—by all counts crucial for normal functioning in the basal ganglia—somehow appears to 'modulate' neural firing patterns in the striatum. Against this background, it is of special interest that much of the neurochemical heterogeneity now being discovered in the basal ganglia relates to the distribution of neuropeptides. Even though the functional range over which these substances may work still has not been adequately specified, certain characteristics of the neuropeptides may be of crucial relevance to their actions in the basal ganglia and to the unusual properties of the striatal dopamine innervations as well. First, some of the peptides act slowly (i.e. their postsynaptic effects peak slowly) and their actions, like those of muscarinic cholinergic and dopaminergic elements, persist for relatively long periods of time (measured at least in minutes). This could be because they act through second-messenger transduction mechanisms, or because they may not be inactivated rapidly, or because they might act at a distance from their sites of release (see, e.g., Krieger et al 1983). Second, there may be special requirements for the release of the peptides so that they function only under certain highly restricted circumstances (see Lundberg et al 1982 for the action of vasoactive intestinal polypeptides [VIP] in autonomic postganglionic neurons). Third, there is evidence that in the peripheral nervous system certain peptides may act at sites other than those immediately postsynaptic to the peptide-containing processes releasing them (Jan & Jan 1983). Fourth, some peptides may have influences that are directed towards the control of metabolic processes rather than (or in addition to) the control of machinery immediately concerned with neural conduction (Krieger et al 1983). Each of these special characteristics of peptides may be reflected in the functional organization of the basal ganglia.

Time scale of neural interactions in the basal ganglia. The prolonged action of the neuropeptides suggests that they may be influential in bringing about changes in the general responsiveness or in the time course of responsiveness of neurons in the basal ganglia. This possibility is worth emphasizing because so many of the clinical symptoms of basal ganglia disease are characterized by episodic or relatively more permanent changes in the quality of movement or affect-state. Examples that come to mind are episodes of involuntary movement (including involuntary vocalizations) in patients with choreas or tics; the on–off phenomena seen in patients treated for parkinsonism; and, on a longer time scale, the fundamental changes in tone and readiness for action of the motor apparatus in individuals suffering from extrapyramidal disorders (Wilson 1929, Martin 1967, Denny-Brown & Yanagisawa 1976). At a different level of analysis, there is physiological evidence that intermittent firing patterns are characteristic of neurons of the external pallidal segment (DeLong & Georgopoulos 1981), and that, among visuo-oculomotor neurons of the pars reticulata, firing sequences can be prolonged enough, and yet specifically event-related enough, to suggest that they have a mnemonic function (Hikosaka & Wurtz 1983c). As both slow muscarinic cholinergic and dopaminergic transmission systems also typify striatal circuits, it seems reasonable to guess that prolonged changes in activity could occur under the impetus of several different conduction lines in the basal ganglia, and that these would be related to one another. In the aggregate, these slow systems must provide the context for the presumably fast actions of glutamatergic/aspartergic cortical afferents and for other rapidly triggered, rapidly terminated transmissions.

Facultative linkages and gain control. Evidence from work on the peripheral nervous system suggests that peptide release may differ from classical neurotransmitter release not only in its duration but also in requiring high rates of stimulation and mobilization of peptide from prohormonal pools in the cytoplasm of the parent neurons (Lundberg et al 1982). If these characteristics hold for central peptidergic transmission, it is possible that peptide-containing pathways could be functionally engaged or disengaged, depending on activity at the sites of synthesis and of release of the peptides in question. The range of firing rates in different nuclei of the basal ganglia may have special significance in this regard because the large number of side-loops associated with the strio-pallido-thalamic and strio-nigro-thalamic pathways—as well as these main pathways themselves—may be open to such fluctuations in activity. According to this suggestion, abnormalities in the pace of discharge along any one of these loops could enhance or decrease the effectiveness of the loop and modify the functioning of the through-paths in an oscillatory as well as in a steady manner. Increase or decrease of loop activity could lead to repetitive bouts of abnormal activity, just as changes in loop activity could lead to difficulties in maintaining a motor stance or in changing from one motor stance to another. These possibilities suggest that part of the apparent change in volitional activity in extrapyramidal disorders could result from abnormalities in the gains of one or more of the loops rather than from changes in initiating signals from regions projecting to the basal ganglia.

Local interactions in striatal neuropil. Every neuropeptide so far localized within the striatum by immunohistochemistry, and also neuropeptide-related ligand binding in the striatum, has been shown to respect the ordering of striatal neuropil into macroscopic striosomal compartments. The fact that the input–output connections of the striatum are broken up into mosaics according to these chemically specified compartments raises the possibility that the compartments represent mechanisms whereby groups of neurons and their processes can be modulated in a coordinated way. Such group modulation could occur through a high density of conventional synaptic contacts established by synchronized inputs. The evidence for electrotonic coupling of nigral neurons projecting to the striatum is of interest in this context (Grace & Bunney 1983a–c). A second possibility is that the striosomes may form sites for local extracellular diffusion of neuroactive substances. There is already evidence in the peripheral nervous system for non-synaptic effects being exerted by luteinizing hormone-releasing hormone at distances approaching the diameter of striosomes (Jan & Jan 1983). It therefore seems plausible that the striosomes themselves could sustain some form of local action by virtue of being within the size range accommodating local diffusion effects, and it may be relevant that they seem to be at least partly rimmed by septa (Graybiel & Ragsdale 1978, 1983). An obvious extension of this hypothesis is that some of the presynaptic effects proposed by Glowinski and his co-workers could be mediated by such local extracellular actions. The dopamine island fibres, being spatially confined to the striosomes, are candidate targets for such effects. Another observation that may be related to compartmentalized regulation is that metabolic activity in the striatum can be non-uniform, as measured by 2-deoxyglucose in the adult (Brown & Wolfson 1983) and by cytochrome oxidase activity in the newborn (Ragsdale & Graybiel 1983), and can be affected non-uniformly in the adult by the application of exogenous VIP (McCulloch et al 1983). Both conventional and non-conventional mechanisms for local control would be consistent with compartmentalization of glia and elements of the microcirculation as well as of neurons and their processes.

Conclusion

Special emphasis has been placed on the neurochemistry of the basal ganglia and their allied nuclei since the dopamine-containing innervation of the striatum was discovered 20 years ago. From this interest have come demonstrations of the high concentrations within the basal ganglia of many of the neurotransmitters and neurotransmitter-related compounds known to act in the brain. We are still at an early stage in attempting to bring together information about these neurotransmitters with information about the pathway anatomy and physiology of the extrapyramidal motor system. New perspectives on the basal ganglia are already beginning to emerge, however, as a result of these efforts. First, there is a growing appreciation that the pathways in the basal ganglia are multiply subdivided according to their chemical specification, and thus are almost certainly subdivided functionally. Second, not only the nigrostriatal loop but also the large number of other loop systems of the basal ganglia can now be viewed as means for modulating these channels and for engaging them or disengaging them according to chemically specified constraints. The functional affiliations of these channels and loops seem as broad as the range of neurotransmitters represented within them. A third realization is that there may be a great range in the action times of different components of the extrapyramidal circuitry. Finally, the basal ganglia, and especially the striatum and the substantia nigra, have become model systems for studying interactions among specified neurotransmitter systems. How these chemically coded systems of the basal ganglia are controlled genetically, and to what extent extrapyramidal disease processes are expressed in terms of their chemical specificities, are among the important questions raised by these new findings. For work on the biochemical organization of the basal ganglia may provide knowledge useful in the clinical situation as well as insights into the logic and cell biology of the brain.

Acknowledgements

It is a pleasure to thank Mr Henry Hall and Ms Linda Hale for their help, and the Seaver Institute for its support.

REFERENCES

Aghajanian GK, Bunney BS 1973 Central dopaminergic neurons: neurophysiological identification and responses to drugs. In: Usdin E, Snyder SH (eds) Frontiers in catecholamine research. Pergamon Press, Oxford, p 643-648

Andén NE, Dahlström A, Fuxe K, Larsson K, Olson L, Ungerstedt U 1966 Ascending monoamine neurons to the telencephalon and diencephalon. Acta Physiol Scand 67:313-326

Bannon MJ, Reinhard JF Jr, Bunney EB, Roth RH 1982 Unique response to antipsychotic drugs is due to absence of terminal autoreceptors in mesocortical dopamine neurones. Nature (Lond) 296:444-446

Beckstead RM 1979 An autoradiographic examination of corticocortical and subcortical projections of the medio-dorsal-projection (prefrontal) cortex in the rat. J Comp Neurol 184:43-62

Beckstead RM, Edwards SB, Frankfurter A 1981 A comparison of the intranigral distribution of nigrotectal neurons labeled with horeradish peroxidase in the monkey, cat and rat. J Neurosci 1:121-125

Björklund A, Lindvall O 1975 Dopamine in dendrites of substantia nigra neurons: suggestions for a role in dendritic terminals. Brain Res 83:531-537

Brown LL, Wolfson LI 1983 [^{14}C]Deoxyglucose studies of the normal rat striatum show heterogeneity and regional differences in glucose utilization. Neurosci Abstr 9:662.

Carpenter MB 1981 Anatomy of the corpus striatum and brain stem integrating systems. In: Brooks VB (ed) Motor control, part 2. American Physiological Society, Bethesda, MD (Handb Physiol sect 1: The nervous system, vol 2) p 947-995

Chéramy A, Leviel V, Glowinski J 1981 Dendritic release of dopamine in the substantia nigra. Nature (Lond) 289:537-542

Chesselet M-F 1984 Presynaptic regulations of neurotransmitter release in the brain: facts and hypothesis. Neuroscience, in press

Chesselet M-F, Graybiel AM 1983a Met-enkephalin-like and dynorphin-like immunoreactivities of the basal ganglia of the cat. Life Sci 33:37-40

Chesselet M-F, Graybiel AM 1983b Subdivisions of the pallidum and the substantia nigra demonstrated by immunohistochemistry. Neurosci Abstr 9:16

Dahlström A, Fuxe K 1964 Evidence for the existence of monoamine containing neurons in the central nervous system. 1. Demonstration of monoamines in the cell bodies of brain stem neurons. Acta Physiol Scand 62 (Suppl 232):1-55

DeLong MR, Georgopoulos AP 1981 Motor functions of the basal ganglia. In: Brooks VB (ed) Motor control, part 2. American Physiological Society, Bethesda, MD (Handb Physiol, sect 1: The nervous system, vol 2), p 1017-1061

Deniau JM, Thierry AM, Feger J 1980 Electrophysiological identification of mesencephalic ventromedial tegmental (VMT) neurons projecting to the frontal cortex, septum and nucleus accumbens. Brain Res 189:315-326

Denny-Brown D, Yanagisawa N 1976 The role of the basal ganglia in the initiation of movement. In: Yahr MD (ed) The basal ganglia. Raven Press, New York (Research publications: Association for Research in Nervous and Mental Disease, vol 55) p 115-149

DeVito JL, Anderson ME 1982 An autoradiographic study of efferent connections of the globus pallidus in Macaca mulatta. Exp Brain Res 46:107-117

Donoghue JP, Herkenham M 1983 Multiple patterns of corticostriatal projections and their relationship to opiate receptor patches in rats. Neurosci Abstr 9:15

Fallon JH 1981 Collateralization of monoamine neurons: mesotelencephalic dopamine projections to caudate, septum, and frontal cortex. J Neurosci 1:1361-1368

Fuxe K, Anderson K, Schwarcz R et al 1979 Studies on different types of dopamine nerve terminals in the forebrain and their possible interactions with hormones and with neurons containing GABA, glutamate, and opioid peptides. Adv Neurol 24:199-215

Geffen LB, Jessell TM, Cuello AC, Iversen LL 1976 Release of dopamine from dendrites in rat substantia nigra. Nature (Lond) 260:258-260

Gerfen CR 1983 Non-topographic order in the basal ganglia: evidence for a second level of organization superimposed upon the topographically ordered striato-nigral projection system. Neurosci Abstr 9:16

Giorguieff-Chesselet M-F, Chéramy A, Glowinski J 1980 In vivo and in vitro studies on the presynaptic control of dopamine release from nerve terminals of the nigrostriatal dopaminergic neuron. In: Littauer UZ et al (eds) Neurotransmitters and their receptors. John Wiley, New York, p 33-47

Goldschmidt RB, Heimer L 1980 The rat olfactory tubercle: its connections and relation to the strio-pallidal system. Neurosci Abstr 6:271

Gonya-Magee T, Anderson ME 1983 An electrophysiological characterization of projections from the pedunculopontine area to entopeduncular nucleus and globus pallidus in the cat. Exp Brain Res 49:269-279

Goedert M, Mantyh PW, Hunt SP, Emson PC 1983 Mosaic distribution of neurotensin-like immunoreactivity in the cat striatum. Brain Res 274:176-179

Grace AA, Bunney BS 1983a Intracellular and extracellular electrophysiology of nigral dopaminergic neurons, 1: Identification and characterization. Neuroscience 10:301-315

Grace AA, Bunney BS 1983b Intracellular and extracellular electrophysiology of nigral dopaminergic neurons. 2: Action potential generating mechanisms and morphological correlates. Neuroscience 10:317-331

Grace AA, Bunney BS 1983c Intracellular and extracellular electrophysiology of nigral dopaminergic neurons. 3: Evidence for electrotonic coupling. Neuroscience 10:333-348

Graybiel AM 1978 Organization of the nigrotectal connection: an experimental tracer study in the cat. Brain Res 143:339-348

Graybiel AM 1979 Some patterns of connectivity in the central nervous system: a tribute to Rafael Lorente de Nó. In: Wilson VJ, Asanuma H (eds) Integration in the nervous system. Igaku-Shoin, Tokyo, p 69-96

Graybiel AM 1982 Correlative studies of histochemistry and fibre connections in the central nervous system. In: Chan-Palay V, Palay S (eds) Cytochemical methods in neuroanatomy. Alan R Liss, New York p 46-67

Graybiel AM 1983 Compartmental organization of the mammalian striatum. Prog Brain Res 58:247-256

Graybiel AM 1984 Modular patterning in the development of the striatum. In: IBRO symposium volume, Raven Press, New York, in press

Graybiel AM, Berson DM 1981 On the relation between transthalamic and transcortical pathways in the visual system. In: Schmitt FO et al (eds) The organization of the cerebral cortex. MIT Press, Cambridge, MA, p 285-319

Graybiel AM, Chesselet M-F 1984 Distribution of cell bodies expressing substance P, enkephalin and dynorphin B in kitten and cat striatum. Anat Rec 208:64A

Graybiel AM, Elde RP 1983 Somatostatin-like immunoreactivity characterizes neurons of the nucleus reticularis thalami in the cat and monkey. J Neurosci 3:1308-1321

Graybiel AM, Ragsdale CW Jr 1978 Histochemically distinct compartments in the striatum of human, monkey and cat demonstrated by acetylcholinesterase staining. Proc Natl Acad Sci USA 75:5723-5726

Graybiel AM, Ragsdale CW Jr 1979 Fiber connections of the basal ganglia. Prog Brain Res 51:239-283

Graybiel AM, Ragsdale CW Jr 1983 Biochemical anatomy of the striatum. In: Emson PC (ed) Chemical neuroanatomy. Raven Press, New York, p 427-504

Graybiel AM, Ragsdale CW Jr, Moon Edley S 1979 Compartments in the striatum of the cat observed by retrograde cell-labeling. Exp Brain Res 34:189-195

Graybiel AM, Ragsdale CW Jr, Yoneoka ES, Elde RP 1981 An immunohistochemical study of enkephalins and other neuropeptides in the striatum of the cat with evidence that the opiate peptides are arranged to form mosaic patterns in register with the striosomal compartments visible by acetylcholinesterase staining. Neuroscience 6:377-397

Graybiel AM, Chesselet M-F, Wu J-Y, Eckenstein F, Joh TE 1983 The relation of striosomes in the caudate nucleus of the cat to the organization of early-developing dopaminergic fibres, GAD-positive neuropil, and CAT-positive neurons. Neurosci Abstr 9:14.

Grofová I, Kita H, Kitai ST 1983 Morphology of nigrostriatal neurons intracellularly stained with HRP. Neurosci Abstr 9:661

Groves PM, Wilson CJ, Young SJ, Rebec GV 1975 Self-inhibition by dopaminergic neurons. An alternative to the "neuronal feedback loop" hypothesis for the mode of action of certain psychotropic drugs. Science (Wash DC) 190:522-529

Haber SN, Elde RP 1981 Correlation between met-enkephalin and substance P immunoreactivity in the primate globus pallidus. Neuroscience 6:1291-1297

Haber, SN, Nauta WJH 1983 Ramifications of the globus pallidus in the rat as indicated by patterns of immunohistochemistry. Neuroscience 9:245-260

Haber SN, Groenewegen HJ Nauta WJH 1982 Efferent connections of the ventral pallidum in the rat. Neurosci Abstr 8:169

Hartmann-von Monakow K, Akert K, Künzle H 1978 Projections of the precentral motor cortex and other cortical areas of the frontal lobe to the subthalamic nucleus in the monkey. Exp Brain Res 33:395-403

Heimer L, Wilson RD 1975 The subcortical projections of the allocortex: similarities in the neural associations of the hippocampus, the piriform cortex, and the neocortex. In: Santini M (ed) Golgi centennial symposium. Raven Press, New York, p 177-193

Hendry SHC, Jones EG, Graham J 1979 Thalamic relay nuclei for cerebellar and certain related fiber systems in the cat. J Comp Neurol 185:679-714

Herkenham M, Nauta WJH 1977 Afferent connections of the habenular nuclei in the rat. A horseradish peroxidase study, with a note on the fiber-of-passage problem. J Comp Neurol 173:123-146

Herkenham M, Nauta WJH 1979 Efferent connections of the habenular nuclei in the rat. J Comp Neurol 187:19-47

Herkenham M, Pert CB 1981 Mosaic distribution of opiate receptors, parafascicular projections and acetylcholinesterase in rat striatum. Nature (Lond) 291:415-418

Hikosaka O, Wurtz RH 1983a Visual and oculomotor functions of monkey substantia nigra pars reticulata. I: Relation of visual and auditory responses to saccades. J Neurophysiol (Bethesda) 49: 1230-1253

Hikosaka O, Wurtz RH 1983b Visual and oculomotor functions of monkey substantia nigra pars reticulata. II: Visual responses related to fixation of gaze. J Neurophysiol (Bethesda) 49:1254-1267

Hikosaka O, Wurtz RH 1983c Visual and oculomotor functions of monkey substantia nigra pars reticulata. III: Memory-contingent visual and saccade responses. J Neurophysiol (Bethesda) 49:1268-1284

Hikosaka O, Wurtz RH 1983d Visual and oculomotor functions of monkey substantia nigra pars reticulata. IV: Relation of substantia nigra to superior colliculus. J Neurophysiol (Bethesda) 49:1285-1301

Hökfelt T, Skirboll L, Rehfeld MF, Goldstein M, Markey K, Dann O 1980 A subpopulation of mesencephalic dopamine neurons projecting to limbic areas contains a cholecystokinin-like peptide: evidence from immunohistochemistry combined with retrograde tracing. Neuroscience 5:2093-2124

Iversen SD 1982 Behavioural effects of substance P through dopaminergic pathways in the brain. In: Substance P in the nervous system. Pitman, London (Ciba Found Symp 91) p 307-324

Jan YN, Jan LY 1983 Some features of peptidergic transmission. Prog Brain Res 58:49-59

Kelley AE, Domesick VB, Nauta WJH 1982 The amygdalostriatal projection in the rat—an anatomical study by anterograde and retrograde tracing methods. Neuroscience 7:615-630

Kievit J, Kuypers HGJM 1977 Organization of the thalamo-cortical connexions to the frontal lobe in the rhesus monkey. Exp Brain Res 29:299-322

Krieger DT, Brownstein MJ, Martin JB 1983 Brain peptides. John Wiley, New York, 1032 pp

Künzle H 1978 An autoradiographic analysis of the efferent connections from premotor and adjacent prefrontal regions (areas 6 and 9) in *Macaca fascicularis*. Brain Behav Evol 15:185-234

Llinás R, Greenfield SA, Jahnsen H 1984 Electrophysiology of pars compacta cells in the in vitro substantia nigra—a possible mechanism for dendritic release. Brain Res, in press

Lundberg JM, Hedlund B, Anggard A et al 1982 In: Bloom SR et al (eds) Systemic role of regulatory peptides. Schattauer, Stuttgart, p 145-168

Martin J Purdon 1967 The basal ganglia and posture. Pitman Medical, London

McCulloch J, Kelly PAT, Uddman R, Edvinsson L 1983 Functional role for vasoactive intestinal polypeptide in the caudate nucleus: a 2-deoxy [^{14}C]glucose investigation. Proc Natl Acad Sci USA 80:1472-1476

Mehler WR 1981 The basal ganglia—circa 1982: a review and commentary. Appl Neurophysiol 44:261-290

Moon Edley S, Herkenham M 1982 Cellular compartments and mosaics of opiate receptors, afferent fiber terminations, and cholinesterase levels in the nucleus accumbens of the rat. Neurosci Abstr 8:173

Moon Edley S, Graybiel AM 1983 The afferent and efferent connections of the feline nucleus tegmenti pedunculopontinus, pars compacta. J Comp Neurol 217:187-215

Moore RY 1981 The anatomy of central serotonin neuron systems in the rat brain. In: Jacobs BL, Gelperin A (eds) Serotonin neurotransmission & behavior. MIT Press, Cambridge, MA, p 35-71

Nastuk MA, Graybiel AM 1983 The distribution of muscarinic binding sites in the feline striatum and its relationship to other histochemical staining patterns. Neurosci Abstr 9:15

Nauta HJW, Cuénod M 1982 Perikaryal cell labeling in the subthalamic nucleus following the injection of [^{3}H]-γ-aminobutyric acid into the pallidal complex: an autoradiographic study in cat. Neuroscience 7:2725-2734

Nauta WJH, Domesick VB 1978 Crossroads of limbic and striatal circuitry: hypothalamo-nigral connections. In: Livingstone KE, Hornykiewicz O (eds) Limbic mechanisms. Plenum, New York, p 75-93

Oertel WH, Tappez ML, Berod A, Mugnaini E 1982 Two color immunohistochemistry for dopamine and GABA neurons in rat substantia nigra and zona incerta. Brain Res Bull 9:463-474

Ragsdale CW Jr, Graybiel AM 1981 The fronto-striatal projection in the cat and monkey and its relationship to inhomogeneities established by acetylcholinesterase histochemistry. Brain Res 208:259-266

Ragsdale CW Jr, Graybiel AM 1983 Butyrylcholinesterase in the dorsal and ventral striatum: observations on histochemical distributions in adult, fetal and neonatal cats. Neurosci Abstr 9:15

Roth RH 1980 Dopamine autoreceptors: pharmacology, function and comparison with postsynaptic dopamine receptors. Commun Psychopharmacol 3:429-445

Schell GR, Strick PL 1983 Origin of thalamic input to the supplementary and arcuate premotor areas. Neurosci Abstr 9: 490

Schiller PH 1982 Central connections of the retinal ON and OFF pathways. Nature (Lond) 297:580-583

Schwyn RC, Fox CA 1974 The primate substantia nigra: a Golgi and electron microscopic study. J Hirnforsch 15:95-126

Skirboll LR, Grace AA, Bunney BS 1979 Dopamine auto- and postsynaptic receptors: electrophysiological evidence for differential sensitivity of dopamine agonists. Science (Wash DC) 206:80-82

Skirboll LR, Grace AA, Hommer DW et al 1981 Peptide-monoamine coexistence: studies of the actions of cholecystokinin-like peptide on the electrical activity of midbrain dopamine neurons. Neuroscience 6:2111-2124

Smith Y, Mackey A, DeBellefeuille L, Parent A 1983 The output organization of the substantia nigra in primate. Neurosci Abstr 9:662

Somogyi P, Bolam JP, Totterdell S, Smith AD 1981 Monosynaptic input from the nucleus accumbens-ventral striatum region to retrogradely labelled nigrostriatal neurones. Brain Res 217:245-263

Steinbusch HWM, Nieuwenhuys R 1983 The raphe nuclei of the rat brainstem: a cytoarchitectonic and immunohistochemical study. In: Emson PC (ed) Chemical neuroanatomy. Raven Press, New York, p 131-207

Swanson LW 1976 An autoradiographic study of the efferent connections of the preoptic region in the rat. J Comp Neurol 167:227-256

Swanson LW 1982 The projections of the ventral tegmental area and adjacent regions: a combined fluorescent retrograde tracer and immunofluorescence study in the rat. Brain Res Bull 9:321-353

Switzer RC, Hill J, Heimer L 1982 The globus pallidus and its rostroventral extension into the olfactory tubercle of the rat: a cyto- and chemoarchitectural study. Neuroscience 7:1891-1904

Thierry AM, Blanc G, Sobel A, Stinus L, Glowinski J 1973 Dopaminergic terminals in the rat cortex. Science (Wash DC) 182:499-501

Tracey DJ, Asanuma C, Jones EG, Porter R 1980 Thalamic relay to motor cortex: afferent pathways from brain stem, cerebellum, and spinal cord in monkeys. J Neurophysiol (Bethesda) 44:532-554

Trulson ME, Jacobs BL 1981 Activity of serotonin-containing neurons in freely moving cats. In: Jacobs BL, Gelperin A (eds) Serotonin neurotransmission and behavior. MIT Press, Cambridge, MA, p 339-365

Ungerstedt U 1971 Stereotaxic mapping of the monoamine pathways in the rat brain. Acta Physiol Scand 197:1-48

Uno M, Yoshida M 1975 Monosynaptic inhibition of thalamic neurons produced by stimulation of the pallidal nucleus in cats. Brain Res 99: 377-380

Wassef M, Berod A, Sotelo C 1981 Dopaminergic dendrites in the pars reticulata of the rat substantia nigra and their striatal input. Combined immunocytochemical localization of tyrosine hydroxylase and anterograde degeneration. Neuroscience 6:2125-2139

Wilson SA Kinnier 1914 An experimental research into the anatomy and physiology of the corpus striatum. Brain 36:427-492

Wilson SA Kinnier 1929 Modern problems in neurology. William Wood & Co, New York, p 120-296

Yoshida M 1974 Functional aspects of, and role of transmitter in, the basal ganglia. Confin Neurol 36:282-291

DISCUSSION

Hornykiewicz: In an attempt to map out, in as much detail as possible, the various neurotransmitter patterns in the human brain, we have recently measured a series of neurotransmitter markers, especially dopamine, noradrenaline and choline acetyltransferase (EC 2.3.1.6), in several subdivisions of the basal ganglia (cf. Hornykiewicz 1981, Hörtnagl et al 1983). Our procedure involves biochemical analysis in fresh-frozen post-mortem material. We divide that part of the hemisphere containing the basal ganglia coronally into 16–18 slices of 3-mm thickness. These slices include all subdivisions of the dorsal and ventral striatum (caudate nucleus, putamen and nucleus accumbens) as well as the globus pallidus and other basal ganglia regions. In each slice, the caudate nucleus, putamen and accumbens are further subdivided into up to nine fields according to dorsoventral and mediolateral coordinates.

The results of our analyses of dopamine and noradrenaline concentrations and choline acetyltransferase activity indicate that all these neurotransmitter systems have specific and highly inhomogeneous distribution patterns within the striatal areas. To date, the following conclusions can be drawn from our studies:

(1) In the head of the caudate nucleus the overall dopamine concentration increases quite significantly in the rostrocaudal direction and falls markedly in the caudate body and tail, roughly giving a bell-shaped distribution curve. The gradient for choline acetyltransferase activity is similar to although not as steep as that for dopamine.

(2) The highest dopamine levels are found in the centre core of the caudate head and the lowest levels in the ventral and medial parts of the nucleus. The latter parts of the caudate also contain the lowest choline acetyltransferase activity. However, the highest choline acetyltransferase activity is found in the dorsolateral segments of the nucleus.

(3) Similar inhomogeneities in dopamine and choline acetyltransferase distribution are also seen in the putamen but here the subregional differences are not as pronounced as in the caudate nucleus, and there is no decline of dopamine and choline acetyltransferase in the caudal putamen subdivisions.

(4) Noradrenaline concentrations are generally very low in the caudate nucleus (less than 0.1 µg/g) and slightly higher in the putamen. In the putamen there is a rather shallow dorsoventral and rostrocaudal gradient, with the ventral and caudal segments containing somewhat more noradrenaline than the dorsal and rostral segments.

(5) A strikingly different pattern of distribution is seen in the area of the ventral (and rostral) striatum constituting the nucleus accumbens, that is the junctional area between the caudate nucleus and the putamen. In contrast to the very low noradrenaline levels in the adjoining parts of the caudate nucleus and putamen (0.1 µg/g or less), the accumbens as a whole contains substantial amounts (around 0.6 µg/g) of this catecholamine. Within the accumbens area, noradrenaline shows a highly inhomogeneous distribution. Whereas the rostral portions contain low noradrenaline levels similar to the levels in the neighbouring caudate nucleus and putamen, the noradrenaline concentration increases drastically in the caudal direction and reaches its highest values in the most dorsomedio-caudal sector, where the noradrenaline level (3–4 µg/g) often exceeds that of dopamine. In contrast, the dopamine and choline acetyltransferase distribution in the accumbens has an inverse pattern, with the lowest levels of dopamine and choline acetyltransferase in the noradrenaline-rich dorsocaudal accumbens subdivisions and high values in the noradrenaline-poor rostral areas.

In summary, it can be said that within the basal ganglia each of the three neurotransmitter systems has its own highly specific and inhomogeneous distribution pattern. Since *in situ* these patterns are projected upon each other, the full picture is like an irregular mosaic, consisting of intricate, from point to point different, networks of interconnected neurotransmitter patches of varying relative composition. It can be assumed that these subtle subregional neurotransmitter patterns impart a strong functional heterogeneity to the various striatal and limbic subdivisions.

Evarts: What does the correspondence between the muscarinic receptors and the dopamine-containing islands mean from the standpoint of the synaptic relations of cells, terminals and inputs, Dr Graybiel?

Graybiel: The hard evidence can only come from electron microscope studies because light microscope autoradiography doesn't give good enough resolution for us to know whether the binding sites we label are presynaptic or postsynaptic. Even so, Mary Nastuk and I have been struck by how detailed the correspondences are between the dopamine islands that we see with fluorescence and the patches of muscarinic ligand binding that we see with autoradiography in serially adjoining sections (Nastuk & Graybiel 1983). One very interesting possibility raised in Dr Glowinski's laboratory is that there may be presynaptic control over the dopamine-containing fibres in the striatum (Giorguieff-Chesselet et al 1980). If so, we may be seeing evidence for muscarinic cholinergic control over the dopaminergic input. Mary and I would very much like to pursue this hypothesis.

Even if the muscarinic ligand is binding to traditional postsynaptic receptors, the results still suggest that the dopamine islands converge with a powerful cholinergic system. We may be seeing part of a mechanism for dopaminergic – cholinergic balance in the striatum.

Evarts: What presynaptic structures might converge to terminate in the same islands? I am thinking now of the muscarinic receptors and wondering about the presynaptic structures that may end on them.

Graybiel: The cholinergic interneurons are obviously one possibility. I don't know whether these might be the big cells you have recorded from. We see the cell bodies of cholinergic neurons more often outside than inside the striosomes but we don't know yet about their axons. As to the other presynaptic structures, there are highly ordered cortical and other projections to these zones, but none of these are known to be cholinergic.

Di Chiara: Before thinking about a presynaptic function of these receptors I would think about a postsynaptic one. As Glowinski and others have shown, the dopaminergic terminals have receptors for GABA and glutamate as well as for acetylcholine, and the reverse is also true. Presynaptic control of the terminals in the striatum can be shown pharmacologically but we don't really know its functional significance. We don't know whether these receptors are simply some kind of embryological remnant which has no real physiological significance in the normal situation.

Somogyi: Is there a stereotyped distribution of striosomes from animal to animal?

Graybiel: Informally, we can say there is great similarity from cat to cat and human to human. We need to do many reconstructions to give you a precise answer, but I have reconstructed two caudate nuclei in the cat. In both of these, the striosomes in rostral sections tend to stretch across the width of the caudate nucleus like the rungs of a ladder, whereas towards the back of the caudate nucleus, they are small and rounded. Most of the patches in any one section are connected with one another out of the plane of that particular section. There may be as few as two branched labyrinths making up the striosomes in a single caudate nucleus.

Somogyi: You mentioned that dynorphin immunoreactivity was strong in the internal segment. Is there a lot in the nigra too?

Graybiel: We find very dense dynorphin-like immunoreactivity spread across almost all of the substantia nigra pars reticulata, except for part of the pars lateralis (Chesselet & Graybiel 1983). That is interesting because the pars lateralis is where somatostatin-like immunoreactivity appears.

Somogyi: What was the marker you used to show the VTA projection selectively?

Graybiel: That picture [not in paper] showed a computer-derived image of the dopamine fluorescence in the caudoputamen of the weaver mutant mouse (Roffler-Tarlov & Graybiel 1983). The residual fluorescence in the weaver looks very much like the innervation derived from the VTA and A8 and we think that much of the nigrostriatal projection is missing or at least greatly diminished in this mouse.

Glowinski: There are problems about the origin of the dopaminergic innervation of the striatum, particularly that of the islands or patches which can be so nicely seen in your picture with tyrosine hydroxylase (TH) immunohistochemical staining. If these patches correspond to dopaminergic nerve terminals originating in the VTA, they should be seen not in the whole striatum but rather in the medial and ventral parts. Another possibility is that there are two types of dopaminergic neuron in the ventral part of the pars compacta (substantia nigra) that innervate the striatum: one giving rise to the 'patches' and the other to the diffuse innervation. Alternatively, the same type of cells could first innervate the 'patches' in the young animal and then progressively innervate the whole structure by collateral branching during development.

A second point concerns the experiments in which you injected labelled amino acids into the prefrontal cortex. Have you done similar studies in the substantia nigra? This could help to determine whether nigral cells are responsible for the non-diffuse dopaminergic innervation in the striatum.

Graybiel: Wright & Arbuthnott (1981) did just that in the rat and described definite patches in the autoradiographically labelled nigrostriatal projection. We are now using retrograde tracers to look at the connection in the reverse direction.

Smith: You said that the striatopallidal neurons lay mainly outside the striosomes. Is that true of the striatonigral neurons as well?

Graybiel: Dr Gerfen in Ed Evarts' laboratory has made injections of retrograde tracer in the substantia nigra in the rat, and he has found clusters of labelled cells which line up with the striosomes (Gerfen 1983). He has also said that he has seen the kind of result that I showed here, with the striosomes standing out as sparsely labelled 'holes' within fields of labelled cells. At MIT we have been doing double tracer experiments to try to figure out these patterns. The problem we all face is that in retrograde labelling experiments, depending on the number of collaterals a neuron has, that neuron may either show heavy labelling or no labelling or an intermediate amount. One fascinating possibility is that some of the striosomal neurons are special by virtue of having particular sets of collateral projections.

Smith: The DNA labelling is a very important observation. Are the neurons in the striosomes special in terms of their Golgi morphology or connectivity as well as their chemistry?

Graybiel: We think they are special. We need to get more animals with thymidine markers in them. Then we can use other techniques to look at the immunohistochemistry of the thymidine-labelled cells. We have just started these experiments.

Smith: Until you gave me the answer about the striatonigral neurons within striosomes I thought that maybe the striosomes were clusters of interneurons.

Graybiel: They are either interneurons or neurons with lots of branches, or neurons with a very special target which sometimes gets hit and sometimes doesn't in the injection experiments.

DeLong: The output groupings are fascinating. Are you saying that there seem to be clusters of neurons that project outwards to the nigra or the pallidum, or maybe to both, and that these are really very different from areas that are receiving inputs?

Graybiel: In fact, although we have mainly studied the striosomes, the entire striatum may have a mosaic organization. My guess is that we simply don't have markers yet for some of the patches that physiologists can find. Dr Rafi Malach and I are trying to record from the striatum, looking for the patches and for topographic mapping in the cortex corticostriatal projection. So far, our results fit very well with yours. At the same time we are working on biochemically identifiable clusters of neurons labelled with neurotransmitter-like compounds and we are working to find out where the clusters project. Even within the striosomal mosaics we have identified so far, there are non-uniformities; for example, dorsal and ventral striosomes in the cat have different amounts of enkephalin-like and substance P-like immunoreactivity.

REFERENCES

Chesselet M-F, Graybiel AM 1983 Subdivisions of the pallidum and the substantia nigra demonstrated by immunohistochemistry. Neurosci Abstr 9:16

Gerfen CR 1983 Non-topographic order in the basal ganglia: evidence for a second level of organization superimposed upon the topographically ordered striato-nigral projection system. Neurosci Abstr 9:16

Giorguieff-Chesselet M-F, Chéramy A, Glowinski J 1980 *In vivo* and *in vitro* studies on the presynaptic control of dopamine release from nerve terminals of the nigrostriatal dopaminergic neuron. In: Littauer UZ et al (eds) Neurotransmitters and their receptors. John Wiley, New York, p 33-47

Hornykiewicz O 1981 Importance of topographic neurochemistry in studying neurotransmitter systems in human brain: critique and new data. In: Riederer P, Usdin E (eds) Transmitter biochemistry of human brain tissue. Macmillan, London, p 9-24

Hörtnagl H, Schlögl E, Sperk G, Hornykiewicz O 1983 The topographical distribution of the monoaminergic innervation in the basal ganglia of the human brain. Prog Brain Res 58:269-274

Nastuk MA, Graybiel AM 1983 The distribution of muscarinic binding sites in the feline striatum and its relationship to other histochemical staining patterns. Neurosci Abstr 9:15

Roffler-Tarlov S, Graybiel AM 1983 Weaver mutation has differential effects on the dopamine innervation of the limbic and non-limbic striatum. Neurosci Abstr 9:661

Wright AK, Arbuthnott GW 1981 The pattern of innervation of the corpus striatum by the substantia nigra. Neuroscience 6:2063-2067

Role of the thalamus in the bilateral regulation of dopaminergic and GABAergic neurons in the basal ganglia

J. GLOWINSKI, M. J. BESSON and A. CHÉRAMY

Groupe NB, INSERM U.114, Collège de France, 11 place Marcelin Berthelot, 75231 Paris Cedex 05, France

Abstract. In halothane-anaesthetized cats implanted with several push–pull cannulae the release of dopamine from nerve terminals and dendrites of the two nigrostriatal dopaminergic pathways is shown to be asymmetrically or symmetrically regulated bilaterally when dopaminergic or GABAergic drugs are infused into one side of the nigra. Nigrothalamic GABAergic neurons may intervene in the asymmetrical regulation, as sagittal section of the thalamic massa intermedia blocks the contralateral effects induced by dopaminergic drugs. The role of thalamic nuclei in the bilateral regulation of dopamine and GABA release in the caudate nucleus and substantia nigra is further demonstrated by studies showing that (1) electrical stimulation of thalamic motor nuclei induces bilateral asymmetrical changes in dopamine release resembling the changes evoked by unilateral sensory stimuli or stimulation of cerebellar nuclei; (2) electrical stimulation of intralaminar or some midline thalamic nuclei leads to bilateral (or contralateral) symmetrical changes in dopamine release, some of these effects being comparable to those induced by unilateral stimulation of the motor cortex; (3) unilateral lesions of motor or intralaminar thalamic nuclei reverse the changes in GABA release in the contralateral caudate nucleus or substantia nigra induced by unilateral infusion of muscimol into the nigra; and (4) unilateral infusion of GABA into thalamic motor nuclei induces bilateral symmetrical regulation of dopamine release in caudate nuclei by means of presynaptic facilitatory influences.

1984 Functions of the basal ganglia. Pitman, London (Ciba Foundation symposium 107) p 150-163

Bilateral regulation of neuronal systems in the basal ganglia was first suspected in our initial studies of the release of newly synthesized [^{3}H]dopamine ([^{3}H]DA) in halothane-anaesthetized cats implanted with a push–pull cannula in each caudate nucleus. Surprisingly, [^{3}H]DA release was markedly increased in the contralateral caudate nucleus after unilateral electrocoagulation of the substantia nigra; but this treatment, as expected, immediately reduced the release of [^{3}H]DA in the ipsilateral structure

(Nieoullon et al 1977a). Unilateral interruption of signals in the substantia nigra therefore seemed to influence the activity of nigrostriatal dopaminergic neurons on the contralateral side.

We then implanted cats with several push–pull cannulae so that we could estimate bilateral release of dopamine from nerve terminals in the caudate nucleus and dendrites in the substantia nigra. Several patterns of response were obtained after potassium or transmitters, or their agonists or antagonists, had been infused into the nigra unilaterally. (1) Dopaminergic drugs induced bilateral asymmetrical changes in [^{3}H]DA release (Table 1) (Leviel et

TABLE 1 Asymmetrical effects of dopaminergic drugs and other stimuli on [^{3}H]DA release in halothane-anesthetized cats

Unilateral treatment	Caudate nucleus		Substantia nigra	
	Ipsi-lateral	Contra-lateral	Ipsi-lateral	Contra-lateral
Dopaminergic drugs				
Amphetamine (10^{-6}M)	↘	↗	↗	↘
α-Methyl-*p*-tyrosine (10^{-4}M)	↗	↘	↘	↗
Sensory stimuli				
Somatic	↗	↘	↘	↗
Visual	↗	↘	↘	↗
Electrical stimulation				
Cerebellar nuclei				
Dentate	↘	↗	↗	↘
Interposate	↘	↗	↗	↘
Motor thalamic nuclei				
VM–VL	↗	↘	↘	↗

Facilitatory (↗) or inhibitory (↘) effects on the release of [^{3}H]DA newly synthesized from [^{3}H]tyrosine.

al 1979a); (2) potassium (30 mM) (Greenfield et al 1980) or GABAergic and glycinergic agonists or antagonists (Leviel et al 1979b) produced bilateral symmetrical responses, with a bilateral increase or decrease in [^{3}H]DA release in the caudate nuclei (Table 2); (3) ipsilateral changes in [^{3}H]DA release were only found with unilateral facilitation or interruption of the transmission of substance P in the nigra (Michelot et al 1979). These treatments also had complex effects on [^{3}H]DA released from dendrites. In most cases, changes in [^{3}H]DA release from nerve terminals were associated with modifications in [^{3}H]DA release from dendrites in the ipsilateral substantia nigra. These changes were either opposite (dopaminergic drugs) or parallel (potassium, GABAergic drugs) to those observed in the caudate nucleus (Tables 1 & 2).

These results raised two main questions. (1) How do messages originating in the substantia nigra on one side reach the contralateral dopaminergic

TABLE 2 Symmetrical effects on [³H]DA release induced by various unilateral treatments in the halothane-anaesthetized cat

Unilateral treatment	Caudate nucleus		Substantia nigra	
	Ipsi-lateral	Contra-lateral	Ipsi-lateral	Contra-lateral
Nigral treatments				
Potassium (30 mM)	↗	↗	↗	↗
Muscimol (10^{-6}M)	↗	↗	↗	n.d
Diazepam (10^{-5}M)	↘	↘	↘	↘
Intralaminar thalamic nuclei[a]				
CM/Pf	↘	↘	↘	↘
CMa	O	O	O	↗
CL/MD	O	↘	O	↘
Midline thalamic nuclei[a]				
NCM	O	O	O	O
RE	O	O	↗	↗
IAM	↗	↗	↗	↗
Cerebral cortex[a]				
Motor	↗	↗	↗	n.d.
Visual	↗	↗	↗	n.d.

Facilitatory (↗) or inhibitory (↘) effects on [³H]DA release. O, no effect; n.d., no data; CM/Pf, n. parafascicularis and adjacent part of n. centrum medianum; CMa, anterior part of n. centrum medianum; CL/MD, n. centralis medialis and adjacent part of n. medialis dorsalis; NCM, n. centralis medialis; RE, n. reuniens; IAM, n. interanteromedialis.)
[a] Electrical stimulation.

neurons? To answer this we need to determine whether different polysynaptic pathways contribute to the bilateral asymmetrical or symmetrical changes in dopamine release from nerve terminals. (2) Why are changes in dopamine release from the dendrites and nerve terminals of corresponding dopaminergic cells either opposite or parallel? According to electrophysiological studies and our own biochemical investigations, dopamine released from dendrites is involved in self-regulation of the activity of dopaminergic cells (Chéramy et al 1981b). Increased availability of dopamine at dopaminergic autoreceptor sites reduces the firing rate of dopaminergic neurons and thus of dopamine release from nerve terminals; the opposite response is seen when nigral transmission of dopamine is reduced. This is why opposite changes in [³H]DA release are sometimes observed in the substantia nigra and the corresponding caudate nucleus, particularly after nigral injection of dopaminergic drugs which facilitate (dopamine, amphetamine, benztropine) or prevent (α-methyl-p-tyrosine, neuroleptics) local transmission of dopamine (Table 1). Other mechanisms must intervene when parallel changes in [³H]DA release occur in the substantia nigra and the ipsilateral caudate nucleus. The activity of dopaminergic cells may no longer be under the control of the dendritic release of dopamine, or there may be no direct relationship between cell firing and the amount of dopamine released from nerve terminals. The latter possibility suggests that potent presynaptic influences are at work, the origin of which has yet to be determined.

Role of the thalamus in bilateral asymmetrical regulation of dopamine release

The role of nigrothalamic neurons and the thalamic massa intermedia in the interhemispheric transfer of messages involved in the bilateral asymmetrical regulation of dopamine release from nerve terminals and dendrites has been demonstrated by experiments in which sagittal sections were made in cats. Sagittal section of the thalamic massa intermedia prevented asymmetrical (when compared with the ipsilateral side) and opposite changes in [^{3}H]DA release in the contralateral caudate nucleus and substantia nigra induced by unilateral infusion of α-methyl-p-tyrosine or amphetamine into the nigra (Chéramy et al 1981a). In contrast, sagittal section of mesencephalic decussations or of fibres passing through the anterior commissure or the anterior part of the corpus callosum did not abolish the contralateral effects of dopaminergic drugs. Bilateral asymmetrical changes in [^{3}H]DA release from nerve terminals and dendrites also occur after unilateral electrical stimulation of cerebellar nuclei (Nieoullon et al 1978a, Nieoullon & Dusticier 1980) or delivery of sensory stimuli (Nieoullon et al 1977b) (Table 1). As observed after infusion of α-methyl-p-tyrosine into the nigra, dopamine release is enhanced in the ipsilateral caudate nucleus and reduced in the contralateral structure; the opposite responses are observed in the corresponding substantia nigra when mild electric shocks are delivered to the right forepaw. These effects of unilateral sensory stimuli may also be mediated by signals reaching the thalamic nuclei, as bilateral changes in [^{3}H]DA release were prevented by section of the thalamic massa intermedia but not by section in other areas (Leviel et al 1981).

A large contingent of nigrothalamic fibres innervate thalamic motor nuclei (ventralis medialis, VM; ventralis lateralis, VL) and electrophysiological studies have indicated either that some of these neurons are activated by dopamine (Ruffieux & Schultz 1980) or that dopamine microiontophoretically injected into the substantia nigra reduces the inhibitory effect of GABA on these cells (Waszczak & Walters 1983). We further investigated the eventual participation of the nigrothalamic GABAergic fibres in transferring messages to the contralateral basal ganglia by examining the effects of unilateral electrical stimulation of VM–VL. Bilateral asymmetrical responses were obtained, [^{3}H]DA release being increased in the ipsilateral caudate nucleus and decreased contralaterally, with opposite responses in the substantia nigra

(Table 1). Complementary experiments in which multi-unit neuronal activity was determined revealed an increased frequency of spikes in the contralateral homologous thalamic nucleus during ipsilateral electrical stimulation of VM–VL (Chéramy et al 1983).

The results just described suggest that nigral neurons projecting to thalamic motor nuclei contribute to the bilateral asymmetrical regulation of nigrostriatal dopamine pathways evoked by unilateral nigral application of dopaminergic drugs. However, on this basis alone, involvement of other thalamic nuclei innervated by nigral cells should not be excluded at present. On the other hand, although multiple connections exist between thalamic nuclei, it remains to be determined how messages reach the contralateral thalamic nuclei and how they are then delivered to the contralateral dopaminergic neurons.

Role of the thalamus and motor cortex in bilateral symmetrical regulation

As well as motor nuclei, intralaminar nuclei (Chéramy et al 1983) and some midline thalamic nuclei (Chéramy et al 1984) that may or may not be connected directly to the substantia nigra are also involved in bilateral or contralateral regulation of dopamine release from nerve terminals and dendrites of nigrostriatal dopaminergic neurons. We have demonstrated this by examining the effects of electrical stimulation of these nuclei. In each case, a different pattern of response(s) was observed which also differed from that obtained by stimulation of VM–VL, in that no bilateral asymmetrical changes in [^{3}H]DA release or the opposite responses in the substantia nigra and the corresponding caudate nucleus were detected. Thus, [^{3}H]DA release was reduced in the four structures, or only in the contralateral substantia nigra and caudate nucleus, after unilateral electrical stimulation of the parafascicularis nucleus and the adjacent posterior part of the nucleus centrum medianum, or of the nucleus centralis lateralis and the adjacent paralaminar part of the nucleus medialis dorsalis, respectively. On the other hand, stimulation of the anterior part of the nucleus centrum medianum—which in contrast to other intralaminar thalamic nuclei receives few nigral inputs—selectively enhanced [^{3}H]DA release in the contralateral substantia nigra (Table 2). Bilateral symmetrical responses were also seen after stimulation of some midline thalamic nuclei, as [^{3}H]DA release was increased in the four structures only after stimulation of the nucleus interanteromedialis, or bilaterally in the substantia nigra after stimulation of the nucleus reuniens (Table 2).

Although electrical stimulation may activate crossing fibres or evoke antidromic responses which could lead to erroneous interpretations, it can nevertheless be concluded that intralaminar and some midline thalamic nuclei exert selective influences on the release of dopamine from nerve terminals and/or dendrites of dopaminergic neurons. Responses obtained after unilateral electrical stimulation of the motor cortex suggest that polysynaptic neuronal loops involving the cerebral cortex may contribute to bilateral symmetrical regulation or to the contralateral parallel changes in [^{3}H]DA release elicited by stimulation of intralaminar nuclei. Indeed, this cortical stimulation markedly enhanced [^{3}H]DA release in the caudate nuclei (Table 2) and the contralateral effect was prevented by sagittal section of the corpus callosum (Nieoullon et al 1978b). According to these observations, the bilateral corticostriatal glutamatergic projection could exert a presynaptic facilitatory influence on dopamine release from nerve terminals of dopaminergic neurons both ipsilaterally and contralaterally. Experiments on rat striatal slices support this statement: L-glutamate stimulates the release of newly synthesized [^{3}H]DA through a tetrodotoxin-insensitive process, suggesting that the receptors involved are located on dopaminergic fibres (Giorguieff et al 1977). Another indication that corticostriatal fibres have a presynaptic facilitatory influence was provided by the parallel changes in [^{3}H]DA release observed in the ipsilateral caudate nucleus and substantia nigra after cortical stimulation (Table 2). Indeed, as dopamine release from dendrites exerts an inhibitory influence on the activity of dopaminergic cells, decreased firing of these cells must be associated with the enhanced release of [^{3}H]DA from dendrites seen in this situation.

As already mentioned, bilateral symmetrical changes in [^{3}H]DA release are also observed in the caudate nuclei or substantia nigra, or both, after unilateral infusion of drugs into the substantia nigra. For instance, infusion of potassium (30 mM), which, as expected, increased local dendritic release of [^{3}H]DA and release of [^{3}H]DA in the ipsilateral caudate nucleus, also enhanced [^{3}H]DA release in contralateral homologous structures (Table 2). Similar observations were made after muscimol (10^{-6}M), a GABAergic agonist, was unilaterally infused into the nigra; at this low concentration muscimol reduces GABA transmission by acting preferentially on presynaptic receptors on nigral afferents (Arbilla et al 1979). Thus [^{3}H]DA release was increased not only locally but also in both caudate nuclei (Table 2). On the other hand, the facilitation of nigral GABAergic transmission induced by unilateral infusion of diazepam (10^{-5}M) resulted in a bilateral symmetrical decrease in the release of [^{3}H]DA in the caudate nuclei and substantia nigra. These responses, comparable to those seen after unilateral electrical stimulation of some intralaminar or midline thalamic nuclei, or of the motor (or visual) cerebral cortex, suggest that common mechanisms exist. These mechanisms must be distinct (partly, at least) from those responsible for asymmetrical bilateral regulation. Recent experiments on the mode of muscimol action on nigral efferent cells and on the role of GABAergic processes in the thalamus have provided us with some new information.

Effects of muscimol on GABAergic processes in the thalamus and basal ganglia

The effects of infusing muscimol (10^{-6}M, 50–60 min) unilaterally into the nigra on the release of [^{3}H]GABA newly synthesized from [^{3}H]glutamine were investigated in the caudate nuclei and substantia nigra or in the

TABLE 3 Effects of a unilateral infusion of muscimol into the nigra on [^{3}H]GABA release in halothane-anaesthetized cats

	Caudate nucleus		Substantia nigra	
	Ipsi-lateral	Contra-lateral	Ipsi-lateral	Contra-lateral
Muscimol (10^{-6}M)				
Unlesioned	↗	↘	↗	↗
VM–VL lesion	↗	↗	↗	↗
IL lesion	↗	↗	↗	↘

Facilitatory (↗) or inhibitory (↘) effects on the release of [^{3}H]GABA newly synthesized from [^{3}H]glutamine. VM–VL, motor thalamic nuclei, ventralis medialis and ventralis lateralis; IL intralaminar thalamic nuclei, centralis lateralis, paracentralis and adjacent part of medialis dorsalis.

ipsilateral thalamic nuclei. Bilateral symmetrical increases in [^{3}H]GABA release were found in the substantia nigra, while [^{3}H]GABA release was enhanced in the ipsilateral caudate nucleus and decreased on the contralateral side (Table 3). This demonstrated another type of complex bilateral regulation of identified neurons in the basal ganglia (Kemel et al 1983).

Concomitant estimations of [^{3}H]GABA release and multi-unit neuronal activity revealed that nigrothalamic GABAergic neurons innervating the VM–VL, the centralis lateralis and the paracentralis nuclei were activated as [^{3}H]GABA release and global multi-unit cellular activity were increased and reduced, respectively, in these thalamic nuclei. Opposite effects were observed in the part of the medialis dorsalis adjacent to the intralaminar nuclei, suggesting either a polysynaptic response or that the nigrothalamic neurons afferent to this thalamic nucleus are non-GABAergic (Gauchy et al 1983).

The contribution of thalamic nuclei to the muscimol-evoked changes in [^{3}H]GABA release in the contralateral basal ganglia was further demonstrated in cats with lesions. Acute lesion of the ipsilateral VM–VL selectively reversed the response in the contralateral caudate nucleus, while lesion of the ipsilateral centralis lateralis and paralaminar zone of the medialis dorsalis reversed the responses in the contralateral caudate nucleus and substantia nigra (Table 3) (Kemel et al 1984).

Further investigations will be needed to establish whether there is any relationship between the effects induced by the infusion of muscimol into the nigra on GABAergic and dopaminergic neurons in contralateral basal ganglia. However, these results reinforce the idea that a better understanding of the variety of messages reaching motor and intralaminar thalamic nuclei and of the relationships between these nuclei is needed before we can more precisely define the mechanisms underlying bilateral symmetrical and asymmetrical regulation of [^{3}H]DA release from the dendrites and nerve terminals of nigrostriatal dopaminergic neurons.

The role of presynaptic influences in the bilateral symmetrical regulation of [^{3}H]DA release in caudate nuclei

Investigations of the effects of unilateral facilitation or interruption of GABAergic transmission in thalamic nuclei on [^{3}H]DA release in the caudate nuclei and substantia nigra should allow some of the problems already discussed in this paper to be resolved. The description of such experiments will be limited here to the effects induced by unilateral infusion of a low concentration of GABA (10^{-5}M) into the VM–VL (Chesselet et al 1983). Since this treatment enhances the local global multi-unit neuronal activity, a reduction in GABA release from GABAergic afferent fibres may occur, due to a preferential action of GABA (10^{-5}M) on presynaptic GABA receptors. In addition, this treatment induces a reduction in global multi-unit neuronal activity in the contralateral homologous nuclei, in contrast to the activity observed after unilateral electrical stimulation of the VM–VL. Infusion of GABA into the VM–VL increased [^{3}H]DA release in the caudate nuclei and the contralateral substantia nigra (Table 4). This pattern of responses, which differs from that evoked by unilateral electrical stimulation of VM–VL, indicates that bilateral symmetrical or asymmetrical regulation of [^{3}H]DA release may depend on the nature of the signals originating from thalamic motor nuclei. This statement also applies to intralaminar nuclei (Romo et al 1983).

Responses evoked by GABA (10^{-5}M) infused into the VM–VL are comparable in some ways to those observed after potassium (30 mM) or muscimol (10^{-6}M) are infused into the nigra. We therefore used this experimental model, with appropriate sagittal sections, in our further investigations of the roles of fibres passing through the corpus callosum or midline thalamic nuclei in transferring inputs to contralateral dopaminergic nerve

TABLE 4 Effects of unilateral thalamic infusion of GABA on [^{3}H]DA release in halothane-anaesthetized cats

	Caudate nucleus		Substantia nigra	
	Ipsi-lateral	Contra-lateral	Ipsi-lateral	Contra-lateral
GABA (10^{-5}M)				
Unlesioned animals	↗	↗	○	↗
Anterior section	↗	↘	○	↗
Thalamic section	↗	↗	○	○

Facilitatory (↗) or inhibitory (↘) effects on the release of newly synthesized [^{3}H]DA. ○, no effect. Anterior section involved the commissura anterior and the rostral part of the corpus callosum; thalamic section involved the massa intermedia and overlying part of the corpus callosum.

terminals and dendrites. Section of the thalamic massa intermedia selectively prevented an increase in the release of [^{3}H]DA in the contralateral substantia nigra, while anterior section of the corpus callosum selectively abolished the GABA-evoked increase in release of [^{3}H]DA in the contralateral caudate nucleus (Table 4) (Romo et al 1984). These findings led to the following conclusions. (1) When parallel changes in [^{3}H]DA release occur in the contralateral caudate nucleus and substantia nigra, the fibres passing through the thalamic massa intermedia seem to be selectively involved in regulating the dendritic release of dopamine, while fibres passing through the anterior part of the corpus callosum appear to be responsible for regulating dopamine release from nerve terminals. (2) The enhanced release of dopamine in the contralateral caudate nucleus results from strong presynaptic facilitatory influences, probably mediated by bilateral corticostriatal fibres. (3) There may be an enhanced release of dopamine from nerve terminals despite a reduction in the firing rate of dopaminergic cells triggered by the enhanced release of dopamine from dendrites (self-inhibition process). The suggestion that GABA infused into VM–VL reduced the firing rate of dopaminergic cells in the contralateral substantia nigra was confirmed (a) by the observation that the release of [^{3}H]DA was reduced in the contralateral caudate nuclei of cats with an anterior sagittal section, and (b) by direct extracellular recording of identified dopaminergic cells.

Concluding remarks

It is now well established that the thalamus plays an important role in bilateral asymmetrical and symmetrical regulation of dopamine release from nerve terminals and dendrites of the nigrostriatal dopaminergic pathways. This seems also to be the case for the changes in GABAergic transmission evoked in the contralateral basal ganglia by unilateral nigral modification of GABA transmission. The effects of dopaminergic and GABAergic drugs on nigral GABAergic neurons innervating motor and intralaminar thalamic nuclei need to be compared before we can define the roles of these nuclei in the bilateral asymmetrical and symmetrical regulation of dopamine release in the caudate nuclei and the substantia nigra. The results reported here have already shown that the cerebral cortex contributes to the symmetrical regulation of dopamine release, and that dopamine transmission in the caudate nucleus is not solely dependent on the activity of dopaminergic cells but is also regulated by potent presynaptic influences.

REFERENCES

Arbilla S, Kamal L, Langer SZ 1979 Presynaptic GABA autoreceptors on GABAergic nerve endings of rat substantia nigra. Eur J Pharmacol 57:211-217

Chéramy A, Chesselet MF, Romo R, Leviel V, Glowinski J 1983 Effects of unilateral electrical stimulation of various thalamic nuclei on the release of dopamine from dendrites and nerve terminals of neurons of the two nigro-striatal dopaminergic pathways. Neuroscience 8:767-780

Chéramy A, Leviel V, Daudet F, Guibert B, Chesselet MF, Glowinski J 1981a Involvement of the thalamus in the reciprocal regulation of the two nigrostriatal dopaminergic pathways. Neuroscience 6:2657-2668

Chéramy A, Leviel V, Glowinski J 1981b Dendritic release of dopamine in the substantia nigra. Nature (Lond) 289:537-542

Chéramy A, Romo R, Godeheu G, Glowinski J 1984 Effects of electrical stimulation of various midline thalamic nuclei on the release of dopamine from dendrites and nerve terminals of neurons of the two nigro-striatal dopaminergic pathways. Neurosci Lett, in press

Chesselet MF, Chéramy A, Romo R, Desban M, Glowinski J 1983 GABA in the thalamic motor nuclei modulates dopamine release from the two dopaminergic nigrostriatal pathways in the cat. Exp Brain Res 51:275-282

Gauchy C, Kemel ML, Romo R, Chéramy A, Glowinski J, Besson MJ 1983 Effects of nigral application of muscimol on [^{3}H]-GABA release and on multi-unit activity in various cat thalamic nuclei. Neuroscience 10:781-788

Giorguieff MF, Kemel ML, Glowinski J 1977 Presynaptic effect of L-glutamic acid on the release of dopamine in rat striatal slices. Neurosci Lett 6:73-77

Greenfield S, Chéramy A, Leviel V, Glowinski J 1980 In vivo release of acetylcholinesterase in cat substantia nigra and caudate nucleus. Nature (Lond) 284:355-357

Kemel ML, Gauchy C, Romo R, Glowinski J, Besson MJ 1983 In vivo release of [^{3}H]-GABA in cat caudate nucleus and substantia nigra. I: Bilateral changes induced by unilateral nigral application of muscimol. Brain Res 272:331-340

Kemel ML, Gauchy C, Glowinski J, Besson MJ 1984 In vivo release of [^{3}H]-GABA in cat caudate nucleus and substantia nigra. II: Involvement of different thalamic nuclei in bilateral changes induced by a nigral application of muscimol. Brain Res, in press

Leviel V, Chéramy A, Glowinski J 1979a Role of the dendritic release of dopamine in the reciprocal control of the two nigrostriatal dopaminergic pathways. Nature (Lond) 280:236-239

Leviel V, Chéramy A, Nieoullon A, Glowinski J 1979b Symmetric bilateral changes in dopamine

release from the caudate nuclei of the cat induced by unilateral nigral application of glycine and GABA-related compounds. Brain Res 175:259-270
Leviel V, Chesselet MF, Glowinski J, Chéramy A 1981 Involvement of the thalamus in asymmetric effects of unilateral sensory stimuli on the two nigrostriatal dopaminergic pathways in the cat. Brain Res 223:257-272
Michelot R, Leviel V, Giorguieff-Chesselet MF, Chéramy A, Glowinski J 1979 Effects of the unilateral nigral modulation of substance P transmission on the activity of the two nigrostriatal dopaminergic pathways. Life Sci 24:715-724
Nieoullon A, Dusticier N 1980 Changes in dopamine release in caudate nuclei and substantia nigrae after electrical stimulation of the posterior interposate nucleus of cat cerebellum. Neurosci Lett 17:167-172
Nieoullon A, Chéramy A, Glowinski J 1977a Interdependence of the nigrostriatal dopaminergic systems on the two sides of the brain in the cat. Science (Wash DC) 198:416-418
Nieoullon A, Chéramy A, Glowinski J 1977b Nigral and striatal dopamine release under sensory stimuli. Nature (Lond) 269:340-342
Nieoullon A, Chéramy A, Glowinski J 1978a Release of dopamine in both caudate nuclei and both substantia nigrae in response to unilateral stimulation of cerebellar nuclei in the cat. Brain Res 148:143-152
Nieoullon A, Chéramy A, Glowinski J 1978b Release of dopamine evoked by electrical stimulation of the motor and visual areas of the cerebral cortex in both caudate nuclei and in the substantia nigra in the cat. Brain Res 145:69-84
Romo R, Chéramy A, Desban M, Godeheu G, Glowinski J 1983 GABA in the intralaminar thalamic nuclei modulates dopamine release from the two dopaminergic nigro-striatal pathways in the cat. Brain Res Bull 11:671-680
Romo R, Chéramy A, Godeheu G, Glowinski J 1984 Distinct commissural pathways are involved in the enhanced release of dopamine induced in the contralateral caudate nucleus and substantia nigra by unilateral application of GABA in the cat thalamic motor nuclei. Brain Res, in press
Ruffieux A, Schultz W 1980 Dopaminergic activation of reticulata neurons in the substantia nigra. Nature (Lond) 285:240-241
Waszczak LB, Walters JR 1983 Dopamine modulation of the effects of γ-aminobutyric acid on substantia nigra pars reticulata neurons. Science (Wash DC) 220:218-222

DISCUSSION

Di Chiara: You concluded that the effects of stimulation of the cerebral cortex were presynaptic because there were changes in dopamine release in both the caudate and the nigra and because section of the corpus callosum abolished the response on the contralateral side. Couldn't these effects be coming from a polysynaptic pathway from a cortico-striato-nigral pathway? Section of the corpus callosum doesn't exclude this possibility.

Glowinski: Do you mean that a signal originating in the motor cortex is transferred to the contralateral substantia nigra?

Di Chiara: The impulses can go from one side of the cortex via the corpus callosum, reach the other side of the cortex and then go to the caudate of both sides and down to the nigra.

Glowinski: This is a possibility, and it could lead to a modification of our interpretation. However, direct presynaptic control of dopamine release in the caudate nucleus is also strongly suggested by experiments in which GABA (10^{-5}M) was infused unilaterally into the thalamic motor nuclei. In this case, dopamine release was increased in both the contralateral caudate nucleus and the contralateral substantia nigra. Only sagittal section of the corpus callosum abolished the effect in the contralateral caudate nucleus; therefore there may be presynaptic regulation by cortical afferents, similar to the type of regulation proposed for the effect of electrical stimulation of the cerebral cortex. This interpretation is supported by the lack of effect of sagittal section on the nigral response. This seems to exclude the involvement of a striatonigral pathway such as you suggest, Dr Di Chiara. In addition, the nigral response, but not that in the caudate nucleus, was abolished by sagittal section of the thalamic massa intermedia, indicating that two distinct pathways are responsible for the increases in dopamine release in the caudate nucleus and substantia nigra. Finally, single-unit recordings of dopaminergic cells in the contralateral substantia nigra showed that the firing rate of the dopaminergic neurons was inhibited. Thus, the enhanced release of dopamine from nerve terminals in the caudate nucleus can only be attributed to facilitatory presynaptic regulation.

Di Chiara: I would be more confident about this presynaptic explanation if you could show that a lesion disconnecting the terminal from the cell body fails to abolish your effects.

Glowinski: In this experiment the lesion was ipsilateral to the site of injection of GABA (10^{-5}M) into thalamic motor nuclei. When the nigrostriatal dopaminergic fibres were completely blocked in this way, GABA still induced an increase in the dopamine released in the ipsilateral caudate nucleus.

Marsden: The dopamine cells are remarkable. With reference to transmitter release, it seems that one end can do the opposite from the other, and that both ends will ignore how fast the neurons are firing and do entirely the wrong thing. Do you think other transmitter neurons using other neurotransmitters work on the same principles?

Glowinski: Nigrostriatal dopaminergic neurons can be regarded as elongated interneurons which, via their nerve terminals and their dendrites, are involved in regulating either local circuit neurons or 'executive' neurons, such as striatonigral GABAergic neurons in the caudate nucleus and nigrothalamic GABAergic neurons in the substantia nigra, respectively. These nigrostriatal neurons can be compared to the dopaminergic amacrine cells in the retina which are involved in local circuits and probably in the regulation of messages passing through bipolar and ganglionic cells. The nerve terminals and dendrites of the nigrostriatal dopaminergic neurons may sometimes, but not always, operate as independent functional units contributing separately to the regulation of events in the caudate nucleus and substantia nigra; and they are themselves subjected to distinct presynaptic regulatory processes controlling the release of dopamine. The efficacy of such processes may partly depend on the basal firing rate of dopaminergic cells. In other situations, depending on the nature of the messages reaching dopaminergic cell bodies, the firing rate of the dopaminergic neurons may predominate and exert a strong influence on the amount of dopamine released from nerve terminals.

Hornykiewicz: It is quite exciting that dopamine release does not seem to be related to the firing rate of the neuron. Does that mean that in your experiments dopamine release is calcium-independent?

Glowinski: No, the release is calcium-dependent. In fact presynaptic receptors participate in the control of calcium flux into nerve terminals. Local calcium currents have been shown to exist at a dendritic level (Llinás & Hess 1976).

Hornykiewicz: So you think it is a local mechanism?

Glowinski: As I have said, I think that local regulatory mechanisms play an important role. In some ways the independent functional units represented by dopaminergic nerve terminals and dendrites are part of local circuits involved in controlling the activity of the main connections in which messages are conducted at high speed.

Smith: We shouldn't get carried away by the dissociation between unit activity and release of dopamine in the caudate. Release of dopamine is a global phenomenon from thousands of nerve endings originating from a few hundred cells. You recorded unit activity in the push–pull cannula where the dose of the drug is high.

Glowinski: In those experiments GABA (10^{-5}M) was delivered into thalamic motor nuclei and single-unit recordings were made in the substantia nigra.

Smith: So you would say that the single units you recorded each represent a population of nigrostriatal dopamine neurons?

Glowinski: I agree that there may be problems in comparing changes in release and in a single-cell activity. First, we are not always comparing events occurring on the same time scale. Dopamine is collected in 10-min fractions in the release studies. However, the average firing rate of the dopaminergic neurons during this period can be calculated, and during this time the rate was reduced. Comparisons are thus valid. Nevertheless, we are not measuring dopamine release from dendrites or nerve terminals from a single identified neuron. This is of course impossible.

Smith: Did you record from the ventral tegmental area?
Glowinski: No.
DeLong: The work you have reviewed is extremely important. We have worked for some time on the substantia nigra of the monkey. Since we found no dramatic changes in pars compacta during limb movements we felt something else must be going on, if dopamine is doing anything in relation to movement.

This system might at first be thought of as having primarily a generalized role in behaviour but from what you say it is not to be viewed in such a simple manner. It looks as if it could instead play an important role in phasic activities. The phasic action of dopamine might be worth discussing later.

The other clue you provided is how the cerebellum and basal ganglia might be interacting. This seems like a very clear mechanism for interactions. Have you any more general thoughts about what these interactions might be?

Glowinski: I cannot give a clear answer on, for example, the functional significance of the interactions between the cerebellum and the nigrostriatal dopaminergic neurons within the basal ganglia. As to the bilateral asymmetrical or symmetrical regulation and the roles of the motor and intralaminar thalamic nuclei in this regulation, there are some methodological difficulties. Pharmacological treatment or electrical stimulation in one thalamic nucleus may well affect the activity of other thalamic nuclei, making the respective roles of these nuclei in the responses observed, namely dopamine release, difficult to define. Similar problems are found with treatment at the substantia nigra level, since nigrothalamic neurons project into different thalamic nuclei which are interconnected. The significance of the changes in GABA release is also uncertain, since GABA may originate from nerve terminals of afferent fibres, collaterals of efferent neurons, or interneurons. This is why we are now simultaneously measuring multi-unit neuronal activity at the site at which GABA release is estimated. This could help at least to differentiate between GABA originating from afferent nerve terminals and from collaterals of efferent cells.

REFERENCE

Llinás R, Hess R 1976 Tetrodotoxin resistant dendritic spikes in avian Purkinje cells. Proc Natl Acad Sci USA 7:2520-2523

γ-Aminobutyric acid and basal ganglia outflow pathways

C. REAVILL, P. JENNER and C. D. MARSDEN

University Department of Neurology, Institute of Psychiatry and *King's College Hospital Medical School, Denmark Hill, London SE5 8AF UK*

Abstract. Neurons containing γ-aminobutyric acid (GABA) are important outflow pathways from the striatum to the pallidal complex and substantia nigra. From these areas GABA-containing neurons pass to the thalamus and to various areas of the brainstem. Manipulation of GABA function in outflow zones in the rat can produce catalepsy, locomotor hyperactivity, stereotypy or circling behaviour, so mimicking the effect of altered dopamine function within basal ganglia. However, the behaviours produced by such manipulation do not form part of the animal's normal activities. Consequently manipulation of GABA action in the outflow zones of the basal ganglia may mimic extrapyramidal movement disorders more closely than the normal functions of these regions of the brain.

1984 Functions of the basal ganglia. Pitman, London (Ciba Foundation symposium 107) p 164-176

Abnormal function of the basal ganglia in humans produces a variety of disorders with abnormal movement including Parkinson's disease, Huntington's chorea, hemiballism, tardive dyskinesia and torsion dystonia. Such diseases may be associated with changes in the neuronal action of dopamine within the striatum. Manipulation of dopamine function in the basal ganglia of rodents produces a variety of changes in motor behaviour. These include catalepsy, stereotypy, locomotor hyperactivity, postural deviation and circling. Recent interest has centred on the pathways by which such behaviours are mediated. In this context, investigation of the effects of altered γ-aminobutyric acid (GABA) function in basal ganglia has been critical. GABA is an important inhibitory neurotransmitter within the basal ganglia, and GABA pathways form important outflows from basal ganglia (see Di Chiara & Gessa 1981). Studies of these GABA pathways have been undertaken to clarify the role of outflow pathways in motor function and to locate GABA-sensitive sites at which GABA-active drugs might be used to treat movement disorders.

One approach to the study of movement disorders is to manipulate GABA function in output zones containing characterized GABA synapses. Unilateral or bilateral focal injections of GABA-active drugs, such as muscimol or picrotoxin, have been used in this kind of manipulation. In this short paper we discuss the results obtained from such experiments in relation to the functions of the basal ganglia. Initially, however, it is necessary to define which outflow pathways from basal ganglia utilize GABA.

GABA-containing outflow pathways from basal ganglia

Many pathways within the basal ganglia utilize GABA (Fig. 1). In brief, the well-characterized strionigral pathway contains GABA, as do collaterals from this pathway to both the internal (entopeduncular nucleus) and external (globus pallidus) segments of the pallidal complex. The dorsal striatum

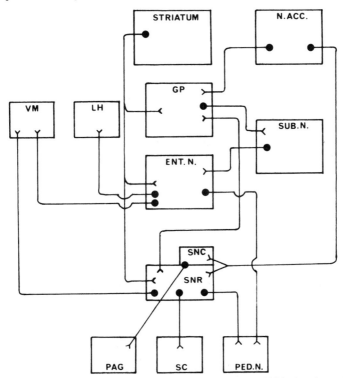

FIG. 1. A summary of the known GABA-containing neurons within basal ganglia and in basal ganglia outflow pathways. N.ACC, nucleus accumbens; GP, globus pallidus; ENT N, entopeduncular nucleus; SUB N, subthalamic nucleus; VM, ventromedial nucleus of the thalamus; LH, lateral habenula; SNC, SNR, substantia nigra zona compacta, zona reticulata; PAG, angular complex (periaqueductal and adjacent mesencephalic reticular formation); SC, lateral and deep superior colliculus; PED. N, nucleus tegmenti pedunculopontinus.

projects to the zona reticulata of the substantia nigra. The nucleus accumbens (ventral striatum) also sends a major projection to the zona reticulata of substantia nigra, with only a small projection to the ventral tegmental area (Walaas & Fonnum 1980). GABA-containing fibres also pass from the nucleus accumbens to the globus pallidus. From the globus pallidus a GABA projection passes to the subthalamic nucleus. The subthalamic nucleus itself sends a large branched projection to the globus pallidus, substantia nigra zona reticulata and entopeduncular nucleus. The pathway from the subthalamic nucleus to the globus pallidus is a GABA-containing pathway, so the projections to zona reticulata and to the entopeduncular nucleus probably also utilize GABA (Rouzaire-Dubois et al 1983).

The major outflow pathways from the striopallidal complex, entopeduncular nucleus and zona reticulata of the substantia nigra also use GABA. From the entopeduncular nucleus GABA projections innervate the ventromedial nucleus of the thalamus and the lateral habenula. The projection from the entopeduncular nucleus to nucleus tegmenti pedunculopontinus does not have a designated transmitter substance but this too is probably GABA. A number of outflow pathways from substantia nigra zona reticulata contain GABA. A branched pathway passes to the ventromedial nucleus of the thalamus and to the deep layers of the superior colliculus (Di Chiara et al 1979b). The zona reticulata may also innervate an area of the mesencephalic reticular formation adjacent to the periaqueductal grey (angular complex). And recently a pathway from substantia nigra to nucleus tegmenti pedunculopontinus has been found to contain GABA (Childs & Gale 1983).

In summary, GABA-containing pathways constitute a major portion of the outflow routes from basal ganglia, presumably mediating motor events. Against this background we will describe the effects produced by GABA-active drugs injected into localized brain areas thought to receive GABA projections.

Focal injection of GABA-active drugs can mimic the effects of basal ganglia lesions

If focal injections of GABA-active drugs into basal ganglia outflow zones in animals are to provide information on the function of the basal ganglia, they should produce local effects. We can illustrate this by focusing on the subthalamic nucleus. In humans, lesions involving the subthalamic nucleus or its connections cause hemiballism (Carpenter & Carpenter 1951). In monkeys, electrolytic or kainic acid lesions damaging the subthalamic nucleus itself, or its connections with the globus pallidus, produce hemiballism (Carpenter et al 1950, Hammond et al 1979). Recently, Crossman and colleagues (1980) have shown that injection of the GABA antagonist picrotoxin into the subthalamic nucleus of baboons also produces contralateral hemiballism. So GABA manipulation of the subthalamic nucleus can produce a movement disorder known to be caused by lesions of this portion of the basal ganglia.

Effects of manipulating GABA function in GABA-sensitive outflow zones from basal ganglia

Attempts have been made to manipulate GABA function in the globus pallidus, entopeduncular nucleus, subthalamic nucleus, substantia nigra, ventromedial nucleus of the thalamus, the deep lateral layers of the superior colliculus and the angular complex (see Di Chiara & Gessa 1981). Bilateral injection of the GABA agonist muscimol into zona reticulata of substantia nigra in low doses induces nodding head movements, with strong stereotyped sniffing, motility and rearing in a few rats (Scheel-Kruger et al 1977). Injection of a higher dose of muscimol induces intense gnawing in all animals. In contrast, the injection of the GABA antagonist picrotoxin into zona reticulata produces a rigid catatonia.

Unilateral injection of muscimol into zona reticulata can induce contraversive rotation, but this is a complex response (Kilpatrick & Starr 1981). Strong contraversive rotational behaviour can be elicited from the central zone of zona reticulata; this is partly haloperidol-sensitive, suggesting some involvement of ascending dopamine pathways. Moderate contraversive circling is elicited from the lateral and ventral regions of zona reticulata; this is not affected by haloperidol so appears to be independent of ascending dopamine systems. As might be expected, injection of the GABA antagonist picrotoxin into zona reticulata induces ipsiversive rotation (Reavill et al 1979).

These results show that GABA manipulation of an area such as zona reticulata can mimic many, if not all, of the effects which are thought to be elicited by changes in dopamine function in the striatum or nucleus accumbens. Similar effects can be observed if GABA function is manipulated in projection areas from substantia nigra zona reticulata.

Bilateral alteration of GABA function in the ventromedial nucleus of the thalamus produces changes in locomotion and induces catalepsy (Di Chiara et al 1979a, Kilpatrick et al 1980). Unilateral manipulation causes postural deviation, although active circling does not occur. Manipulation of GABA function with picrotoxin or muscimol in the superior colliculus/angular complex can cause changes in motility and produce stereotyped behaviour, or elicit circling responses (Redgrave et al 1981, Kilpatrick et al 1982, Di Chiara et al 1981, Reavill et al 1982). However, if all the available information on manipulation of outflow areas by GABA-active drugs is considered, the conclusion must be that specific behaviours are not elicited from different areas of outflow. This point is illustrated in Table 1. With some exceptions, and with some differences in the degree of intensity of effect, changes in locomotion, stereotypy, catalepsy and circling can be induced from many regions. Clearly, there is no simple pattern of one motor behaviour being mediated through one particular outflow system.

TABLE 1 Behaviours induced by bilateral or unilateral injection of GABA agonist or antagonist drugs into basal ganglia outflow zones in the rat

Area	Bilateral injection			Unilateral injection	
	Stereotypy	Locomotion	Catalepsy	Posture	Locomotion
Globus pallidus	++[a]	++[a,b]	+[a]	++ N[a]	0[a]
Entopeduncular nucleus	+++[a]	++[a]	++[b]	++ T[a,b]	++[a,b]
Subthalamic nucleus	+++[a]	++[a]	++[b]	++ T[a,b]	++[a,b]
Substantia nigra zona reticulata	+++[a]	+[a]	+++[b]	+++ N[a,b]	+++[a,b]
Ventromedial thalamus	0	++[b]	++[a]	+ N[a,b]	+[a,b]
Superior colliculus/ angular complex	+++[b]	+[b]	?++[a]	++ N[a,b]	++[a,b]

0, absent; +, weak; ++, moderate; +++, strong.
N, neck; T, trunk.
a, agonist drugs, e.g. muscimol.
b, antagonist drugs, e.g. picrotoxin, bicuculline.

Other approaches to solving the role of outflow stations in basal ganglia function have also been used but will not be discussed here. These have included lesioning of basal ganglia outflow structures to determine their effect on dopamine-mediated behaviours and the production of circling behaviour in rodents by unilateral 6-hydroxydopamine lesions of the medial forebrain bundle, with subsequent lesioning of outflow systems to remove this response (see for example Leigh et al 1983).

What does GABA manipulation of basal ganglia outflow reveal?

Do the behavioural effects of manipulation of these GABA-dependent basal ganglia systems reveal the functions of these regions of the brain? Probably not.

One potential confusion lies in the concept developed by Hughlings Jackson more than a century ago. Hughlings Jackson suggested that the positive effects resulting from damage to a particular brain area may not reveal anything about the function of that part of the brain. It is more likely that such positive effects are caused by the release of distant brain regions from the control normally imposed by the damaged area. The negative symptoms caused by damage, namely those functions that are lost, are more likely to reveal the normal role of a particular brain region.

GABA manipulation of basal ganglia outflow produces mainly positive changes, that is, unusual behavioural effects which are not normally observed in the intact animal. Consider again the behaviours which are studied in such experiments. Locomotor hyperactivity, involving incessant movement of the animal from place to place as though compulsively driven, cannot be considered a normal phenomenon. Locomotion in rats is normally directed towards some purposeful activity such as exploration of the surroundings. But within an enclosed space these animals rapidly habituate and exhibit only occasional brief bursts of locomotor activity. The compulsive, repetitive and purposeless oral movements that occur in stereotypy in humans are not normally observed in rats. All normal rats show some licking, gnawing or biting but this is usually goal-directed, being a component of normal behaviours such as feeding or grooming. Most normal rats also show some apparently purposeless perioral movements, but frantic licking or biting of the cage walls is never observed. Nor is it normal for rats to maintain themselves in the abnormal positions and postures that characterize catalepsy. Finally, it is unknown for normal rats to run compulsively in tight circles. These driven behaviours do not constitute part of the normal repertoire of rat behaviour and they constantly interfere with normal motor activity.

GABA manipulation of basal ganglia outflow zones thus probably produces positive phenomena, more closely related to the production of abnormal movement disorders than to the normal functions of these regions. Are there then any negative effects of such animal experiments which might provide a better clue to normal function of the basal ganglia? The only one that might help would be a decrease in normal locomotor activity. A true decrease in locomotion has been reported only from studies of the entopeduncular nucleus, the subthalamic nucleus and the superior colliculus/angular complex. Perhaps it is the normal behaviour of the animal that should be studied rather than drug-induced driven behaviours. More subtle tests of animal activity may reveal the role of GABA-containing outflow pathways in basal ganglia functioning.

In the manipulation technique high concentrations of potent GABA agonist and antagonist drugs are injected into single areas of basal ganglia outflow. Yet many of the GABA output pathways branch to innervate a number of regions. Filion & Herbert (1983) have applied unilateral electrical stimulation to the entopeduncular nucleus to induce contraversive head-turning. Lesioning the outflow from the entopeduncular nucleus to the ventro-anterior/ventrolateral nuclei of the thalamus had no effect on such head-turning. Similarly, lesioning the other major outflow route from the entopeduncular nucleus to the nucleus tegmenti pedunculopontinus had no effect on head-turning. However, when both these pathways were lesioned simultaneously head-turning was markedly reduced. Such findings imply that there is considerable functional overlap in the pathways arising from any individual area, such that destruction of one pathway alone will have no effect on the final behavioural response. To understand the role of GABA-dependent outflow systems from the zona reticulata of substantia nigra it may be necessary to simultaneously lesion or manipulate all target areas, i.e. those to the ventromedial thalamus, the lateral superior colliculus, the angular complex and the nucleus tegmenti pedunculopontinus.

Conclusions

GABA neuronal systems form important intrinsic links within outflow systems from basal ganglia. Manipulation of GABA function at various sites will mimic the effects of behaviours which can be elicited from dopamine-containing areas of basal ganglia. Such studies have identified GABA-sensitive sites within basal ganglia, and the regions to which they project, at which new therapeutic agents might act. However, this approach appears to reveal little of the normal function of the basal ganglia or, indeed, of the involvement of any individual outflow pathway in the mediation of specific motor behaviours. More refined behavioural techniques will be needed to determine how such systems operate in the normal animal.

REFERENCES

Carpenter MB, Carpenter CS 1951 Analysis of somatotopic relations of the corpus Luysi in man and monkey. Relation between the site of dyskinesia and distribution of lesion within the subthalamic nucleus. J Comp Neurol 95:349-370

Carpenter MB, Whittier JR, Mettler FA 1950 Analysis of choreoid hyperkinesia in the rhesus monkey. Surgical and pharmacological analysis of hyperkinesia resulting from lesion in the subthalamic nucleus of Luysi. J Comp Neurol 92:293-331

Childs JA, Gale K 1983 Neurochemical evidence for a nigrotegmental GABAergic projection. Brain Res 258:109-114

Crossman AR, Sambrook MA, Jackson A 1980 Experimental hemiballism in the baboon produced by injection of a gamma-aminobutyric acid antagonist into the basal ganglia. Neurosci Lett 20:369-372

Di Chiara G, Gessa GL (eds) 1981 GABA and the basal ganglia. Raven Press, New York (Adv Biochem Psychopharmacol 30)

Di Chiara G, Morelli M, Porceddu ML, Gessa GL 1979a Role of thalamic γ-aminobutyric in

motor functions: catalepsy and ipsiversive turning after intrathalamic muscimol. Neuroscience 4:1453-1465
Di Chiara G, Porceddu ML, Morelli M, Mulas ML, Gessa GL 1979b Evidence for a GABAergic projection from the substantia nigra to the ventromedial thalamus and to the superior colliculus of the rat. Brain Res 176:273-284
Di Chiara G, Porceddu ML, Imperato A, Morelli M 1981 Role of GABA neurons in the expression of striatal motor functions. In: Di Chiara G, Gessa GL (eds) GABA and the basal ganglia. Raven Press, New York (Adv Biochem Psycopharmacol 30) p 129-163
Filion M, Herbert R 1983 Redundancy in ascending and descending pathways mediating head turning elicited by entopeduncular stimulation in the cat. Neuroscience 10:169-176
Hammond C, Feger J, Bioulac B, Soutegrand JP 1979 Experimental hemiballism in the monkey produced by unilateral kainic acid lesion in corpus Luysii. Brain Res 171:577-580
Kilpatrick IC, Starr MS 1981 Involvement of dopamine in circling responses to muscimol depends on intranigral site of injection. Eur J Pharmacol 69:407-419
Kilpatrick IC, Starr MS, Fletcher A, James TA, MacLeod NK 1980 Evidence for a GABAergic nigrothalamic pathway in the rat. 1. Behavioural and biochemical studies. Exp Brain Res 40:45-54
Kilpatrick IC, Collinridge GL, Starr MS 1982 Evidence for the participation of nigrotectal γ-aminobutyric-containing neurones in striatal and nigral-derived circling in the rat. Neuroscience 7:207-222
Leigh PN, Reavill C, Jenner P, Marsden CD 1983 Basal ganglia outflow pathways and circling behaviour in the rat. J Neural Transm 58:1-41
Reavill C, Jenner P, Leigh N, Marsden CD 1979 Turning behaviour induced by injection of muscimol or picrotoxin into the substantia nigra demonstrates dual GABA components. Neurosci Lett 12:323-328
Reavill C, Leigh PN, Muscatt S, Jenner P, Marsden CD 1982 The involvement of the superior colliculus and midbrain reticular formation in the expression of circling behaviour. In: Kohsaka M et al (eds) Advances in dopamine research. Pergamon, Oxford (Adv Biosci 37) p 201-217
Redgrave P, Dean P, Souki W, Lewis G 1981 Gnawing and changes in reactivity produced by microinjection of picrotoxin into the superior colliculus of rats. Psychopharmacology 75:198-203
Rouzaire-Dubois B, Scarnatti E, Hammond C, Crossman AR, Shibazaki T 1983 Microiontophoretic studies on the nature of the neurotransmitter in the subthalamo-entopeduncular pathway of the rat. Brain Res 271:11-20
Scheel-Kruger J, Arnt J, Magelund G 1977 Behavioural stimulation induced by muscimol and other GABA agonists injected into the substantia nigra. Neurosci Lett 4:351-356
Walaas I, Fonnum F 1980 Biochemical evidence for γ-aminobutyrate containing fibres from the nucleus accumbens to the substantia nigra and ventral tegmental area in the rat. Neuroscience 5:63-72

DISCUSSION

Di Chiara: I don't agree that by manipulating different subdivisions of the basal ganglia with lesions or local injection of drugs one only gets information on abnormal behaviour. The methodological approach we use simply establishes neurotransmitter mechanisms in the neural pathways that subserve the transmission of information from the basal ganglia, and can be applied to abnormal as well as normal motor behaviour. Studying spontaneous or conditioned behaviour is not necessarily the best model. We prefer to see how lesions or drug manipulations can modify a pharmacological syndrome, such as the apomorphine syndrome, which we attribute to the basal ganglia.

So we know that if we manipulate a certain system at a certain place and modify the apomorphine syndrome, we are likely to interfere with the expression of basal ganglia function. The answer to your question about normal behaviour would be to do the same manipulations but to look at the spontaneous behaviour of the animal, conditioned or not. The problem then would be to find a behaviour that can be ascribed with certainty to the basal ganglia.

Smith: I agree with you to some extent. You and others are revealing sites where chemicals act but you are not telling us much about specific neural pathways. You are telling us that neural pathways might exist which use GABA and that the target area is sensitive to GABA, but we don't know exactly what the pathways are doing. There is so much GABA around and so many GABA synapses. Information processing in any part of the brain involves neurons and not just brain areas. If GABA or muscimol is injected into the globus pallidus the drug will presumably act at all sites where there are natural GABA synapses. In addition it may act at extrasynaptic sites. So exactly where are the GABA synapses on a typical neuron? I have chosen a pyramidal neuron for the purposes of illustration (Fig. 1). The GABA boutons in this figure are repre-

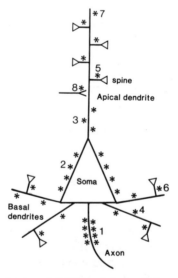

FIG. 1 (*Smith*). Sites of termination of GABAergic synaptic boutons on a 'typical' pyramidal neuron (see text for explanation).

sented by asterisks and I have tried to give a rough indication of the likely density of these boutons on different parts of the neuron.

Several workers have shown that glutamate decarboxylase (GAD)-immunoreactive boutons can be found in synaptic contact with the initial segments of axons (site 1), cell bodies (site 2), the proximal shafts of apical (site 3) or basal (site 4) dendrites, the spines of apical (site 5) or basal (site 6) dendrites, distal dendritic shafts (site 7) and on afferent nerve terminals (site 8) (for recent studies see Somogyi et al 1983, Freund et al 1983). If one injects a GABA agonist or antagonist into a brain area one has no idea whether it is acting at one of these particular sites or at all of them. That is important because the different sites do different things in information processing. For example, if the GABA boutons on a proximal dendritic shaft are active and dense enough in their innervation they would probably cut off all the excitatory input from the more distal parts of that particular dendrite. If the GABA boutons on basal dendrites are active but not those on apical dendrites, the excitatory input from the apical dendrities will continue to come into the cell body. The resulting selectivity of input will affect the information that this neuron transfers through its axon to other regions.

Particularly interesting is the presence of GAD-immunoreactive boutons on the necks of dendritic spines in the striatum (J. P. Bolam et al, unpublished work). These spines frequently receive excitatory input on their heads. So the action of GABA at this site can be even more selective, cutting off input at the level of individual spines (Diamond et al 1970, Kemp & Powell 1971). Incidentally, one would not necessarily see this by recording from the cell body. Since many spines occur on distal regions of the dendrite, GABA acting on the necks of these spines might not cause any detectable change in membrane properties recorded at the soma (Jack et al 1975). You would, of course, see changes in cell firing if there was tonic excitatory input into these spines, since the action of GABA would be to cut this off selectively.

The presynaptic action of GABA (No. 8 on Fig. 1) in the spinal cord is well known since the classical studies of Eccles and his colleagues, and there is now morphological evidence of GAD-immunoreactive boutons in synaptic contact with primary afferent terminals (Barber et al 1978). Whether similar types of axo-axonic contacts between GABA boutons and other axon terminals exist elsewhere, including the basal ganglia, is less certain and further morphological studies are needed.

In the striatum of the rat occasional GAD-immunoreactive boutons have been found on the initial segments of axons (J.P. Bolam et al, unpublished work), but the density of such boutons is very much lower than is found in the hippocampus and cortex (Somogyi et al 1983, Freund et al 1983). The identity of the neurons in the striatum that receive such input to their initial segment remains to be established, but it can be predicted that GABA could influence the output of these striatal neurons, either to distant regions or to local targets in the striatum, by modifying the membrane conductance at the level of the axon initial segment.

The next step is clearly to identify the sources of the different GABA boutons that innervate different regions of an individual neuron. A start has been made for the cerebral cortex (Freund et al 1983) and hippocampus (Somogyi et al 1983), where it has been shown that site 1 (the axon initial segment) receives almost all its GABAergic input from a single type of local circuit neuron, the axo-axonic cell. What I have tried to show is that it is only by analysis of this kind that we shall be able to account for the actions of GABA when it is applied to a particular brain area.

Yoshida: Dr Smith described a complicated scheme for the GABAergic synapses on a neuron. In 1971 and 1972 we published papers on the inhibitory and GABAergic nature of the caudatonigral and caudatopallidal fibres (Precht & Yoshida 1971, Yoshida & Precht 1971, Yoshida et al 1971, 1972). In 1970, Rinvik & Grofová published an important paper on substantia nigra neurons. They said that the caudatonigral fibre terminal made mosaic-type synapses around the dendrite of nigral neurons and that these contained pleomorphic vesicles. Similar types of synapses were observed in caudatopallidal terminals. So far as the caudatonigral or caudatopallidal inhibitory terminals are concerned, the situation is not as complicated as you showed.

Calne: Dr Jenner, would you comment on the initial reports of GAD depletion in Huntington's disease and its relevance, if any, to deficits in this disorder.

Jenner: Obviously if the inhibitory control is removed from basal ganglia there must be, through the loop system, an enhancement of the outgoing motor information. Dr Hornykiewicz proposed that dopaminergic information continues to reach the striatum through the intact nigrostriatal system but the loss of GABA-containing and cholinergic-containing cells in Huntington's disease leads to a deficit in information processing. There is very little else we can say from the type of study that we are doing.

Yoshida: Since we have shown the inhibitory and GABAergic nature of the pallidothalamic and nigrothalamic pathways in cats, I wanted to see whether the situation was similar in human beings. I have had a rare opportunity of examining at autopsy the brain of a woman who died of postencephalitic parkinsonism. In this particular case the cells of the substantia nigra, including the zona compacta and zona reticulata, were completely lost. Since the nigrothalamic pathway is GABAergic, we measured the GABA content in the thalamus of this patient. We cut frontal sections 150-μm thick caudal to the middle of the thalamus in the rostrocaudal direction, made microdissections of 2×2 mm^2 by Dr Kanazawa's technique, and measured the GABA content in each sample using gas chromatography–mass spectrometry (Yoshida et al

1981). The GABA concentration was remarkably low compared to that in the normal control, paralysis agitans and Huntington's chorea. So the human being would also have GABAergic nigrothalamic fibres.

Iversen: I accept the anatomical facts that David Smith described but we have no tools at the moment for selectively manipulating GABA receptors in one part of the nigral complex rather than the other, or for knowing whether there is more than one kind of GABA receptor at a given CNS site. Maybe with the development of pharmacological tools we shall have better and more selective antagonists. We have to defend the lesion approach for the time being, because there is no other way of tackling these problems. It is acceptable to interfere with an anatomical system to accentuate a behaviour but it is dangerous then to go backwards and infer normal function from the nature of that abnormal behaviour. But at least the 'lesion' is a tool for tracing functional links in anatomical circuits in the way Dr Di Chiara uses it. I think Peter Jenner's point regarding positive and negative symptoms is an important one. In parallel in such studies we should be exploring a wider range of naturalistic normal behaviours involving motor responses with the sophisticated observational methods now available.

Smith: I entirely agree. I was just saying that we must develop the next step forward.

Kitai: We have been studying the intracellularly labelled subthalamic neurons and their axon terminals. Their terminals are quite different from well-established GABAergic synapses in the various areas in the CNS. When you manipulate the subthalamus with GABA agonists or antagonists, you might be manipulating not the subthalamic system but the pallido-subthalamic GABAergic inputs. You might indeed be manipulating the pallidal rather than the subthalamic operations. We would all be wise to keep up with anatomy and physiology.

Glowinski: The debate between reductionists and integrationists is an old one. The approaches are complementary and we are well aware of the difficulty of interpreting observations on behaviour induced by locally injected drugs. Dr Jenner, what do you think of the approach of inducing abnormal behaviour by local application of a pharmacological substance and then restoring normal behaviour with another local treatment in a second structure which is under the control of the first one?

Jenner: Dr Di Chiara has used it extensively and we have used it too. Many of the behaviours which we initiate in this way I would call driven behaviours. They are not part of the normal repertoire of the animal and they are much more compulsive than I would have liked when trying to examine the normal functioning of basal ganglia.

Iversen: The approach of using drugs to restore normality to a behaviour distorted with drugs or CNS damage has been widely used, for example by

Randrup in Denmark studying the effect of dopamine-blocking drugs on amphetamine stereotypy. It is a very positive approach that finds a place in clinical psychiatry. For example, in considering how neuroleptics control schizophrenic behaviour, the ability of the drugs to restructure complex behaviour rather than merely eliminate unwanted responses has been questioned. In experimental work in animals efforts have been made to distinguish the effects of neuroleptics on motor output from their ability to reinstate organized behaviour.

Wing: Although your main point is that one should study lesions that have negative rather than positive effects on behavioural output, it is important to remember that we have to lesion many places to make sure that the deficits one sees with a lesion at a particular site aren't seen with a lesion elsewhere. This tells us about the localization of function.

REFERENCES

Barber RP, Vaughn JE, Saito K, McLaughlin B, Roberts E 1978 GABAergic terminals are presynaptic to primary afferent terminals in the substantia gelatinosa of the rat spinal cord Brain Res 141:35-55

Diamond J, Gray EG, Yarsargil GM 1970 The function of the dendritic spine: an hypothesis. In: Anderson P, Jansen JKS (eds) Excitatory synaptic mechanisms. Universitets Forlaget, Oslo, p 213-222

Freund TF, Martin KAC, Smith AD, Somogyi P 1983 Glutamate decarboxylase-immunoreactive terminals of Golgi-impregnated axo-axonic cells and of presumed basket cells in synaptic contact with pyramidal cells of the cat's visual cortex. J Comp Neurol 221(3):263-278

Jack JJB, Noble D, Tsien RW 1975 Electric current flow in excitable cells. Clarendon Press, Oxford, p 218-224

Kemp JM, Powell TPS 1971 The site of termination of afferent fibres in the caudate nucleus. Philos Trans R Soc Lond B Biol Sci 262:413-427

Precht W, Yoshida M 1971 Blockage of caudate-evoked inhibition of neurons in the substantia nigra by picrotoxin. Brain Res 32:229-233

Rinvik E, Grofová I 1970 Observations on the fine structure of the substantia nigra in the cat. Exp Brain Res 11:229-248

Somogyi P, Smith AD, Nunzi MG, Gorio A, Takagi H, Wu J-Y 1983 Glutamate decarboxylase immunoreactivity in the hippocampus of the cat. Distribution of immunoreactive terminals with special reference to the axon initial segment of pyramidal neurons. J Neurosci 3:1450-1468

Yoshida M, Precht W 1971 Monosynaptic inhibition of neurons of the substantia nigra by caudate-nigral fibers. Brain Res 32:225-228

Yoshida M, Rabin A, Anderson ME 1971 Two types of monosynaptic inhibition of pallidal neurons produced by stimulation of the diencephalon and substantia nigra. Brain Res 30:235-239

Yoshida M, Rabin A, Anderson ME 1972 Monosynaptic inhibition of pallidal neurons by axon collaterals of caudate-nigral fibers. Exp Brain Res 15:333-347

Yoshida M, Ariga T, Kanazawa I, Miyatake T 1981 Topographical distribution of γ-aminobutyric acid within the cat thalamus in relation to the basal ganglia, as determined by mass fragmentometry. J Neurochem 37:670-676

General discussion 2

The striatal mosaic and local control of dopamine

Evarts: Ann Graybiel gave us a beautiful demonstration of the striking overlap between muscarinic receptors and the dopamine islands. It used to be said that the striatum was like the liver: when you had seen one piece you had seen it all. A tremendous degree of differentiation has now emerged, both at the micro-level that Ann has demonstrated and at the macro-level that Dr Hornykiewicz has described. The change that Ann Graybiel found in the weaver mutant is a fascinating indication of the extent to which disease may affect one part of the striatum and not another.

Jacques Glowinski then pointed out that there are mechanisms for dopamine release that do not depend on the impulse frequencies of the dopaminergic cells. Mahlon DeLong noted that this lack of dependence on impulse frequency pointed to the existence of local regulation of dopamine availability, and we might now discuss this local regulation. What approaches could be used to reveal the mechanisms for local regulation that are indicated by the phenomena that Glowinski has described?

Graybiel: If there are local mechanisms for presynaptic control at the level of the caudate nucleus, striosomes could play some role because of the correspondence between dopamine islands and ligand binding. Probably it will be necessary to do immunohistochemistry of receptors, with antibodies raised to receptors, to settle this. There is very little evidence from electron microscopy for boutons terminating on other boutons or on axons, so something even less conventional may be needed as the physical mechanism. One possibility is that the axons release peptides and that these are degraded slowly enough to have local effects. It is interesting that so many peptides are present in high concentrations in the striatum, but could they diffuse across 200 or 300 μm? If so, this might be the key to understanding that compartments set up local metabolic pools.

Bolam: If there is communication between terminals in the striatum then it is not via classical synapses, which we have never seen.

Arbuthnott: There is already in Dr Smith's diagram (p 172) a presynaptic action that doesn't need presynaptic morphology. If there is an excitatory input to the head of a dendritic spine and an inhibitory one to the neck of the spine, the excitation won't be transmitted to the soma. The result will look like presynaptic inhibition in all the conventional neurophysiological tests, except perhaps in transmitter release. Such an arrangement could explain, for instance, the effect of dopamine on corticostriatal connections which we recently described (Brown & Arbuthnott 1983). We don't need to have the dopamine terminal on an axonal process in order to inhibit the postsynaptic effect of individual terminals. But all this doesn't help with the release story.

Somogyi: The presynaptic action of transmitters raises another question, namely whether the message mediated by the transmitter is addressed to a particular target or is received diffusely by nearby neuronal processes. I would like to outline a hypothesis which unfortunately has no facts to support it yet. Nevertheless it could resolve the paradox that the brain goes to all the trouble of making precise synaptic wiring yet almost any substance can affect the release of others without any apparent specific structural relationship between the nerve terminals. Let us assume that two terminals containing two different transmitters, say dopamine and glutamate, terminate on the same dendrite in the striatum. We know that these are very precisely placed synapses on the postsynaptic neuron. Therefore there is already a precisely wired channel for interaction between the two terminals through the postsynaptic dendrite, besides the possibility of a diffuse interaction through receptors on the presynaptic terminal. The action of one of the transmitters will change the properties of the dendrite, which can in turn send a message to the other terminal, changing its properties. Professor Glowinski has just demonstrated that dopamine presumably released from dendrites can affect neurotransmission in the substantia nigra.

Glowinski: There are different levels at which this problem should be discussed. First, *in vitro* experiments with striatal slices have shown that several types of transmitters or their agonists modify the spontaneous or evoked release of dopamine. Presynaptic receptors involved in this kind of regulation may or may not be located on dopaminergic nerve terminals. This presynaptic regulation, which may be either direct or indirect, is nevertheless quite specific for a given type of nerve terminal. Indeed, differences can be seen in the presynaptic regulation of dopamine or 5HT release in this kind of pharmacological *in vitro* experiment.

At the second level, interactions between identified nerve terminals should be demonstrated morphologically. Recently, Pickel et al (1981) and Descarries et al (1980), who have used different approaches to identify dopaminergic boutons in the electron microscope, have described axo-axonic contacts or 'appositions' which may represent the morphological basis for the interactions discovered in biochemical studies. Appositions between dopaminergic dendrites or between dendrites and nerve terminals or even glial cells have also been observed in the substantia nigra (Wassef et al 1981). These types of contacts, which differ from classical synapses, should be taken into consideration.

At the third level, presynaptic regulation should be demonstrated *in vivo*, to confirm its physiological significance. The demonstration of physiological presynaptic regulation of dopamine release by striatal cholinergic interneurons is difficult, but it can be done for striatal afferent pathways, for instance those originating in the cerebral cortex. I hope that some of the results presented in my paper show that this type of experiment is possible and that conclusions on the involvement of physiological presynaptic regulation can be reached.

The last point concerns the relationships between nerve activity and presynaptic regulatory mechanisms in the control of dopamine release. The two processes may influence each other. The potency of presynaptic facilitatory or inhibitory regulation depends on the level of nerve activity. For example, presynaptic facilitatory regulation can hardly be seen when the dopaminergic cells are activated.

Kitai: Dr C.J. Wilson is intracellularly recording from the cortical neurons and intracellularly labelling the recorded neurons in the sensorimotor cortex. Stimulating electrodes are placed bilaterally in the striatum. His findings indicate that cortical neurons can be activated bilaterally and the labelled neurons have their axons projecting to both striata. These findings support Dr Glowinski's hypothesis. Dr Hornykiewicz, you showed us that there was more dopamine concentrated in the centre of your caudate slice. How do you explain that in light of the known anatomy of the caudate?

Hornykiewicz: First, the dopamine islands Dr Graybiel described are not unlike what we see. Second, as to the known anatomy of the caudate, Fuxe and his colleagues (Olson et al 1972) have observed two types of dopamine fluorescence in the rat striatum, a 'diffuse' and a 'dotted' system. They found the classical diffuse dopamine system mainly in the central parts of the striatum, whereas the dotted type of dopamine system formed islands of fluorescence mainly along the medial and lateral borders of the nucleus. It is possible that the especially high dopamine levels which we found in the central parts of the human caudate and, to a lesser extent, in the putamen, correspond anatomically to the distribution of the diffuse dopamine system. In this case, the caudate parts with lower dopamine might correspond to the dotted ('islandic') dopamine system. It would be interesting to study the distribution pattern of cholecystokinin in the human basal ganglia. Hökfelt and his colleagues (1980) showed that in the rat some dopamine neurons of the dotted type contain cholecystokinin as a possible co-transmitter.

Graybiel: From what is known about peptides, the action of dopamine and muscarinic acetylcholine effects, we suspect that different pathways may be using different time scales—some slow, some fast. The coordination of these various timings is probably very important. Abnormal behaviours in basal ganglia disease could result from very small changes in the timing in the loop circuits of the basal ganglia, and there are many of these.

Another point we should remember in thinking about non-conventional effects is that there are large numbers of glial cells in the basal ganglia. New evidence suggests that glial cells may have many different receptors for many neurotransmitters, at least *in vitro*. Also, many years ago, Palay (1966) pointed out that glial cells can, on a very local scale, isolate groups of synapses along what otherwise seem to be continuous fields of postsynaptic space. If the glial cells, in addition to or by virtue of having receptors, can alter the metabolism of adjoining neurons, then that would be another way of controlling the circuits.

Nauta: I have a question that concerns the efferent rather than the afferent side of the striatum: what is the relationship between the compartments of the striatum as we begin to see them unravelled by studies such as Dr Graybiel's, and the close diadic and triadic clustering of small and medium-sized striatal neurons first recognized by Patricia Mensah (1977) in the light microscope? Electron microscope studies have shown large areas of nearly uninterrupted soma-to-soma contact between the members of such clusters. The intervening cleft has been variously measured as 25 nm (Adinolfi & Pappas 1968) and 3–9 nm in width (Domesick 1981). The lower of these estimates could raise the suspicion of electrotonic couplings and, with it, the question whether the striatal output might, in part at least, be generated by functional units having more than one axon.

Graybiel: In a very interesting set of recent papers Grace and Bunney have suggested that in the pars compacta there are groups of dopamine-containing neurons that are electrotonically coupled (Grace & Bunney 1983). Steve Bunney and I have been talking about the possibility that those couplings could be reflected in the caudate nucleus, maybe even in the form of the dopamine island–striosomal system.

Calne: Jacques Glowinski referred to the cross-talk between the two sides. Mahlon DeLong's cells and Ed Evarts' cells were contralateral to movement, and certainly in parkinsonian patients one may see substantial motor deficits on one side with a clinically normal hand on the other side. Although there is bilateral dopamine depletion, Professor Hornykiewicz would probably confirm that the depletion is asymmetrical. There are clearly significant lateralizations in terms of output correlating with contralateral physical damage. What is the biological significance of the strong connections between the two sides of the basal ganglia?

Glowinski: Our experiments were on halothane-anaesthetized cats, and this can be considered an artificial situation. I agree that experiments in conscious animals are needed and we are now trying to do such experiments in the rat. Bilateral regulation has been observed in preliminary experiments on these animals, but more work should be done.

As to the significance of bilateral regulation of nigrostriatal dopaminergic neurons in relation to behaviour, I think it may have something to do with the coordination of motor behaviour and posture, in particular. However, there is no experimental proof yet for such an interpretation.

Smith: There isn't really a dichotomy between reductionists and integrationists. It is just the way we all do our experiments. The reductionist approach is essential for understanding the information processing going on at the neuronal level. This approach can be combined with the more global approach of the physiologists and behaviourists, but obviously it will take a long time. One thing we can now do is describe the precise chemical and anatomical framework so that you know that if you apply a certain drug by microiontophoresis to a certain area or layer or striosome it is likely to produce one or more effects on information transfer. This can be done by combining the electron microscope and immunohistochemical analysis of individual synapses with the demonstration of the connections of the neurons that give or receive these synaptic boutons.

The information from electron microscope immunocytochemistry alone is almost useless—it just tells us about possibilities. But it is now technically feasible to show, for example, for identified striatal output neurons, whether their input comes from another spiny neuron inside or outside a striosome. We can show whether their afferent GABA boutons come from the local circuit GABA neurons that Paul Bolam described earlier in the striatum. We can show that some of the other inputs of the same identified neuron come from the cortex whereas others come from the thalamus. We can put the dimension that is missing from the reductionist approach on the reductionist map by showing the distant connections to the other brain areas of individual synapses that are chemically identified with respect to their transmitter.

Finally, we can bring in the dimension of 'function' by trying to record the activity of neurons in relation to specific types of behaviour, as Rod Porter, Mahlon DeLong and Edmund Rolls have done in the basal ganglia. If the technical difficulties of recording intracellularly in the basal ganglia of behaving animals can be overcome, these 'functionally' characterized neurons can be injected with horseradish peroxidase, as Steven Kitai does, and then their chemistry and synaptic connections can be studied by techniques that are at an advanced stage of development (Freund & Somogyi 1983, Somogyi et al 1983a,b).

REFERENCES

Adinolfi AM, Pappas GD 1968 The fine structure of the caudate nucleus of the cat. J Comp Neurol 133:167-184

Brown JR, Arbuthnott GW 1983 The electrophysiology of dopamine (D_2) receptors: a study of the actions of dopamine in corticostriatal transmission. Neuroscience 10:349-355

Descarries L, Bosler O, Berthelet F, Des Rosiers M 1980 Dopaminergic nerve endings visualized by high resolution autoradiography in adult rat neostriatum. Nature (Lond) 284:620-622

Domesick VB 1981 Further observations on the anatomy of nucleus accumbens and caudatoputamen in the rat: similarities and contrasts. In: Chronister RB, DeFrance JF (eds) The neurobiology of the nucleus accumbens. Haer Institute for Electrophysiological Research, Brunswick, Maine

Freund TF, Somogyi P 1983 The section Golgi impregnation procedure. 1. Description of the method and its combination with histochemistry after intracellular iontophoresis or retrograde transport of horseradish peroxidase. Neuroscience 9:463-474

Grace AA, Bunney BS 1983 Intracellular and extracellular electrophysiology of nigral dopaminergic neurons—3. Evidence for electronic coupling. Neuroscience 10:333-348

Hökfelt T, Rehfeld JF, Skirboll L, Ivemark B, Goldstein M, Markey K 1980 Evidence for coexistence of dopamine and CCK in meso-limbic neurones. Nature (Lond) 285:476-478

Mensah PL 1977 The internal organization of the mouse caudate nucleus: evidence for cell clustering and regional variation. Brain Res 137:53-66

Olson L, Seiger A, Fuxe K 1972 Heterogeneity of striatal and limbic dopamine innervation: highly fluorescent islands in developing and adult rats. Brain Res 44:283-288

Palay SL 1966 The role of neuroglia in the organization of the central nervous system. In: Rodahl K, Issekutz B (eds) Nerve as a tissue. Hoeber-Harper, New York, p 3-10

Pickel VM, Beckley SC, Joh TH, Reis DJ 1981 Ultrastructural immunocytochemical localization of tyrosine hydroxylase in the neostriatum. Brain Res 225:373-385

Somogyi P, Freund TF, Wu J-Y, Smith AD 1983a The section-Golgi procedure. 2. Immunocytochemical demonstration of glutamate decarboxylase in Golgi-impregnated neurons and in their afferent synaptic boutons in the visual cortex of the cat. Neuroscience 9:475-490

Somogyi P, Kisvarday ZF, Martin KAC, Whitteridge D 1983b Synaptic connections of morphologically identified and physiologically characterised large basket cells the striate cortex of cat. Neuroscience 10:261-294

Wassef M, Berod A, Sotello C 1981 Dopaminergic dendrites in the pars reticulata of the rat substantia nigra and their striatal input. Combined immunocytochemical localization of tyrosine hydroxylase and anterograde degeneration. Neuroscience 7:2125-2139

Behavioural effects of manipulation of basal ganglia neurotransmitters

SUSAN D. IVERSEN

Neuroscience Research Centre, Merck Sharp and Dohme Ltd, Hertford Road, Hoddesdon, Herts EN11 9BU, UK

Abstract. Topographically organized dopaminergic projections from the extrapyramidal structures of the ventral mesencephalon (substantia nigra and ventral tegmental area) to the dorsal (body of caudate–putamen) and ventral (anterior-ventral caudate, nucleus accumbens, tuberculum olfactorium) striatum subserve sensorimotor integration in the rat. Selective depletion of DA impairs the animal's ability to integrate sensory input with motor output; in the dorsal striatum the exteroceptive sensory input and in the ventral or limbic striatum the interoceptive input principally related to motivation and affect. Grafts of fetal DA neurons to the damaged dorsal striatum reverse sensorimotor asymmetry and sensory neglect. A large number of other excitatory and inhibitory neurotransmitters, including recently discovered neuropeptides, contribute to the functional balance afforded by the DA neurons. This chemical heterogeneity of the basal ganglia offers the possibility that novel therapeutic approaches with drugs could be used to control the chemical imbalances in basal ganglia that are associated with a number of neurological and psychiatric conditions.

1984 Functions of the basal ganglia. Pitman, London (Ciba Foundation symposium 107) p 183-200

The term 'basal ganglia' has carried different connotations in the course of time. Anatomists originally tended to use it to include most, if not all, of the large nuclei in the interior of the brain and although a number of structures have now been excluded from the list, there is still no generally accepted definition. Clinicians tend to include the caudate nucleus and lentiform nuclei (putamen and globus pallidus), usually realigned in terminology as *striatum* (caudate nucleus and putamen) and *pallidum* (globus pallidus). Since these elements of the corpus striatum have clearly defined anatomical and functional relationships with the motor system, this restricted use of the term has been widely accepted. It is becoming clear that the corpus striatum has strong relationships with other telencephalic grey matter, suggesting that its role may not be strictly motor. In this paper attention will be focused on the striatum and globus pallidus and on their efferent and afferent relations.

Chemical morphology of the striatum

Modern anatomical techniques have allowed substantial advances to be made in describing the cell types of the striatum, their detailed morphology and reciprocal connectivity (Pasik et al 1980). The chemical architecture of the striatum matches its anatomical complexity. In a recent review Graybiel & Ragsdale (1983) cite 189 references to neuroactive compounds and related substances, including synthetic and degrading enzymes, in striatum. In addition to glutamate, γ-aminobutyric acid (GABA) and the monoamines, striatum is reported to contain a wide range of the more recently discovered neuropeptides. These chemical systems are in some cases intrinsic to striatum and in others extrinsic, arising from cell bodies located some distance away.

To determine the origin of chemical components in a structure two techniques are commonly used. (1) Lesions to the intrinsic neurons are made and if the neurotransmitter is reduced it is assumed to have been located in cell bodies of the destroyed neurons. Terminals are resistant to toxins such as kainic or ibotenic acid. (2) Anatomically defined input/output pathways are severed and the structure is investigated for decreases in specific neurotransmitters. Summarizing such approaches, Fonnum & Walaas (1979) reported that acetylcholine, GABA and enkephalin are endogenous to striatum whereas dopamine (DA), 5-hydroxytryptamine (5HT), noradrenaline and glutamate are extrinsic.

The relationships of these chemical systems to the intrinsic neurons of the striatum are not fully understood but recent advances in immunohistochemistry, combined with modern Golgi and tracing techniques, continue to define them (Pasik et al 1980). The chemically coded efferent projections of striatum to lower sites in the extrapyramidal motor system are also receiving detailed attention, typified by the elegant studies of Somogyi and Brown and their colleagues (Somogyi et al 1979, Somogyi et al 1982).

The detailed distribution of neurotransmitters in striatum and globus pallidus has been the focus of studies using a wide range of anatomical techniques of increasingly fine resolution. The striatum has been divided, on the basis of anatomical (Heimer & Wilson 1975) and functional studies, into dorsal and ventral sectors (Iversen & Fray 1982). The dorsal striatum refers to the head and body of the caudate–putamen and the ventral striatum includes the nucleus accumbens, olfactory tubercle and closely related parts of the limbic system. These two sectors of striatum were originally differentiated on the basis of their DA input from the substantia nigra on the one hand and the ventral tegmental area on the other. The distinction is borne out by the observation that whereas neocortex projects topographically to dorsal striatum, the limbic system (hippocampus, amygdala and septum) projects topographically to ventral striatum. Although the two sectors are defined by

DA and 'cortical' afferents, the distinction provides a useful framework for describing the distribution of the many other chemically coded components of the striatum.

Three broad categories of chemical messenger can be identified (L. L. Iversen 1984) and all are found within striatum: (1) fast excitatory or inhibitory transmitters, typified by glutamate and GABA; (2) slow excitatory or inhibitory transmitters, typified by the monoamines noradrenaline, DA, 5HT and acetylcholine; (3) perhaps even slower transmitters or long-term modulators typified by the neuropeptides; more than 30 of these have now been identified in brain, all with peripheral physiological functions in addition to any effects they may have in the central nervous system.

Fonnum & Walaas (1979) provided an excellent review of the chemical innervations extrinsic and intrinsic to striatum. Neocortical projections to dorsal striatum and allocortical projections to ventral striatum utilize glutamate; lesions of the cortex or of limbic structures (kainic acid in the hippocampus or surgical section of the fornix reduced glutamate uptake or glutamate levels in the relevant target areas in striatum (Divac et al 1977, Walaas & Fonnum 1979). Electrophysiological experiments confirm the excitatory nature of the pathway between hippocampus (Groenewegen et al 1981), amygdala (Yim & Mogenson 1982) and ventral striatum. Glutamate agonists such as kainic or ibotenic acid excite neurons bearing glutamate receptors to such an extent that degeneration is induced. Consequently, these drugs have been widely used as toxins to induce selective damage of intrinsic striatal neurons bearing glutamate receptors and to determine the distribution of other neurotransmitters in cell bodies or terminal innervations. With such an approach it has been demonstrated that GABA is almost, if not totally, confined to terminals derived from local neurons which are evenly distributed in the rostrocaudal plane. These neurons also give rise to long extrinsic projections to globus pallidus and substantia nigra.

The striatum is rich in monoamines. The catecholamine dopamine is found in high concentrations throughout dorsal and ventral striatum. The DA innervation arises from neurons in the ventral mesencephalon, with the A9 DA neurons (substantia nigra, zona compacta) innervating the dorsal striatum topographically and the A10 DA neurons (ventral tegmental area) innervating the ventral striatal complex.

The distribution of 5HT neurons and their projections was originally described using the formaldehyde fluorescence method but more recently with immunofluorescence (Steinbusch 1981). The dorsal striatum has a relatively sparse 5HT innervation but the ventrolateral striatum and nucleus accumbens have been shown to have a much denser, although patchy, 5HT innervation. The noradrenaline innervation of striatum is even more restricted. Finally, acetylcholine occurs in high concentrations in striatum, and lesion studies indicate that, like GABA, acetylcholine is intrinsic to striatum. Recent studies using acetylcholinesterase as a marker reveal that this enzyme is heterogeneously distributed in striatum in association with the DA innervation. However, acetylcholinesterase does not appear to be associated with DA innervation of the limbic structures (Marshall et al 1983).

The neuropeptide substance P and the opioid peptide enkephalin are found in neurons located in the posterior and anterior striatum respectively. In both cases these neurons innervate locally within striatum but form long extrinsic projections to the substantia nigra (substance P neurons) and globus pallidus (enkephalin neurons). Immunohistochemical staining of peptides and specific receptor binding assays have provided information on the distribution of many other neuropeptides in the brain: melanocyte-stimulating hormone, Met-enkephalin and Leu-enkephalin, dynorphin, angiotensin II, avian polypeptide, bradykinin, cholecystokinin, neurotensin, secretin, somatostatin, substance P, thyrotropin-releasing hormone, vasoactive intestinal polypeptide and vasopressin (cited in Graybiel & Ragsdale 1983) have all been reported to exist in striatum. One has the impression from this survey that neuropeptides occur in great density and variety in striatum, particularly in the limbic ventral striatum. It is also interesting that cholecystokinin coexists with DA in a subpopulation of A10 DA neurons innervating ventral striatum (Hökfelt et al 1980). Thus cholecystokinin is likely to be released together with DA in these striatal terminal areas.

Functional studies of neurotransmitters in the striatum

How does one go about studying such a mosaic of neurotransmitters or modulators within a brain structure? The striatum has provided in many ways a model system for the development of experimental techniques for answering such questions.

Is the striatum functionally homogeneous?

In terms of its intrinsic neuronal elements and its extrinsic and intrinsic neurotransmitter content, striatum appears at first sight to be homogeneous. However, the neocortex and limbic system provide the information input to striatum and the organization of these pathways strongly suggests that striatum is not functionally homogeneous (S. D. Iversen 1984). Neocortex projects topographically to dorsal striatum. For example, medial frontal cortex in the rat projects to the anteromedial part of dorsal striatum, whereas suprarhinal or orbital cortex projects to a more ventrolateral sector (Beckstead 1979). In a range of species in which cortical lesions and silver staining of degeneration in striatum were used, these projection fields from different areas of cortex appeared discrete and little overlap was observed.

In the monkey, autoradiographic staining of amino acids transported from cortex to striatum has revealed more widespread and overlapping areas of influence in striatum from given cortical foci (Goldman & Nauta 1977, Jones et al 1977, Yeterian & Van Hoesen 1978). Yeterian & Van Hoesen state 'we find ourselves in substantial agreement with Divac's (1972) statement that "there is ground to believe that regions of both the caudate nucleus and putamen differ only as much as their cortical and thalamic inputs differ".'

Manipulation of neurotransmitters within a functionally distinct area of striatum

Starting from the same interpretation as Yeterian & Van Hoesen (1978), Stephen Dunnett and I investigated the effect on behaviour of selective chemical manipulations within anatomically defined sectors of rat striatum that received specified cortical input. Earlier lesion studies had defined the behavioural impairments associated with selective lesions to sectors of rat neocortex: spatial learning impairments after medial frontal cortex lesions (Wikmark et al 1973, Kolb et al 1974), deficiencies in the control of responses after orbital lesions (Kolb et al 1974, Neill et al 1974), and regulatory abnormalities after orbital lesions (Kolb & Nonneman 1975, Kolb et al 1977).

We sought further evidence of dissociation of the functions in the head of the caudate between the foci receiving input from medial and frontal cortex and orbital cortex. Using kainic acid to destroy intrinsic neurons we compared the effects of anteromedial and ventrolateral lesions in rats in spatial habit learning and reversal and on the withholding of responses during go/no-go performance and its extinction in a straight runway. A double dissociation was found: anteromedial lesions impaired spatial habit reversal but not the withholding of responses in the go/no-go task or during extinction; ventrolateral lesions resulted in the opposite pattern (Dunnett & Iversen 1981). Furthermore, selective removal of DA in the anteromedial focus or damage to the intrinsic neurons at this site results in spatial learning impairments (Dunnett & Iversen 1979). Equally, kainic acid and 6-hydroxydopamine lesions to ventrolateral striatum impaired the control of body weight and the acquisition of performance in a schedule involving differential reinforcement of low rates of responding. Since this schedule requires a slow rate of lever pressing, the impairment is viewed as one of response disinhibition. In this study kainic and 6-hydroxydopamine lesions to anteromedial cortex impaired neither aspect of behaviour (Dunnett & Iversen 1982a).

Selective damage to 5HT terminals in these striatal sites did not result in significant impairment of behaviour.

As we shall see later, the striatum is conceived of as an interface between cortical levels of processing and the generation of motor responses. In addition to 'cognitive' impairments, damage or pharmacological manipulation of the striatum or its afferent pathways impairs sensorimotor integration. Unilateral lesions to striatum or substantia nigra result in unilateral sensorimotor neglect as assessed from sensory responsiveness (Marshall et al 1980) or rotational behaviour (Ungerstedt 1971b). Dunnett & Iversen (1982b) have localized this impairment to an area of dorsal striatum (the mid-ventral sector) which receives input from lateral cortical areas (Webster 1961). Anterior cortical lesions in rat, including damage to this lateral cortical area (Whishaw et al 1981), result in sensorimotor orientation impairments. Both kainic acid and 6-hydroxydopamine lesions to midventral striatum produce the same sensorimotor deficit. Bilateral lesions of substantia nigra which disrupt afferent and efferent striatal connections result in global sensorimotor impairment; rats are totally unresponsive—they do not move, eat or drink (Ungerstedt 1971a).

Unilateral disruption of the neurotransmitters found in dorsal striatum or substantia nigra results in rotational behaviour, the direction depending on whether the manipulation is agonistic or antagonistic to striatal output. Experiments in which DA, GABA, acetylcholine, enkephalins and substance P were manipulated in striatum or in the substantia nigra all lead to motoric asymmetries (see Pycock 1980 for review).

Further work is needed to determine the role of transmitter candidates other than DA in local areas of striatum. Differential effects of cholinergic manipulations in dorsal and ventral sectors of head of caudate on shock avoidance behaviour were reported by Neill & Grossman in 1970. Subsequently, Neill & Herndon (1978) focused on cholinergic/catecholamine interactions in the ventral aspect of head of caudate where lesions were known to result in response disinhibition. Dopamine itself, the dopamine-releasing drug amphetamine, and the anticholinergic scopolamine, all disrupted response control on a schedule involving differential reinforcement of low rates of responding, strengthening the view that DA and acetylcholine are functionally antagonist in striatum.

Neuronal grafting as an approach to localization of function within striatum

Selective lesioning of specific neurotransmitter candidates or their inactivation with pharmacological antagonists are well tried techniques. Further support for functional involvement of transmitters is afforded in experiments where lesion-induced deficits can be reversed with drugs acting as agonists to the specific transmitter system. For example, apomorphine, the DA agonist, given systemically reverses the sensorimotor neglect after unilateral nigrostriatal lesions.

In collaboration with Björklund and Stenevi of the University of Lund, Stephen Dunnett and I have evaluated how well fetal neurotransmitter neurons grafted to damaged transmitter pathways in adult rats can reverse the behavioural deficits associated with those lesions. The rotational behaviour (Björklund et al 1980, Dunnett et al 1981a) and sensory neglect (Dunnett et al 1981b) induced with 6-hydroxydopamine lesions in the nigrostriatal DA pathway are reversed by grafting fetal DA neurons locally to the damaged sector of striatum. These grafted neurons survive in the host, form fibre outgrowth and show normal levels of biochemical activity (Björklund et al 1981) despite being isolated from their normal site (substantia nigra) and connections in the CNS.

Grafts in the dorsal striatum reverse rotational behaviour but not the sensory neglect; by contrast lateral striatal grafts which innervate the midventral sector of striatum reverse the sensory neglect most effectively (Fig. 1). It remains to be seen whether similar grafts can reverse the more cognitive impairments after bilateral 6-hydroxydopamine lesions to head of caudate.

It seems unlikely that release of DA from these isolated neurons can be under precise neurophysiological control. In a number of instances it has been suggested that monoamine release in CNS is diffuse in both space and time, serving to modulate nervous integration in terminal areas rather than conveying specific signals. If this is so, it may explain how diffuse release of DA from fetal neurons can effectively contribute to striatal integration.

Parallel organization of the mesolimbic DA neurons in ventral striatum

The small size and relative inaccessibility of the ventral striatum has made experimental studies more difficult. Anatomical studies reveal that the focal structure of this complex, the nucleus accumbens, receives organized limbic input utilizing glutamate as a transmitter, just as the dorsal striatum receives cortical input. The DA neurons of the ventral tegmental area (VTA) provide a topographically organized input. It seems likely that the additional transmitters important in the nigrostriatal circuitry are also components of this parallel limbic system.

In pursuing the parallel organization of the limbic and non-limbic DA systems to the functional level we suggested (Iversen & Fray 1982) that the ventral striatum serves sensorimotor integration of interoceptive rather than of exteroceptive information and is important for species-specific motivation-

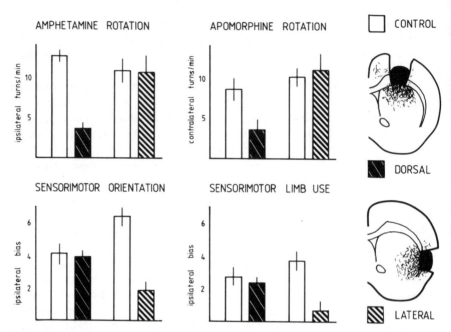

FIG. 1. Summary of results described in Dunnett et al (1981a,b). The open bars indicate the drug-induced rotational behaviour (above) and sensorimotor bias to the ipsilateral side (below) after unilateral 6-hydroxydopamine lesions to the substantia nigra. Dorsal grafts (black bars with white stripes) of fetal dopamine neurons reduce amphetamine and apomorphine rotation but have no effect on sensorimotor orientation or limb use. By contrast, lateral grafts (black and white striped bars) have no effect on rotation but reduce sensorimotor bias and reinstate symmetry.

al behaviour. Damage to the ventral striatum or the VTA impairs organized motivational behaviour such as exploration of novel environments (Iversen & Fray 1982), maternal behaviour (Gaffori & Le Moal 1979), psychogenic polydipsia (Robbins & Koob 1980) and food hoarding (Stinus et al 1978).

Jonathan Alpert and I have recently completed a detailed study of hoarding behaviour in the rat after selective 6-hydroxydopamine lesions of the nucleus accumbens or dorsal striatum. When rats on a restricted feeding schedule were given unlimited food, they ate but they also hoarded pellets in the home cage. We studied hoarding behaviour in the two groups of lesioned rats and in a control group, using the apparatus illustrated in Fig. 2. Sixteen to eighteen hours after a meal the home cage was attached to the open field and the behaviour of the animals was observed for 30 min. The 6-hydroxydopamine lesion in VTA significantly impaired the latency to the beginning of hoarding and reduced the amount of food hoarded (Fig. 3), whereas the caudate-putamen lesion did not. On a number of other measures, feeding in the group

- HOARDING FIELD -

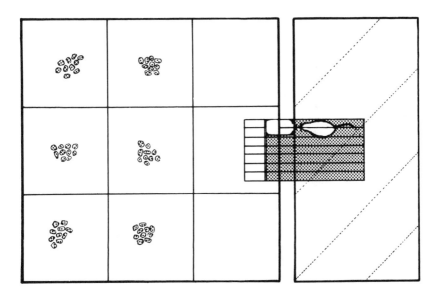

68·5 x 68·5 x 45·0 cm

FIG. 2. View from above of the hoarding apparatus, illustrating attachment of home cage to the field. Food pellets were available in the centre of six marked squares. A fluorescent strip light (180 W) illuminated the field.

with VTA lesions was observed to be abnormal: more time was devoted to eating but due to the reduced rate of feeding there was no increase in food intake—the same number of meals were taken but feeding was ineffective since it occurred in fragmented bouts, more often in the open field than in the home cage. The VTA group also showed less organized exploration of the open field. Subsequently all groups were treated with 62.5 µg apomorphine i.p./kg body weight and re-tested in the hoarding field. Apomorphine produced a significant reversal of the hoarding deficit in the rats with VTA lesions (Fig. 3).

Direct manipulation of dopaminergic neurons in VTA by the application of neurotransmitters in the vicinity of DA neurons stimulates the motivational behaviours impaired by DA lesions or enhanced by direct application of DA or its agonists in terminal areas. SP (Eison et al 1982), enkephalin (Kelley et al 1980), neurotensin (Kalivas et al 1981) and GABA antagonists (Mogenson et al 1979, Stinus et al 1982) stimulate locomotor activity and exploration when injected into VTA. In the terminal areas neurotensin (Ervin et al 1981)

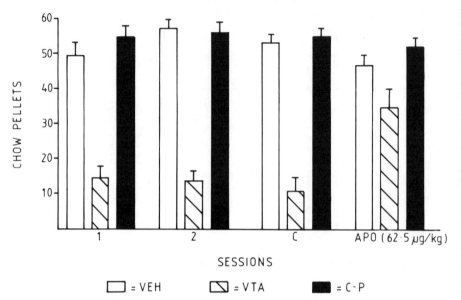

FIG. 3. Effect of 6-hydroxydopamine lesion in the ventral tegmental area (VTA) or caudate putamen (C-P) or of a control lesion (VEH) on hoarding behaviour during test sessions 1 and 2. In session 3 the lesion groups and VEH controls were treated with ascorbic acid (0.2 mg/kg) and the results are shown above C(= drug control session). In session 4 apomorphine (62.5 μg/kg; final three columns) was administered and hoarding increased in the impaired VTA-lesioned rats.

and cholecystokinin (Schneider et al 1983) appear to inhibit DA-mediated functions.

The importance of dopamine innervation of the striatum

It is clear that the DA innervation of the striatum is fundamental to the nervous integration achieved by that structure. Abnormalities of dopamine transmission result in Parkinson's disease, are a component in Huntington's disease, and are implicated in schizophrenia and aspects of depressive illness. The importance of the forebrain DA neurons is reflected in the number of regulatory mechanisms endogenous in these neurons and afforded by the wide range of other transmitters or modulators found in DA terminal and cell body regions.

In the dopamine circuitry itself there is reciprocal release of DA from terminal areas and cell body dendrites, there are autoreceptors on terminal and cell body regions that regulate local release, there is enhanced activity of intact DA neurons in partially damaged DA pathways, and the cortical DA pathway plays a role in the corticostriatal control of dopamine receptor activity in nucleus accumbens. As to controls extrinsic to the DA neurons, acetylcholine modulation in terminal areas is clearly very important, as is GABA feedback to the cell body regions. Preliminary experiments suggest that, of the recorded neuropeptides, substance P, neurotensin, somatostatin, cholecystokinin, thyrotropin-releasing hormone and enkephalin modulate dopaminergic functions in the substantia nigra and VTA.

REFERENCES

Beckstead RM 1979 An autoradiography examination of cortico-cortical and subcortical projections of the mediodorsal projection (prefrontal) cortex in the rat. J Comp Neurol 184:43-62

Björklund A, Dunnett SB, Stenevi U, Lewis ME, Iversen SD 1980 Reinnervation of the denervated striatum by substantia nigra transplants: functional consequences as revealed by pharmacological and sensorimotor testing. Brain Res 199:307-333

Björklund A, Stenevi U, Dunnett SB, Iversen SD 1981 Functional reactivation of the deafferented neostriatum by nigral transplants. Nature (Lond) 289:497-499

Divac I 1972 Neostriatum and functions of prefrontal cortex. Acta Neurobiol Exp 32:461-477

Divac I, Fonnum F, Storm-Mathisen J 1977 High affinity uptake of glutamate in terminals of corticostriatal axons. Nature (Lond) 266:377-378

Dunnett SB, Iversen SD 1979 Selective kainic acid and 6-hydroxydopamine induced caudate lesions: some behavioural consequences. Neurosci Lett Suppl 3:S207

Dunnett SB, Iversen SD 1981 Learning impairments following selective kainic acid-induced lesions within the neostriatum of rats. Behav Brain Res 2:189-209

Dunnett SB, Iversen SD 1982a Neurotoxic lesions of ventrolateral but not anteromedial neostriatum in rats impair differential reinforcement of low rates (DRL) performance. Behav Brain Res 6:213-226

Dunnett SB, Iversen SD 1982b Sensorimotor impairments following localised kainic acid and 6-hydroxydopamine lesions of the neostriatum. Brain Res 248:121-127

Dunnett SB, Björklund A, Stenevi U, Iversen SD 1981a Behavioural recovery following transplantation of substantia nigra in rats subjected to 60HDA lesions of the nigrostriatal pathway 1. Unilateral lesions. Brain Res 215:147-161

Dunnett SB, Björklund A, Stenevi U, Iversen SD 1981b Grafts of embryonic substantia nigra reinnervating the ventrolateral striatum ameliorate sensorimotor impairments and akinesia in rats with 60HDA lesions of the nigrostriatal pathway. Brain Res 229:209-217

Eison AS, Eison MS, Iversen SD 1982 The behavioural consequences of a novel substance P analogue following infusion into the ventral tegmental area or substantia nigra of rat brain. Brain Res 238:137-152

Ervin GN, Birkemo LS, Nemeroff CB, Prange AJ 1981 Neurotensin blocks certain amphetamine-induced behaviours. Nature (Lond) 291:73-76

Fonnum F, Walaas I 1979 Localisation of neurotransmitter candidates in the neostriatum. In: Divac I, Öberg RGE (eds) The neostriatum. Pergamon, Oxford, p 53-69

Gaffori O, Le Moal M 1979 Disruption of maternal behaviour and appearance of cannibalism after ventral mesencephalic tegmentum lesions. Physiol Behav 23:317-323

Goldman PS, Nauta WJH 1977 An intricately patterned prefronto-caudate projection in the rhesus monkey. J Comp Neurol 172:369-386

Graybiel AM, Ragsdale CW 1983 Biochemical anatomy of the striatum. In: Emson PC (ed) Chemical neuroanatomy. Raven Press, New York, p 427-504

Groenewegen HJ, Arnolds DEAT, Lopes de Silva FH 1981 Afferent connections of the nucleus accumbens in the cat, with special emphasis on the projections from the hippocampal region—an anatomical and neurophysiological study. In: Chronister RB, DeFrance JF (eds) The neurobiology of the nucleus accumbens. Haer Institute Press, Brunswick, Maine, p 41-74

Heimer L, Wilson RD 1975 The subcortical projections of the allo-cortex: similarities in the neural associations of the hippocampus, the piriform cortex and the neocortex. In: Santini M (ed) Golgi centennial symposium: perspectives in neurobiology. Raven Press, New York, p 177-193

Hökfelt T, Skirboll L, Rehfeld JF, Goldstein M, Markey K, Dann O 1980 A subpopulation of mesencephalic dopamine neurones projecting to limbic areas contains a cholecystokinin-like peptide: evidence from immunochemistry combined with retrograde tracing. Neuroscience 5: 2093-2124

Iversen LL 1984 Amino acids and peptides: fast and slow chemical signals in the nervous system. Proc R Soc Lond B Biol Sci, in press

Iversen SD 1984 Behavioural aspects of the cortico-subcortical interaction with special reference to the striatum. In: Reinoso-Suarez F (ed) Neural integration at basic and cortical brain levels. Raven Press, New York, in press

Iversen SD, Fray PJ 1982 Brain catecholamines in relation to affect. In: Beckman A (ed) The neural basis of behaviour. Spectrum, New York, p 229-269

Jones EG, Coulter JD, Burton H, Porter R 1977 Cells of origin and terminal distribution of corticostriatal fibers arising in the sensorimotor cortex of monkeys. J Comp Neurol 173:53-80

Kalivas PW, Nemeroff CB, Prange AJ Jr 1981 Increase in spontaneous motor activity following infusion of neurotensin into the ventral tegmental area. Brain Res 229:525-529

Kelley AE, Stinus L, Iversen SD 1980 Interactions between D-Ala-Met enkephalin, A10 dopaminergic neurones and spontaneous behaviour in the rat. Behav Brain Res 1:3-24

Kolb B, Nonneman AJ 1975 Prefrontal cortex and the regulation of food intake in the rat. J Comp Physiol Psychol 88:806-815

Kolb B, Nonneman AJ, Singh RK 1974 Double dissociation of spatial impairments and perseveration following selective prefrontal lesions in rats. J Comp Physiol Psychol 87:772-780

Kolb B, Whislaw IQ, Schaller TT 1977 Aphagia, behaviour sequencing and body weight set point following orbital frontal lesions in rats. Physiol Behav 19:93-103

Marshall JF, Berrios N, Sawyer S 1980 Neostriatal dopamine and sensory inattention. J Comp Physiol Psychol 94:833-846

Marshall JF, Van Oordt K, Kozowski MR 1983 Acetylcholinesterase associated with dopaminergic innervation of the neostriatum: histochemical observations of a heterogenous distribution. Brain Res 274:283-289

Mogenson GJ, Wu M, Manchanda SK 1979 Locomotor activity initiated by microinfusions of picrotoxin into the ventral tegmental area. Brain Res 161:311-319

Neill DB, Grossman SP 1970 Behavioural effects of lesions or cholinergic blockade of the dorsal and ventral caudate of rats. J Comp Physiol Psychol 71:311-317

Neill DB, Herndon JG 1978 Anatomical specificity within rat striatum for the dopaminergic modulations of DRL responding and activity. Brain Res 153:529-538

Neill DB, Ross JF, Grossman SP 1974 Comparison of the effects of frontal striatal and septal lesions in paradigms thought to measure incentive motivation on behavioural inhibition. Physiol Behav 13:297-305

Pasik T, Pasik P, DiFiglia M 1980 Synaptic organisation of the striatum and pallidum in the monkey. In: Szentágothai J et al (eds) Regulatory functions of the CNS subsystems. Pergamon, Oxford/Akadémiai Kiádó, Budapest (Adv Physiol Sci 2), p 161-174
Pycock CJ 1980 Turning behaviour in animals. Neuroscience 5:461-514
Robbins TW, Koob GF 1980 Selective disruption of displacement behaviour by lesions of the mesolimbic dopamine system. Nature (Lond) 278:409-412
Schneider LH, Alpert JE, Iversen SD 1983 CCK-8 modulation of mesolimbic dopamine: antagonism of amphetamine-stimulated behaviours. Peptides (NY) 4:749-753
Somogyi P, Hodson AJ, Smith AD 1979 An approach to tracing neuron networks in the cerebral cortex and basal ganglion. Combination of Golgi staining, retrograde transport of horse radish peroxidase and anterograde degeneration of synaptic boutons in the same material. Neuroscience 4:1805-1852
Somogyi P, Priestley JV, Cuello AC, Smith AD, Bolam JP 1982 Synaptic connections of substance P-immunoreactive nerve terminals in the substantia nigra of the rat. Cell Tissue Res 223:469-486
Steinbusch HWM 1981 Distribution of serotonin-immunoreactivity in the central nervous system of the rat—cell bodies and terminals. Neuroscience 6:557-618
Stinus L, Gaffori O, Simon H, Le Moal M 1978 Disappearance of hoarding and disorganisation of eating behaviour after ventral mesencephalic tegmentum lesions in rats. J Comp Physiol Psychol 92:289-329
Stinus L, Herman JP, Le Moal M 1982 Gabaergic mechanisms within the ventral tegmental area: involvement of dopaminergic (A10) and non-dopaminergic neurones. Psychopharmacology 77:186-192
Ungerstedt U 1971a Adipsia and aphagia after 6-hydroxydopamine induced degeneration of the nigro-striatal dopamine system. Acta Physiol Scand 367:95-122
Ungerstedt U 1971b Postsynaptic supersensitivity after 6-hydroxydopamine induced degeneration of the nigro-striatal dopamine system. Acta Physiol Scand Suppl 367:69-93
Walaas I, Fonnum F 1979 The effects of surgical and chemical lesions on neurotransmitter candidates in the nucleus accumbens. Neuroscience 4:209-216
Webster KE 1961 Cortico-striate interrelations in the albino rat. J Anat 95:532-544
Whishaw IQ, Schallert T, Kolb B 1981 An analysis of feeding and sensorimotor abilities of rats after decortication. J Comp Physiol 95:85-103
Wikmark RGE, Divac I, Weiss R 1973 Retention of spatial delayed alteration in rats with lesions in the frontal lobes: implications for a comparative neuropsychology of prefrontal system. Brain Behav Evol 8:329-339
Yeterian EH, Van Hoesen GW 1978 Cortico-striate projections in the rhesus monkey: the organisation of certain cortico-caudate connections. Brain Res 139:43-63
Yim CY, Mogenson GJ 1982 Response of nucleus accumbens neurones to amygdala stimulation and its modification by dopamine. Brain Res 239:401-415

DISCUSSION

Glowinski: All the cells you see may be related to the dopaminergic neurons, but there are other cells as well.

Iversen: There must be, since the entire ventral tegmental area of the fetus is used to prepare graft material. The development of these grafting methods will depend on neurobiologists working out methods of cell-sorting to produce purer classes of neurons for transplantation, but this is a very difficult task. When the solid tissue grafts are used the ventral tegmental area within the block is disconnected from all normal CNS connections. The dopamine-rich ventral tegmental areas are taken from the ventral mesencephalon and placed in a cortical site where they receive connections from neurons in surrounding cortical tissue, which grow into the graft. The functional significance of these abnormal inputs has not been explored in any detail yet.

Rizzolatti: It is very interesting that you have found sensory neglect after lesions of that part of the striatum which receives its input from sensorimotor cortex. We have recently described neglect after unilateral ablation of the postarcuate cortex in the monkey (Rizzolatti et al 1983). Thus, it appears that in rat and monkey neglect may occur after lesions in areas related to the motor system. A second interesting observation is the presence of different types of neglect after different lesions. Conceptually it is important to stress that besides the 'classical' neglect caused in humans by right parietal lesions, there are other phenomenologically different types of neglect. For example in the monkey after lesions of area 6 the hemi-inattention concerns mostly the contralateral hemisoma and the contralateral peripersonal space, while after a lesion of area 8 it concerns mostly the contralateral 'far' space, without any obvious impairment in the animal's responsiveness to somatosensory stimuli (Rizzolatti et al 1983). Thus hemi-inattention may concern different modalities and different parts of the space.

Iversen: I agree. It would be interesting to study neglect in the monkey and explore grafting potential in a higher animal. We need to know far more about the organization of rat cortex with respect to sensory attention and other integrative functions before we make further selective lesions and focal striatal grafts.

Marsden: In work on animals, 'neglect' of necessity has to be judged from the failure to make a motor response. In human patients with dopamine deficiency, neglect is quite different from that seen in patients with parietal lobe lesions. In clinical usage neglect means the inability to perceive in an area of space, or in an area of the body, whereas in Parkinson's disease it means the inability to make a motor response to an external stimulus.

Rizzolatti: It is true that in animals sensory deficits are judged by the absence of motor behaviour, but in the examples I mentioned before, the motor deficits could not explain the lack of reactivity to certain stimuli. For example, the animals after an area 6 lesion do not blink in response to menace and do not jump away from frightening stimuli presented in the peripersonal space, although they have no paralysis which prevents them from doing so.

Marsden: That may be true for the frontal lobes, but in Parkinson's disease spontaneous blinking is already impaired.

Do these experiments, which beautifully show that isolated behavioural deficits can be produced by destroying or interfering with synaptic regions of the caudate–putamen, indicate that the *same* deficit can be produced by lesions in the caudate–putamen as by lesions in the cerebral cortex? Are these two parts of the brain doing the same thing?

Iversen: The end-product or response is exactly the same. In these behavioural tasks we have to go on to try and specify the particular contribution of the caudate–putamen to what we call the functional interface between cortex and striatum. We haven't begun to do that. We use the word cognition sometimes but that doesn't tell us what the mechanism is. By cognition we mean complicated associations that we don't understand, and that really isn't adequate. We have to try to move forward in animal studies, as cognitive psychologists have done, to specify in humans the processes involved in higher CNS integration. That is not going to be an easy task. I have no behavioural tasks yet that I could use to differentiate the special contribution of an area of cortex and its related striatal area to some aspect of sensorimotor integration.

Rolls: I am concerned by your use of the term sensorimotor when you are dealing with what you have also referred to as cognitive tasks. For example you are dealing with spatial reversal and with extinction, as well as with orientation produced by somatosensory stimuli. The first two of those don't seem to me to be sensorimotor functions.

There is also the question of the relative contributions of the cortex and striatum when a similar effect is produced by a lesion in either structure. We have tried to analyse that by recording from neurons in the cortex and in the corresponding part of the striatum in the monkey. Damage to either the orbito-frontal cortex or the head of the caudate nucleus produces the sort of extinction impairment that you described (see Thorpe et al 1983, Rolls et al 1983). The sensory information necessary for this computation of whether to extinguish was present in the cortex. This included visual and gustatory information, both of which the monkey needs for computing whether it should continue responding to a visual stimulus, depending on the taste previously associated with it (Thorpe et al 1983). When we recorded from the corresponding part of the striatum, the sensory information did not appear to be represented. Instead a signal in the striatum effectively reflected whether the monkey should go or not on a particular trial (Rolls et al 1983). Thus the actual cognitive computation may be done in the cortex. The role of the striatum may be to choose or switch between different cortical inputs reaching it (the striatum) to produce perhaps one behavioural output (see Rolls et al 1983, Rolls 1984).

Iversen: I think the concept of switching is important when thinking about striatum. In any situation usually more than one stimulus can control behaviour but only one eventually controls it at a particular time. In behavioural experiments in which rats have food, water or wood chips available, the hypothalamus is electrically stimulated and one specific response is seen, usually eating. If food is removed, however, the animal instantly switches to one of the other appropriate responses. When one stimulus has lost its priority another takes over immediately in selecting the relevant behaviour. I propose that dopamine in striatum is essential for both switching on eating and switching to the subsequent response.

Turning to your first point, I use the term sensory in the loosest possible way. It is synonymous with information anywhere in the cortex. Your implication is that I should use the term for processing behind the central fissure, whereas I include highly integrated information in the frontal lobe, for example, as sensory in my terminology.

Rolls: Sensory information may be needed to deal with a complex computation about whether to do a spatial reversal or to extinguish, but that computation would not normally be called a simple sensorimotor function input.

Iversen: I agree and I would accept different terms for different levels of sensorimotor integration if acceptable ones could be found.

Hornykiewicz: It is very important to stress the possibility that dopamine, rather than transmitting specific information, serves an enabling function in the striatum. This concept was developed especially in connection with noradrenergic effects on the cerebellum and hippocampus by B.J. Hoffer, F.E. Bloom and their colleagues (cf. Moore & Bloom 1979). They showed that the catecholamines, rather than influencing the neurotransmitter processes in a distinct way, served a kind of bias-adjusting or gain-setting function. When applied iontophoretically or released synaptically, catecholamines such as noradrenaline improved both inhibitory and excitatory neurotransmission processes in given target systems (cf. Moises et al 1979). Dopamine may well serve such a 'quality improving' function in the striatum. The situation in Parkinson's disease would support this idea. It seems that it is not a specific and narrowly circumscribed function of the striatum that is affected by the nigrostriatal dopamine deficit. Rather, the whole processing system in the striatum seems to be deranged. That would also make it understandable why dopamine substitution is so effective in restoring striatal function. Dr Iversen, would you agree that dopamine has this enabling function?

Iversen: Yes, I agree with everything you have said. But we are only talking about one transmitter and we have to do equivalent experiments on the other components of the isodendritic core to determine whether its enabling function is peculiar to dopamine. It may turn out that dopamine is not unique in this respect but serves to modulate the striatal complex only. Some of Ann Graybiel's developmental studies show that dopamine also plays a crucial role in the early stages of anatomical development in striatum which ultimately result in the striosome organization of the adult striatum. It will be fascinating to learn whether the same or different dopamine neurons are involved in development and adult functioning of the striatum.

Calne: One problem is in distinguishing between what is primarily an information-processing phenomenon and what is an information transfer phenomenon. David Marsden referred to different parts doing the same thing. If there is a lesion of the corticospinal neuron, the lower motor neuron, the peripheral nerve, the neuromuscular junction or the muscle, the output (movement of the limb) is affected in a similar way. We may be dealing predominantly with transfer phenomena rather than processing phenomena.

To follow up what Dr Hornykiewicz said, if there is just an 'enabling' effect of dopamine, does it make any difference whether you put A9 cells into ventral striatum, A10 cells into dorsal striatum, or adrenal cells into either?

Iversen: Grafting experiments have been done with adrenal cells. Chromaffin cells injected into the ventricle of rats with unilateral nigrostriatal lesions sufficed to give some compensation of the rotational behaviour induced by the lesion. The compensation of turning in those experiments is not as complete as we see with grafting of dopaminergic neurons to the damaged striatum. As to the involvement of A9 versus A10 dopaminergic neurons, the ventral tegmental area and the substantia nigra of the fetal rat are so small that we haven't yet been able to do that. The tissue for grafting is prepared from a block of A9 and A10 neurons.

Hornykiewicz: I don't think it should matter where you take the dopamine cells from as long as they are capable of producing dopamine. After all, if you inject L-dopa you get restoration of function.

Iversen: There is always a rather global output in these systems we measure. If we looked more carefully at the behaviour we might see that the compensation is not as good or not as complete with different cell types, but that would need very careful investigation with refined behavioural measures.

DeLong: These are fascinating findings. In the parkinsonian patient a problem is that there is a great richness of behavioural disturbance. The overriding motor impairments such as rigidity and akinesia make testing in a similar manner almost impossible. Have any parallels been found in humans for this type of disturbance in the processing or transfer of information?

Iversen: The question of whether or not parkinsonian patients have a cognitive disturbance has been addressed in a number of studies. There seem to be mixed results, with some people finding little evidence for cognitive impairments and others claiming, as we shall hear later, that there are impairments on specific tasks. There is, however, always a danger in expecting to see perfect parallels between clinical syndromes in neurology or psychiatry and those in animal models with supposedly similar CNS dysfunction. In the clinical syndromes degeneration has progressed slowly over many years and in the animal model it is induced acutely.

REFERENCES

Moises HC, Woodward DJ, Hoffer BJ, Freedman R 1979 Interactions of norepinephrine with Purkinje cell responses to putative amino acid neurotransmitters applied by microiontophoresis. Exp Neurol 64:493-515

Moore RY, Bloom FE 1979 Central catecholamine neuron systems: anatomy and physiology of the norepinephrine and epinephrine systems. Annu Rev Neurosci 2:113-168

Rizzolatti G, Matelli M, Pavesi G 1983 Deficits in attention and movement following the removal of postarcuate (area 6) and prearcuate (area 8) cortex in macaque monkeys. Brain 106:655-673

Rolls ET 1984 Activity of neurons in different regions of the striatum of the monkey. In: McKenzie J, Wilcox L (eds) The basal ganglia: structure and function. Plenum, New York

Rolls ET, Thorpe SJ, Maddison SP 1983 Responses of striatal neurons in the behaving monkey. 1. Head of the caudate nucleus. Behav Brain Res 7:179-210

Thorpe SJ, Rolls ET, Maddison S 1983 Neuronal activity in the orbito-frontal cortex of the behaving monkey. Exp Brain Res 49:93-115

The neostriatum viewed orthogonally

IVAN DIVAC

Institute of Neurophysiology, University of Copenhagen School of Medicine, Blegdamsvej 3C, Copenhagen N, Denmark

Abstract. The work described in this paper suggests that: (1) the neostriatum (a tier of the forebrain) is functionally heterogeneous, i.e. different neostriatal regions mediate responses in different situations; (2) regional specialization may be attributed to selective cortico-neostriatal connections, as indicated by anatomical and neurobehavioural evidence and studies with 2-deoxyglucose; (3) cortical areas and their corresponding principal neostriatal regions are organized 'in series'; (4) small lesions which cause severe but specific 'cognitive' effects do not induce movement disorders; (5) the transmitter of the cortico-neostriatal connections is glutamate. This work and other evidence suggests that the forebrain consists of systems that each contain a 'specific' thalamic nucleus and its associated cortical area and neostriatal region. This view, with three additional assumptions, can tentatively explain not only experimental neurobehavioural results but also certain symptoms of diseases involving the neostriatum.

1984 Functions of the basal ganglia. Pitman, London (Ciba Foundation symposium 107) p 201-215

At the suggestion of the organizers of this symposium, this paper summarizes my work on the neostriatum. This work has focused on relations between the neostriatum and the cerebral cortex. In neurobehavioural experiments I used tasks involving choice. This approach differs from the more frequently encountered approach which studies neostriatal relationships with the substantia nigra, where turning behaviour is the dependent variable. As expected, these two approaches have suggested different concepts about neostriatal functions.

My interest in cortico-striatal relations was triggered in the early 1960s by the discovery of the prefrontal syndrome in monkeys (*Macaca mulatta*) with lesions in the head of the caudate nucleus (Rosvold & Szwarcbart 1964), and by the description of topographic projections of cortical areas to different regions of the neostriatum (W. J. H. Nauta 1964). This seminal work was the basis for an experiment which attempted to demonstrate that the connections between specific cortical areas and associated neostriatal regions had a functional significance. Specific behavioural tests were available for detecting damage to three distinct cortical areas in the monkey, and these areas were known to project to different striatal regions. Small lesions in each of these neostriatal regions impaired performance essentially only in the task which had been found sensitive to damage of the anatomically associated cortical area (Divac et al 1967).

This study generated further questions which were experimentally studied in my subsequent work discussed below.

The problem of accidental damage to cortical connections

The behavioural consequences of neostriatal lesions could be attributed to accidental damage to connections of the closest cortical area. This possibility was tested in a number of ways. In the most conclusive study of this kind, lesions in the neostriatum of rats were produced by kainic acid. These lesions impaired the performance in a task sensitive to ablation of the prefrontal cortical area. In the same animals we demonstrated that cortical afferents passing through the damaged neostriatal region were able to transport horseradish peroxidase (Divac et al 1978). Injections of kainic acid into another brain region did not induce prefrontal impairment (Divac et al 1978, note added in proof), nor did injections into other neostriatal regions (Dunnett & Iversen 1981). It no longer seems likely that the apparently close functional relationship between cortical areas and neighbouring neostriatal regions is a spurious phenomenon produced by accidental damage to passing fibres. (See Divac 1972a, Divac & Diemer 1980, Divac et al 1982, and Öberg & Divac 1979 for additional evidence.)

The question of species generality

Was the observed cortico-neostriatal relation a unique species specialization or could it be demonstrated also in species other than the monkey? The problem is not trivial. Firstly, across-species comparisons throw light on the phylogenetic distribution of a given trait and thus ultimately on its evolutionary history. Secondly, a trait seen in many different species is more likely to be present in humans too than a trait seen in only one species, even if that species is a primate. Anatomical studies (reviewed in Divac 1972a) have shown topographic cortico-neostriatal projections in many species of different lines of descent. Our studies in cats (Divac 1968b, 1972b, Öberg & Divac 1975, Rosenkilde & Divac 1976) and rats (Divac 1971, Divac et al 1978, Öberg & Divac 1975, Wikmark et al 1973) convincingly demonstrated that in representatives of the rodent and carnivore orders, as in the primate line, the effect of a small neostriatal lesion depends on the localization of the lesion and replicates the effect of ablation of the anatomically associated cortical area. Thus, it seems probable that the close cortico-neostriatal linkage is a property of all mammals, inherited from a common ancestor and retained in the human brain.

Similarity of lesion effects (cortical versus neostriatal)

We have searched for, but not found, tasks involving choice in which lesions of the neostriatum would produce effects different from those found after ablation of the related association cortical area (reviewed in Divac 1972a, Öberg & Divac 1979). Thus, both our own work and that of others (see the reviews just mentioned) indicates a remarkable degree of functional similarity between an association cortical area and its neostriatal principal target. In studies where other claims have been made, it has not been shown that the cortical and neostriatal ablations involve only the principally interrelated members of the cortico-neostriatal pair.

In contrast, neostriatal damage does not seem to replicate the motor and sensory deficits seen after lesions of the respective cortical areas. One may assume that some of the functions of motor cortical areas are in fact mediated via the cortico-striatal route, but that these functions cannot be detected with lesioning techniques since the relevant deficits are unavoidably obscured by the more conspicuous consequences of loss of function of the corticospinal path. The primary visual and probably the auditory sensory areas seem to innervate only very small neostriatal regions (e.g. Collins & Divac 1984). Hence, the contribution of the neostriatal regions related to these sensory areas at this stage of sensory processing may be relatively small.

Sequential or parallel organization of cortico-neostriatal pairs

The unidirectionality of cortico-neostriatal connections suggests that there is a hierarchical, in series, relationship between cortex and neostriatum (Divac 1972a, 1977a). If this is so, the impairment seen after a lesion of either of the two formations (e.g. the prefrontal cortical area or its principal neostriatal region) should not reappear when a lesion of the other formation is added in animals which had relearnt a task after the first lesion. (The assumption is that the task is now solved in a different way, with the participation of other brain mechanisms.) In addition, there should be no additive effect of lesions of the two levels of the same cortico-neostriatal pair. In experiments with cats, lesions were made in one of the structures and the animals were retrained. The first lesion, whether cortical or neostriatal, always produced impairment in sensitive prefrontal tasks, whereas the second lesion (made after relearning) never did (Wikmark & Divac 1973, and unpublished data). In addition, there was no evidence for additivity of the prefrontal cortical and neostriatal lesions (Divac 1968b), nor was there any sign that additional lesions of the critical neostriatal region delayed relearning of a prefrontal task in cats with large frontal ablations (Divac 1974). These results strongly suggest an 'in series' relationship between the prefrontal cortical area and its principal neostriatal target in the cat. The species generality of this organization, however, has been insufficiently explored and some evidence suggests a different relationship in rats and young monkeys (see Divac 1980).

Morphological substrate of neostriatal heterogeneity

In our studies the neostriatum has been subdivided on the basis of topographical cortico-neostriatal projections. These connections were demonstrated by silver impregnation of degenerating axons. More recently, however, anterograde transport of radioactive amino acids demonstrated that association areas of the cortex project not only to the known neostriatal target but also to other parts of the neostriatum. Directly connected cortical areas seem to project also to each others' neostriatal target sites (Yeterian & Van Hoesen 1978). Thus, even a small neostriatal lesion destroys the target of more than one cortical area. Why, then, are the behavioural effects of such small lesions so specific? In recent studies with 2-deoxyglucose, strong epileptic activation of one cortical area coactivated only the 'traditional' projection area in the neostriatum, i.e. the region so designated by the silver impregnation techniques (Collins & Divac 1984, Divac 1983, Divac & Diemer 1980, Divac et al 1982). The activation was not seen in other regions of the neostriatum, suggesting that projections from a cortical area to different neostriatal regions are quantitatively or qualitatively different. We have therefore suggested that cortico-neostriatal connections fall into two categories, 'principal' and 'auxiliary'. Principal projections are seen with the silver impregnation and autoradiographic tracing techniques. These projections mediate activation of the neostriatum from a cortical epileptic focus and transmit information essential for a given behaviour. The auxiliary connections have thus far been documented only by the very sensitive autoradiographic anatomical technique. Their functional significance is at present obscure and should probably be sought in the way neostriatal input and output are organized and by means of behavioural studies in which sensitive tasks are used (further discussion in Divac & Diemer 1980).

Levels of neostriatal heterogeneity

The work described thus far suggests that the neostriatum consists of regions related to specific cortical areas. Many recent experiments have shown that cortico-neostriatal connections and some other constituents of the neostriatum have a patchy distribution (references in Divac 1983). Recently I observed that after epileptic discharges in a cortical area, a neostriatal region sometimes appears to be patchily activated. Thus, the patches are a functional as well as a morphological property of the neostriatum. Neostriatal regions may therefore be further subdivided into smaller units, possibly comparable to cortical layers (Graybiel 1983) or columns, or both.

Neurotransmitter of cortico-neostriatal connections

Pharmacological manipulations of the cortico-neostriatal path might further contribute to an understanding of cortico-neostriatal relations. We provided evidence that the transmitter in this path is probably glutamate (Divac et al 1977). This result has been confirmed in several other laboratories and has become potentially relevant for human neuropathology because high concentrations of acidic amino acids, including glutamate, are neurocytotoxic (see references in Divac 1977b). Like others, I have suggested that an abnormality of this transmission may be responsible for the lesions seen in Huntington's chorea (Divac 1977b). Recently, Kuhl et al (1982) obtained evidence that even at early stages of this disease, before changes in the volume of the neostriatum can be detected, the neostriatum is hypoactive rather than hyperactive as would be expected from the hypothesis of glutamatergic hyperexcitation. A non-excitatory cytotoxic effect related to glutamate transmission is, however, not excluded.

Discussion

The evidence reviewed here suggests that the forebrain is divisible not only into tiers such as the thalamus, cortex, striatum and pallidum (e.g. H. J. W. Nauta 1979) but also into 'vertical' functional 'systems', each consisting of a thalamic nucleus, a cortical area, a neostriatal region and quite possibly a pallidal region. This vertical organization was emphasized 15 years ago (Divac 1968a). The prosencephalic systems may be further divisible into modular subsystems consisting of cortical columns, thalamic 'barreloids' and neostriatal modules (Mountcastle 1979, Divac 1983) (see Fig. 1). While forebrain tiers may process different inputs in each of their system-

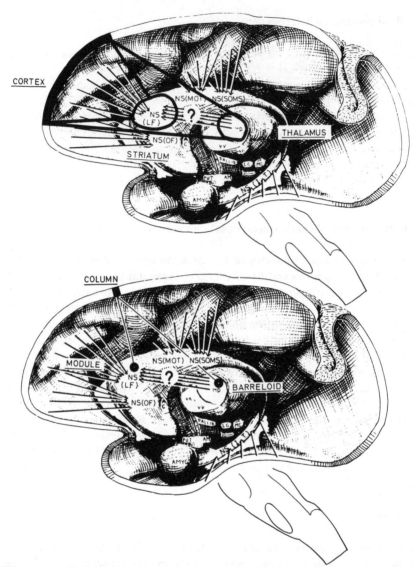

FIG. 1. Schematic illustration of systems consisting of regions (above) and possibly also modules (below) of the three tiers of the forebrain. Amyg, amygdala; C, caudate nucleus; LG, lateral geniculate nucleus; MD, mediodorsal nucleus thalami; MG, medial geniculate nucleus; NS (IT), neostriatum related to the inferotemporal cortex; NS (LF), neostriatum related to the lateral prefrontal cortex; NS (OF), neostriatum related to the orbital frontal cortex; NS (MOT), neostriatum related to the motor cortex; NS (soms), neostriatum related to the somatosensory cortex; Pallid, globus pallidus; Put, putamen; VA, ventral anterior nucleus thalami; VP, ventral posterior nucleus thalami. (Reproduced, with permission, from Krieg W.J.S. 1963 Connections of the cerebral cortex. Brain Books, Evanston, IL. Fig. 230)

subdivisions in the same way, all the components of a vertical system, regardless of the tier they belong to, may process the information relevant for a particular behaviour (task) in different ways. In other words, neostriatal regions differ with regard to *what* information is processed rather than *how* it is processed (Öberg & Divac 1981).

Horizontal and vertical perspectives on forebrain organization allow us to reconcile the two apparently incompatible sets of data emerging from two rarely interacting research traditions. I am referring to clinical observations and experiments primarily aimed at understanding the symptoms of basal ganglia diseases on the one hand, and the more cognitively oriented line of animal research as described in this paper on the other hand (see also Divac & Öberg 1979). Three assumptions are needed for reconciliation of the two traditions. (1) Each neostriatal region (at least the regions related to association cortical areas) has ultimate access to the peripheral motor neurons which mediate movements of *all* body parts (Fig. 2). In this way one can understand both the control of the same movements by different neostriatal regions (Divac et al 1967) and the control of different movements by the same system (i.e. a given behaviour is impaired by appropriate neostriatal lesions regardless of the kind of movement required in the test; see Öberg & Divac 1979). This empirically founded assumption helps us to understand why the intensity of clinical symptoms may depend on the extent of the pathological process in the neostriatum. (2) The neostriatum regulates different behaviours by virtue of patterned neuronal activity (see Rolls 1983). The patterns are region-specific and probably imposed by the associated cortical areas. Therefore any treatment (including faradic electrical stimulation and local application of drugs; see Table 1 in Öberg & Divac 1979) confined to a neostriatal region can only impair (but not elicit or improve) behaviour that is under the control of the forebrain system to which the region belongs. It follows also that, in humans, small areas of destruction in the neostriatum that are often considered to produce no effects should in fact induce impairments detectable by appropriate neuropsychological methods. (3) Neurological symptoms (such as rigidity and tremor) of diseases affecting the neostriatum may be in part a consequence of a tonic bias originating in pathologically changed neostriatal tissue. In humans, the bias seems to involve the chain of neurons from the neostriatum to the peripheral motor neurons (Divac 1977a). The bias is stronger when the pathological process involves one transmitter or a population of neurons in the neostriatum than after total destruction of the same amount of tissue. The present and the previous assumption explain why surgical lesions in the neostriatum or anywhere along the path between the neostriatum and the peripheral motor neurons reduce or eliminate tremor and rigidity while further increasing akinesia (i.e. neurobehavioural deficits).

The sketch of neostriatal functioning presented here (Figs. 1 and 2), broad as it is, still leaves a number of important aspects untouched. Some of them have already been discussed (e.g. Divac 1977a). Yet the present line of reasoning seems to embrace a wider spectrum of data than other current views, and it offers a large number of testable hypotheses for clinical as well as laboratory research.

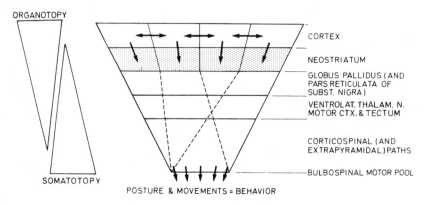

FIG. 2. Hypothetical representation of some relationships of the neostriatum. The endbrain is 'organotopically' organized (different systems mediate different functions) but shows no somatotopicity except in the sensorimotor system. At the other end, the final common path is entirely somatotopic. (For other features of this scheme, see text.)

Acknowledgements

The work described in this paper owes much to L. Mihailovic, H. E. Rosvold, J. Konorski, M. Mishkin, J. M. Warren and A. M. Laursen, who enabled me to work at the Institute of Pathophysiology of the Beograd School of Medicine; the Laboratory of Neuropsychology, NIMH, Bethesda; the Nencki Institute of the Polish Academy of Science: the Animal Behavior Laboratory of the Pennsylvania State University; and the Institute of Neurophysiology of the University of Copenhagen. Many colleagues at these and other institutions have contributed in various ways and degrees to the outcome of my work. The most profound influence has come from R. G. E. Öberg (previously Wikmark) who also made valuable comments on the present paper. The work described (and my upkeep) was supported by grants from the US Agency of International Development to Yugoslavia, from the Department of Health Education and Welfare (PL 480) to Poland, and from the National Institute of Mental Health (MH 04726) to J. M. Warren. Since

1970 the work described here has been supported mainly by the University of Copenhagen and the Danish Medical Research Council. Some recent experiments received support from Grete Jensen's Studielegat, Director Jacob Madsen og Hustru Olga Madsens Fond and NOVO's Fond.

REFERENCES

Collins RC, Divac I 1984 Neostriatal participation in prosencephalic systems, evidence from deoxyglucose autoradiography. Adv Neurol 40:117-122

Divac I 1968a Functions of the caudate nucleus. Acta Biol Exp (Warsaw) 28:107-120

Divac I 1968b Effects of prefrontal and caudate lesions on delayed response in cats. Acta Biol Exp (Warsaw) 28:149-167

Divac I 1971 Frontal lobe system and spatial reversal in the rat. Neuropsychologia 9:175-183

Divac I 1972a Neostriatum and functions of prefrontal cortex. Acta Neurobiol Exp 32:461-477

Divac I 1972b Delayed alternation in cats with lesions of the prefrontal cortex and the caudate nucleus. Physiol Behav 8:519-522

Divac I 1974 Caudate nucleus and relearning of delayed alternation in cats. Physiol Psychol 2:104-106

Divac I 1977a Does the neostriatum operate as a functional entity? In: Cools AR et al (eds) Psychobiology of the striatum. Elsevier, Amsterdam, p 21-30

Divac I 1977b Possible pathogenesis of Huntington's chorea and a new approach to treatment. Acta Neurol Scand 56:352-360

Divac I 1980 Functions of the neostriatum: cortex-dependent or autonomous. Acta Neurobiol Exp 40:85-94

Divac I 1983 Two levels of functional heterogeneity of the neostriatum. Neuroscience 10:1151-1155

Divac I, Diemer NH 1980 The prefrontal system in the rat visualized by means of labelled deoxyglucose. Further evidence for functional heterogeneity of the neostriatum. J Comp Neurol 190:1-13

Divac I, Öberg RGE 1979 Current conceptions of neostriatal functions. History and an evaluation. In: Divac I, Öberg RGE (eds) The neostriatum. Pergamon, Oxford, p 215-230

Divac I, Rosvold HE, Szwarcbart MK 1967 Behavioural effects of selective ablation of the caudate nucleus. J Comp Physiol Psychol 63:184-190

Divac I, Fonnum F, Storm-Mathisen J 1977 High affinity uptake of glutamate in the terminals of cortico-striatal axons. Nature (Lond) 266:377-378

Divac I, Markowitsch HJ, Pritzel M 1978 Behavioural and anatomical consequences of small intrastriatal injections of kainic acid in the rat. Brain Res 151:523-532

Divac I, Milivojevic Z, Krivokuca Z 1982 Activation pattern in the prefrontal system of kainic acid-pretreated rats. Acta Neurobiol Exp 43:223-225

Dunnett SB, Iversen SD 1981 Learning impairments following selective kainic acid-induced lesions within the neostriatum of rats. Behav Brain Res 2:189-209

Graybiel AM 1983 Compartmental organization of the mammalian striatum. Prog Brain Res 58:247-256

Kuhl DE, Phelps ME, Markham CH, Metter EJ, Riege WH, Winter J 1982 Cerebral metabolism and atrophy in Huntington's disease determined by ^{18}FDG and computed tomographic scan. Ann Neurol 12:425-434

Mountcastle VB 1979 An organizing principle for cerebral function: the unit module and the distributed system. In: Schmitt FO, Worden FG (eds) The neurosciences fourth study program. MIT Press, Cambridge, MA p 21-42

Nauta HJW 1979 A proposed conceptual reorganization of the basal ganglia and telencephalon. Neuroscience 4:1875-1881

Nauta WJH 1964 Some efferent connections of the prefrontal cortex in the monkey. In: Warren JM, Akert K (eds) The frontal granular cortex and behavior. McGraw-Hill, New York, p 397-407

Öberg RGE, Divac I 1975 Dissociative effects of selective lesions in the caudate nucleus of cats and rats. Acta Neurobiol Exp 35:675-689

Öberg RGE, Divac I 1979 'Cognitive' functions of the neostriatum. In: Divac I, Öberg RGE (eds) The neostriatum. Pergamon, Oxford, p 291-313

Öberg RGE, Divac I 1981 Levels of motor planning. Cognition and the control of movement. Trends Neurosci 4:122-124

Rolls ET 1983 The initiation of movements. In: Massion J et al (eds) Neural coding of motor performance. Springer, Berlin, p 97-113

Rosenkilde CE, Divac I 1976 Time discrimination performance in cats with lesions in the prefrontal cortex and the caudate nucleus. J Comp Physiol Psychol 90:343-352

Rosvold HE, Szwarcbart MK 1964 Neural structures involved in delayed response performance. In: Warren JM, Akert K (eds) The frontal granular cortex and behavior. McGraw-Hill, New York, p 1-15

Wikmark RGE, Divac I 1973 Absence of effects of caudate lesions on delayed responses acquired after large frontal ablations in cats. Isr J Med Sci 9:92-97

Wikmark RGE, Divac I, Weiss R 1973 Retention of spatial delayed alternation in rats with lesions in the frontal lobes. Implications for a comparative neuropsychology of the prefrontal system. Brain Behav Evol 8:329-339

Yeterian EH, Van Hoesen GW 1978 Cortico-striate projections in the rhesus monkey; the organization of certain cortico-caudate connections. Brain Res 139:43-63

DISCUSSION

Carpenter: Most of the evidence suggests that the corticostriatal pathways come from the upper part of the fifth lamina but some investigators say there may be a unique cortical projection from supragranular layers of the cortex to the striatum. This may represent a different set of the corticofugal fibres (Kitai et al 1976, Oka 1980, Royce 1982).

Divac: May I discuss the reverse path? I have seen labelled cells after cortical injections of retrograde tracers only in the nucleus basalis of Meynert, which in the monkey enters the lamina between globus pallidus and the putamen, never far away from the striatopallidal junction. Maybe some neostriatal cells project so widely that I have failed to inject a sufficiently large cortical area to label them conspicuously. Alternatively, the neurons of the nucleus basalis might become displaced into the neostriatum in some species.

Graybiel: In cats Dr Chesselet and I may have seen the cells that you are speaking of. They are giant cholinergic (CAT-positive) neurons that lie in the putamen, with a few also sprinkled in the cell bridges of the internal capsule and at the very lateral margins of the caudate nucleus. But they seem to form a special group, perhaps of displaced nucleus basalis neurons.

Marsden: You said that lesions in the caudate nucleus produced no motor effects, Dr Divac, but the sort of lesions that Sue Iversen and you make must produce profound effects on motor behaviour. You used the word cognitive to describe motor behaviour of a high order. This is somewhat confusing because clinical psychologists use 'cognitive' to describe intellectual processes such as perception and intellect. We need a term such as perceptual motor function or behavioural motor control. 'Sensorimotor' isn't quite right.

Divac: How does one separate the 'lower motor', 'higher motor' and 'intellectual' functions? Even the highest cognitive functions are inferences from movements producing speech, written text, art, computer chips or a wink at an appropriate moment. Notice that our monkeys were unable to choose correctly in delayed alternation, yet they made the needed movements adequately in object reversals and visual discrimination. In all three tasks used in that experiment, the 'motor plans' seem to be identical; yet each lesion affected not particular movements but the correct use of the movements in one of the tasks only. Is this a 'motor' or 'cognitive' impairment? I agree, however, that we do not have good terminology. We used the term 'cognitive' only because we couldn't find a better alternative to emphasize that we were not referring to motor functions in the narrow sense of the term. We do not believe that the terms 'motor plan or program' or 'sensorimotor integration' are any better (Divac & Öberg 1979, Öberg & Divac 1981).

Calne: Obviously you can't study cognition in animals and therefore you are making inferences from what you can see, i.e. motor responses. But in view of the subtlety of some of the motor phenomena, would you infer that there may be a truly cognitive deficit in the animals if you could get at it?

Divac: To my mind, delayed alternation is a task which reveals the existence of cognitive processes in animals. An animal can show, by making a series of appropriate responses, that it remembers the rather complex rule: 'disregard all earlier trials; if remember where you went on the previous trial, go to the other side for reward, and remember where you went'. This paradigm can be motorically accomplished in many ways. Thus, the motor programs can vary i.e. are irrelevant.

Parkinsonism alone cannot critically influence the issue of involvement of the neostriatum in cognitive functions. In addition, we must not assume without direct evidence that in parkinsonism the striatum is devoid of any function. Instead, it is safe to assume that it functions abnormally. This malfunctioning probably contributes to the triad of motor symptoms of the disease, and to cognitive impairments. The fascination with the most striking features of this illness may be why the latter were not discovered earlier. Certain abnormalities are not found unless we have tools to discover them and unless we make use of these tools.

The parkinsonian symptom which seems to be most directly related to the dysfunction of the neostriatum is akinesia. Whether akinesia deserves to be considered a purely motor symptom is debatable. Consider the following interpretation: if a particular cortical area or the related neostriatal region is inactivated, e.g. destroyed, the organism is unable to resolve adequately certain behavioural situations. If a different neostriatal region is damaged, the organism will be unable to do something else. These losses could be called 'partial akinesias': the animal or the patient is able to engage successfully in many other activities besides those for which the damaged systems are normally responsible. If the entire isocortex or the related neostriatum is dysfunctioning or inactivated, the result is akinesia for all movements (behaviours) which originate in the isoprosencephalic systems. This is not to say that all movements have become impossible. For example those originating in alloprosencephalon, the brainstem or spinal cord may still take place; the preserved movement-initiating mechanisms may even be 'released'. In parkinsonism the dysfunction of the alloprosencephalon (to which nucleus accumbens belongs) may be responsible for a significant additional element of akinesia. A further active component of parkinsonian akinesia may be attributable to hypofunction of noradrenergic transmission (Narabayashi et al 1984).

Wing: The brain is working towards integrated sensorimotor function: action involves both perception and motor control, as the term perceptual–motor skill implies. Thus psychologists think the most fruitful approach is to try and characterize the relative contributions of logically distinct processes and how these relate to the basal ganglia rather than to say that basal ganglia function is all sensory or all motor.

Di Chiara: This question of semantics could be resolved if we did the same experiments in humans and animals. Have you applied delayed alternation tasks to humans and investigated whether patients with Parkinson's or Huntington's disease show deficits in them?

Divac: Patients with Huntington's disease are impaired in delayed alternation, like primates with prefrontal cortical or neostriatal lesions (Oscar-Berman et al 1982). I don't know what happens with parkinsonian patients in this situation. Let me say something else: clinicians know that in humans a lesion restricted to the neostriatum does not produce the classical motor symptoms of basal ganglia disease.

Calne: Clinicians don't all agree on that.

Divac: I would predict that if one knew which cortical area projected to the destroyed part of the striatum in a patient, one could use *human* neuropsychological tests to demonstrate the specific deficit associated with that lesion.

Marsden: The difficulty is that the tests used for animals are too simple for humans. Humans can use intellectual processes to get over the deficit. So you can't show alternation defects in patients with frontal lobe lesions.

Evarts: Would you say that patients with prefrontal lesions have cognitive disorders?

Marsden: The neuropsychologists with whom I work would say that. But the situation is different with lesions of the premotor cortex.

Evarts: But don't those patients have sensorimotor disturbances?

Marsden: They have a perceptual–motor disturbance.

Evarts: We would certainly agree that prefrontal cortex is important in cognition and conscious experience. Ivan Divac's point is that lesions of the basal ganglia may produce cognitive impairments without movement impairments.

Divac: There is another important point. Patients with Huntington's or Parkinson's disease have no aphasia as far as I know. Parkinsonian patients are very unevenly impaired in the language and performance parts of general intelligence tests: the tests show no language impairments. Could it be that in humans the cortical areas involved in language have become independent of the path through the basal ganglia (Öberg & Divac 1981)? Maybe these areas are much more highly developed in humans than the comparable areas in animals. Maybe, therefore, language *per se* is not affected by the basal ganglia damage. The cases with apparent 'subcortical aphasia' have lesions extending into the white matter and have demonstrably impaired cortical blood flow. The issue, however, is not settled.

Rolls: I am a little disturbed by the statement that frontal lobe damage does not produce complex or cognitive dysfunctions. Perhaps one can link this point back to the issue of whether striatal damage produces something like a cognitive deficit. In the Wisconsin card-sorting task, cards have to be sorted by colour, number of symbols, or shape, depending on whether the examiner says 'right' or 'wrong' to a particular placement. The switching from one category to another is impaired in patients with lesions of the frontal lobe (see Goodglass & Kaplan 1979). The link with striatal work is that Cools et al (1984) claim to get a similar deficit in parkinsonian patients, although there is of course the possibility that the effect in parkinsonian patients is due to frontal cortex dysfunction.

DeLong: It could also be the basal forebrain cholinergic system, since this is also damaged in some parkinsonian patients.

Rolls: At least it provides one with a task which could be used in patients who may have lesions in the head of the caudate nucleus. Perhaps one could therefore study the cognitive dysfunctions in such patients.

Marsden: I agree entirely about the prefrontal cortex. The question I was asked was whether the defects caused by lesions of the premotor cortex were cognitive.

I would like to return to the question of the subthalamic nucleus. The effects of subthalamic lesions are dramatically motor and there appear to be no cognitive deficits when that nucleus is damaged, at least unilaterally. If one accepts that, to what extent does the subthalamic nucleus, which modulates the output of the globus pallidus and substantia nigra pars reticulata, also regulate the ventral pallidum and its output by the medial dorsal nucleus to the frontal cortex? Is the subthalamic nucleus purely concerned with the dorsal 'motor' striatum and 'motor' pallidum and its output, or does it also interact with the ventral striatum, ventral pallidum and projections to frontal cortex?

Rolls: Anatomically the ventral pallidum projects to the subthalamic nucleus. One might expect to see a deficit in cognitive functions if there were bilateral damage to the subthalamic nucleus, which of course is unlikely to occur. But even given the anatomy, the information coming out of, say, the ventral pallidum or the head of the caudate nucleus does not necessarily have to go through the subthalamic nucleus. I am not sure you would predict a cognitive deficit, but if you found one perhaps you would not be surprised, if there was bilateral damage to the subthalamic nucleus.

Calne: We are at an interesting stage with clinicopathological correlations now. With computerized tomography and nuclear magnetic resonance scanning we can define and locate lesions with greater precision than ever before. Most of the lesions I have seen in the basal ganglia, using these techniques, lead to motor deficits that may be either florid or subtle. They are usually secondary to vascular disease because that is the commonest cause of focal lesions. They are sometimes masked or modified because the internal capsule is very close and may also be damaged. I have a patient with classical dystonia from a lesion that is almost confined to the putamen and is undoubtedly caused by a vascular abnormality. What may be misleading is that there are certain pseudo-lesions such as calcification of blood vessels which need not affect neuronal function.

Marsden: I agree that dystonia is perhaps more obvious with lesions of the putamen. We have a single clinicopathological case of a tumour restricted to the ventrolateral portion of the putamen, in the area where the motor cortex projects. Occasionally, however, one sees infarcts or haemorrhages in the head of the caudate nucleus, without clinical symptoms as far as one can tell.

Calne: But you may see that in the motor cortex.

Marsden: We may have looked at the wrong thing.

DeLong: One of the problems is that the cognitive deficits Ivan Divac described are produced by bilateral symmetrical lesions of the caudate and these rarely occur in humans. It is therefore not surprising that these disturbances are not seen. Motor deficits on the other hand can be observed after unilateral lesions of the putamen.

A dramatic disorder that has not received enough attention is hemiballismus. This disorder results from a tiny lesion in the subthalamic nucleus, which seems to regulate the outflow from the pallidum. Malcolm Carpenter did some of the pioneer work on that. This disorder clearly demonstrates how important the basal ganglia are for motor function in primates. In the rat and cat I understand there is no such syndrome after lesions of the subthalamic nucleus. That again clearly tells us that primate organization is different from the rat. People working in different species or in different parts of the basal ganglia don't need to argue about whether the basal ganglia are primarily 'motor' or 'cognitive'. If you lesion the caudate you will not disturb motor function in the sense that we understand it, and vice versa. Teuber once said that the basal ganglia are 'more than motor', which is a useful expression that doesn't commit one to something others would disagree with.

Divac: I don't want to leave the impression that I am saying that basal ganglia do not have motor functions. The entire brain does. The very position of the neostriatum on the way from the cortex towards the final common path indicates some motor role. The neostriatum mediates different aspects of behaviour, which of course involves movements. Possibly one part of the striatum is 'more motor' than the rest. This part is the target of the motor cortex. An important problem at present is how to understand coexistence of the 'motor' and 'association' parts of the neostriatum.

REFERENCES

Cools AR, Van Den Bercken JHL, Horstink MWI, Van Spaedonck MWI, Berger HJC 1984 Basal ganglia disorders in man: a shifting aptitude disorder manifested in motor and cognitive modalities. In: McKenzie, Wilcox L (eds) The basal ganglia: structure and function. Plenum, New York

Divac I, Öberg RGE 1979 Current conceptions of neostriatal functions. History and an evaluation. In: Divac I, Öberg RGE (eds) The neostriatum. Pergamon, Oxford, p 215-230

Goodglass H, Kaplan E 1979 Assessment of cognitive impairment in the brain-injured patient. In: Gazzaniga MS (ed) Neuropsychology. Plenum, New York (Handb Behav Neurobiol 2) p 3-22

Kitai ST, Kocsis JD, Wood J 1976 Origin and characteristics of the cortico-caudate afferents: an anatomical and electrophysiological study. Brain Res 118:137-141

Narabayashi H, Kondo T, Nagatsu T, Hayashi A, Suzuki T 1984 DL-Threo-3,4 dihydroxyphenylserine for freezing symptom in parkinsonism. Adv Neurol 40:117-122

Öberg RGE, Divac I 1981 Levels of motor planning. Cognition and the control of movement. Trends Neurosci 4:122-124

Oka H 1980 Organization of the cortico-caudate projections. A horseradish peroxidase study in the cat. Exp Brain Res 40:203-208

Oscar-Berman M, Zola-Morgan SM, Öberg RGE, Bonner RT 1982 Comparative neuropsychology and Korsakoff's syndrome. III. Delayed response, delayed alteration and DRL performance. Neuropsychologia 20:187-202

Royce GJ 1982 Laminar origin of cortical neurons which project upon the caudate nucleus: a horseradish peroxidase investigation in the cat. J Comp Neurol 205:8-29

General discussion 3

Dopamine depletion and replacement

Evarts: What can we learn about the operation of aminergic mechanisms from the improvements that Sue Iversen has described? This question brings us back to our discussion of mechanisms for local regulation of dopamine availability. Transplanted aminergic cells do not have the feedback pathways that would normally be available to control the activity of dopaminergic neurons. This fact is especially obvious when the implanted cells are not neuronal at all, but adrenal. The techniques that Sue has described have many ramifications.

Iversen: As I mentioned briefly in my paper, grafting compensation is apparently not unique to dopaminergic neurons. It also occurs in neurons of the isodendritic core of brain, which runs from brainstem to the basal forebrain and includes other neurotransmitter systems such as noradrenaline, acetylcholine and 5HT. We have done similar experiments in rats with fetal cholinergic grafts to the cholinergically denervated hippocampus. A lesion is made in the fornix fimbria severing cholinergic input to hippocampus. A cholinergic graft in the fornix lesion cavity survives, grows into the hippocampus and reinnervates in what appears to be a normal acetylcholine pattern. Such grafts reverse the deficits on memory tasks that are seen after hippocampal lesions or section of the fornix. That is even more remarkable because we had generally supposed that with its very specific morphology the hippocampus is one of the most highly wired and tightly coupled neurophysiological structures in the brain. Yet apparently acetylcholine from grafted neurons is able to influence neuronal integration sufficiently to restore hippocampal functioning.

Arbuthnott: Isn't there a risk in the assumption that these cells no longer 'know' what they would have 'known' in their original position? What we don't know is what kind of new inputs they collect in the place to which they have been transplanted.

Iversen: That is true. We have looked at this in dorsal–striatal dopamine grafts, and they do pick up extraneous connections. For example, substance P-containing neurons in cortex grow out and innervate dorsal dopamine grafts. Presumably the longer the grafts were left, the more connections they would spuriously pick up, resulting perhaps eventually in neuronal chaos and even more dysfunction than that created with the original lesions. The initial improvement might be lost as the graft acquired all sorts of spurious connections by growth in the central nervous system.

DeLong: Have any of these experiments been done in primates?

Iversen: They have been done in humans. I am not aware of any results in other primates yet.

Smith: For most of the behaviour deficits in the nigrostriatal system that had been reduced or abolished by the grafts in your system you had to give amphetamine to produce the reversal. In other words you induced release of dopamine artificially, by pharmacological means. But you also did some experiments without amphetamine, didn't you?

Iversen: We used amphetamine or apomorphine to induce pronounced asymmetry so that we could evaluate compensation. Spontaneous asymmetry is not very great after a unilateral nigrostriatal lesion unless one stresses the animal in some way. But the experiments on sensorimotor neglect didn't involve any pharmacological manipulation.

Smith: You concluded from this that the function of dopamine was to enhance behaviour globally. You were testing various specific behaviours but are you able to test behaviours involving choice or motivation?

Iversen: If we had used other paradigms which were sensitive to caudate lesions we could have looked at some of the more complex tasks involving what I call sensorimotor integration at the cognition level. We don't know yet whether we can reverse and normalize more complex kinds of performance. That would involve a bilateral lesion and a bilateral graft, and this work is being done now. If we were unable to reverse deficits on complex tasks, that would have important implications for what dopamine is doing in the striatum and would suggest that perhaps the garden-sprinkler hypothesis is not the correct one.

Smith: You need to control the power of the sprinkler.

Iversen: Yes, there may be another mechanism for controlling the gain in the dopamine system.

Hornykiewicz: I still venture to predict that you will get reversal in more complex tasks, simply from what we know in Parkinson's disease. It is not only L-dopa that is effective but also direct-acting dopamine agonists. Whereas the activity of L-dopa may depend on impulse-related release of newly formed dopamine, direct-acting dopamine agonists do not, as far as we know, depend on neuronal activity at all. They are there and that seems to be sufficient to restore more or less all the deficits in Parkinson's disease.

Iversen: That is why I am encouraged to think in terms of the non-specific or garden-sprinkler hypothesis.

Evarts: Dr Hornykiewicz mentioned earlier (page 110) that in parkinsonism losses of dopamine are greater in the putamen than in the accumbens or the caudate. Might it be that impairments of higher brain functions are absent in parkinsonian patients because the areas of the basal ganglia that are projected upon by prefrontal and temporal cortex are not those with the maximum loss of dopamine?

Hornykiewicz: The putamen is the more severely affected basal ganglia region in the idiopathic variety of Parkinson's disease. In patients with postencephalitic Parkinson's disease dopamine deficiency in the caudate is just as severe as in the putamen. This is so because in the postencephalitic condition cell loss in the substantia nigra is very severe and diffusely distributed, so the dopamine in the caudate is as severely reduced as in the putamen. The question is: what do clinicians know about cognitive or higher brain function disturbances in postencephalitic cases compared with idiopathic cases? That may give us a clue to caudate function.

Marsden: Many more behavioural disturbances are seen in patients with postencephalitic parkinsonism. Some such patients may exhibit only minor motor deficits.

Evarts: What do you mean by behavioural?

Marsden: Patients with compulsive behaviour disorders, schizophrenic, psychotic or antisocial tendencies of any kind. This doesn't solve the problem because unfortunately the abnormalities of postencephalitic parkinsonism are not only more diffuse within the basal ganglia, but also extend well beyond the basal ganglia to many other structures.

Divac: Is there a good human model for basal ganglia lesions? Parkinsonism is not the best model, because the pathological changes involve dopamine not only in the striatum but also outside it. There are also decreased amounts of noradrenaline, serotonin and some peptides. Occasionally there is a loss of cholinergic cells of the basal forebrain. Parkinsonian patients are known to have an excess of somatostatin in the cerebrospinal fluid. These changes cannot exist without some functional consequences. Kuhl and his collaborators (1984) examined parkinsonian patients with the deoxyglucose–PET scan. They found hypometabolism throughout the brain, in the cortex as well as in the basal ganglia. These widespread changes argue against the common belief that all symptoms of Parkinson's disease should be attributed only to the dysfunctioning neostriatum. Kuhl et al (1982) have also reported that in patients in the early stages of Huntington's disease, even in the offspring who have not yet developed the disease, there is hypometabolism restricted to the striatum. These changes are seen before a CT scan shows any shrinkage of the striatum or cortex. Normal metabolism does not necessarily mean normal function but perhaps Huntington's disease is a better model than Parkinson's.

Calne: I disagree. It depends very much on which basal ganglia you are talking about. In Huntington's disease damage is much more extensive in the cerebral cortex, for example, than it is in parkinsonism. For the dopaminergic components of the basal ganglia, as Professor Hornykiewicz has often pointed out, Parkinson's disease is the model *par excellence* . If one puts in drugs which are relatively selective dopamine-blocking agents one will get clinical parkinsonism. Many of the other changes that occur could be construed as secondary.

For example late parkinsonian patients who die have been subjected to an enormous barrage of pharmacological agents, such as L-dopa and dopamine agonists. It is difficult to interpret the basic bradykinesia, rigidity, tremor and postural disturbances of the parkinsonian syndrome as being the result of anything other than a specific dopaminergic problem within the basal ganglia. The one category of drug that consistently reverses the various components of the syndrome (to differing extents) is the dopaminomimetic group. Dopamine-blocking drugs produce the syndrome and dopaminomimetic drugs reverse it. Nature has given us a superb model of a lesion of the dopamine components of the basal ganglia, specifically the nigrostriatal pathway. I don't think any other disease can compete.

Evarts: It is true that dopamine is critically involved. However, the loss of dopamine may be least in the nucleus accumbens, and most in the putamen. [But see p 265.]

Marsden: I don't think we can take Huntington's disease as a model of basal ganglia function, because the cortical changes of Huntington's disease are extensive early in the disease. This makes it exceedingly difficult to interpret any change whatsoever in Huntington's disease. The hypometabolism in the apparently anatomically intact caudate shown by Kuhl and his colleagues in Huntington's disease was seen in symptomatic and presymptomatic cases. If the caudate can be abnormal by this test, without symptoms, that doesn't tell us anything useful either. In Parkinson's disease the dilemma is whether such patients have a selective functional striatal dopamine deficit confined to the motor striatum. This becomes particularly important, for one has to examine patients with early Parkinson's disease, before they develop dementia, to assess the impact of restricted basal ganglia dysfunction. A third of these patients become demented as the disease progresses. Present thinking is that this dementia is a consequence of changes in the nucleus basalis, associated with Lewy body degeneration, causing secondary depletion of acetylcholine throughout the cortex. The human model of abnormal basal ganglia function has to be parkinsonian patients before they become demented. The question is whether, at that stage of the illness, one is dealing with a selective *putamenal* motor striatal depletion of dopamine. There are parkinsonian patients who are severely affected but not demented, in whom L-dopa works only intermittently. They swing from normal mobility with apparently normal striatal dopamine function to gross immobility at intervals throughout the day. But for L-dopa, they would be dead. At that stage of motor disability, at their worst, such patients may be bedridden. What would you expect their caudate and accumbens dopamine concentrations to be?

Hornykiewicz: Those very severely affected patients would have a heavier dopamine deficiency in the caudate and accumbens than the early cases. We have been able to show this by correlating cell losses in the substantia nigra with the decrease in dopamine in the caudate and putamen (Bernheimer et al 1973). It is very dangerous to say that Parkinson's disease is due to a purely putamenal deficiency of dopamine. It is true that the caudate is less affected but it is still severely affected when one looks at cell losses in those substantia nigra segments that project to the caudate nucleus.

Marsden: What are the numbers for regional dopamine losses in advanced parkinsonism?

Hornykiewicz: One can extrapolate from the loss of striatal dopamine to nigral cell loss. In advanced parkinsonism the loss of dopamine in the putamen is at least 95% and often more. In the caudate and nucleus accumbens it is more than 80%. The loss in the caudate nucleus and accumbens is high enough for one to expect symptoms.

Marsden: How much depletion were you producing with your local 6-hydroxydopamine lesions in various areas of the striatum, Dr Iversen?

Iversen: I can't answer that from the local lesion studies, since we did not do regional biochemical assays but rather fluorescence histochemistry. In earlier studies we attempted to make larger caudate lesions and found that caudate is always more difficult to damage than the nucleus accumbens. In caudate we get something like 80% loss, at best, and in nucleus accumbens usually over 90%. That is probably a reflection of the relative size of the structures and with focal striatal lesions the depletion should be greater.

Marsden: The conclusion I would draw from that is that the depletion in the caudate and accumbens in severe Parkinson's disease is sufficient to cause dysfunction of the caudate and accumbens.

Evarts: What level of depletion in the putamen is usually associated with the initial symptomatology?

Hornykiewicz: We have done extensive correlations in that respect (Bernheimer et al 1973). The surprising result was that the parkinsonian patient had to have an 80% loss of caudate dopamine before clinicians saw any symptoms. At this stage of illness the dopamine loss in the putamen was just over 85%.

Iversen: There is quite a striking parallel with the experimental work in animals. With depletions of 65 or 75% of striatal dopamine after 6-hydroxydopamine we rarely saw effects on behaviour. Only at the magic figure of about 80% depletion did we begin to see any deficits.

Rolls: Is there sufficient depletion of dopamine in the frontal cortex of these patients to produce some of the symptoms that are seen?

Hornykiewicz: Recently, Scatton et al (1983) found a 60% depletion of dopamine in frontal cortex (Brodmann area 9) in non-dopa treated patients with Parkinson's disease. Earlier, we observed a dopamine reduction of more than 90% in Brodmann area 25, that is the parolfactory gyrus (Farley et al 1977), an area which may have something to do with cognitive functions.

I am glad to hear from Dr Iversen that in rats, too, less than 80% striatal dopamine depletion does not result in any symptoms. What is it that makes the figure of 80% dopamine loss such a magic number? I think that basically it is the large capacity of the brain dopamine system for functional compensation in the face of partial damage that makes up for smaller dopamine losses. When we calculated homovanillic acid:dopamine ratio in order to obtain a measure of dopamine turnover in the human basal ganglia we observed that in Parkinson's disease this ratio was shifted markedly in favour of homovanillic acid, indicating an increased dopamine turnover in both the caudate nucleus and in the putamen (Bernheimer & Hornykiewicz 1965, Bernheimer et al 1973). It seems that when dopamine losses are lower than 80%, this compensatory overactivity of the remaining dopamine neurons is sufficient to keep the dopamine reductions asymptomatic. However, when the loss of dopamine neurons was 90% or more, marked symptoms were present. Obviously overactivity in the remaining 10% or so of dopamine neurons was simply not enough to maintain striatal function near normal levels. All these figures apply to the caudate nucleus dopamine. As mentioned, the putamenal dopamine levels at all stages of clinically overt Parkinson's disease are noticeably lower than the caudate values.

Marsden: The crucial question is whether that degree of dopamine depletion in the caudate in the human being means that the caudate isn't working.

Models for basal ganglia disease

Calne: I want to comment again on the quest for the best model for basal ganglia disease. I think that is an improper question. It is like saying that the best model of spinal cord disease is poliomyelitis, or subacute degeneration of the cord, or motor neuron disease. The pathways are all intermingled so one can't say that any disease is a model of function unless one is much more specific. We may have a new handle for the problem in relation to the dopamine system. Stan Burns has reported from studies with the selective neurotoxin, NMPTP (*N*-methyl-4-phenyl-1,2,3,6-tetrahydropyridine), that the degeneration in primates is confined to the nigrostriatal–dopamine system. It does not affect other dopamine neurons. That is very important because NMPTP produces a disorder in human subjects that is clinically indistinguishable from Parkinson's disease. In one patient who died after exposure to NMPTP there was selective degeneration of the zona compacta of the substantia nigra. If we are precise and take one component of the basal ganglia, the nigrostriatal pathway from the zona compacta, everything we know is in accord with Parkinson's disease being a condition in which this specific region is disturbed. It is very difficult to prove that all the other changes we have been discussing are anything other than secondary phenomena.

Di Chiara: Another possibility is that in human beings dopamine depletion in the basal ganglia is able to produce more pronounced 'motor' than 'cognitive' deficits, while in animals the reverse is true. Rats show a rapid recovery of motor function after bilateral massive lesions of dopaminergic neurons induced by 6-hydroxydopamine. This might underline the possibility that when one goes from lower to higher mammals the caudate–putamen tends to specialize as a 'motor' centre at the expense of its 'cognitive' functions. One might then suggest that to get a cognitive deficit in humans so much dopamine would have to be lost as to be incompatible with life, because of the motor impairment it produces. So perhaps we shall never be able to answer the problem of the 'cognitive' functions of basal ganglia by using parkinsonian patients.

Marsden: There is no best model in human beings. Every model suffers from some difficulty. We have discussed Huntington's disease and Parkinson's disease. Vascular lesions or tumours are exceedingly rare and nearly always unilateral, so they don't tell us anything. The whole system is bilaterally organized so it doesn't help to have unilateral regions. There are a number of degenerative diseases that hit the basal ganglia quite hard, but all of them produce damage in other parts of the brain. At present there is no pure basal ganglia disease in humans. It is not a question of what is best but what is the most informative of a bad bunch.

Evarts: We should accept that. Ann Graybiel's observations on the weaver mutant, for example, provide a good animal model but the weaver is not a model for all disorders of the basal ganglia. Ultimately we may discover genetic disorders with selective effects in the other parts of the basal ganglia.

Graybiel: A very large study is being done by Drs von Sattel and Richardson at the Massachusetts General Hospital in Boston. They have collected well over 200 post-mortem brains of patients who died with the diagnosis of Huntington's disease. They staged these brains in terms of severity of loss of striatal mass, and according to cytoarchitectural changes. The loss of volume starts medially in the head of the caudate nucleus, progresses laterally into the putamen and then spreads ventrally. The nucleus accumbens seems to be selectively spared until the latest stages of the disease process. It is curious that in the weaver mouse, the nucleus accumbens again is spared, although there is only 15% atrophy, by weight, of the striatum. There may be topographic differences among different parts of the striatum that will be picked up genetically. If so, there is every reason to think that in another mutant the nucleus accumbens might be differentially affected.

Carpenter: Another clinical way of looking at this is that parkinsonism differs a great deal from the other forms of dyskinesia, such as athetosis, chorea and ballismus, with common features. These other forms of dyskinesia vary in amplitude and in force, they occur in both immediate and delayed sequence, and they form a spectrum of activity which bears a relationship to muscle tone.

Athetosis at one end of the spectrum almost always occurs in association with some degree of spasticity, whereas ballism is associated with marked hypotonus. These are two very different classifications but they cover most forms of dyskinesia.

Divac: The 'best' disease for those interested in functions of a brain formation is the disease that least affects structures outside that formation and definitely affects the formation (in this instance the neostriatum). I still believe that at present Huntington's disease in its early stages is a 'better' disease for understanding functions of the neostriatum than is parkinsonism in a similarly early stage of its clinical evolution. I am not aware of any pathological evidence obtained in *early* cases of Huntington's chorea which demonstrated any abnormality outside the neostriatum. The cases we know are patients who died in an advanced stage of the disease. It is worth mentioning that Lyle & Gottesman (1977) in a longitudinal study of genetic candidates for Huntington's disease demonstrated premorbid (i.e. before the manifestation of classical symptoms of the disease) deficits in intellectual abilities. The degree of these deficits grows with proximity of the clinical onset of the disease.

Marsden: You are quite right that there is not much information on the extent of cortical changes in Huntington's disease. Pathologists can only quantitate the cortical changes in terms of gross obvious atrophy. Cell counts are only just beginning to be made in otherwise homogeneous areas of the brain. No one knows what the drop-out of neurons in the cerebral cortex of patients with Huntington's disease is at the early stages of the disease. At the time of diagnosis of this disease about three-quarters of the patients show clear cortical atrophy on computerized tomography scans, whereas at the time of diagnosis of Parkinson's disease less than a quarter of the patients show atrophy of their cerebral cortex on CT scans. That is the evidence that makes me think that, at present, Huntington's disease is not the best of a bad bunch.

The point about localized damage in Huntington's disease is intriguing. It goes some way towards resolving one of the dilemmas of clinical neurology. That is, how can damage to the same part of the brain, the basal ganglia, produce on the one hand a range of involuntary movement disorders like chorea and, on the other hand, no movement, as in parkinsonian akinesia? It is possible to conceive of a scenario in which the early brunt of Huntington's disease falls on the caudate nucleus while there is a relatively intact motor putamen, so allowing the presentation of chorea. But as the putamen gets more involved the chorea disappears and eventually most patients end up rigid and parkinsonian. This relative distribution of damage may resolve the paradox of the two extremes of movement disorders that we see with basal ganglia disease.

Calne: There is clearly a close relationship between the two. In patients with postencephalitic Parkinson's disease it is common to see chorea with parkinsonism. And we can drive patients between chorea and parkinsonism with drugs that stimulate or block dopamine receptors.

REFERENCES

Bernheimer H, Hornykiewicz O 1965 Herabgesetzte Konzentration der Homovanillinsäure im Gehirn von Parkinsonkranken Menschen als Ausdruck der Störung des zentralen Dopaminstoffwechsels. Klin Wschr 43:711-715

Bernheimer H, Birkmayer W, Hornykiewicz O, Jellinger K, Seitelberger F 1973 Brain dopamine and the syndromes of Parkinson and Huntington. J Neurol Sci 20:415-455

Farley IJ, Price KS, Hornykiewicz O 1977 Dopamine in the limbic regions of the human brain: normal and abnormal. Adv Biochem Psychopharmacol 16:57-64

Kuhl DE, Phelps ME, Markham CH, Metter EJ, Riege WH, Winter J 1982 Cerebral metabolism and atrophy in Huntington's disease determined by ^{18}FDG and computed tomographic scan. Ann Neurol 12:425-434

Kuhl DE, Metter EJ, Riege WH 1984 Patterns of local cerebral glucose utilization determined in Parkinson's disease by the [^{18}F]fluorodeoxyglucose method. Ann Neurol, in press

Lyle OE, Gottesman II 1977 Premorbid psychometric indicators of the gene for Huntington's disease. J Consult Clin Psychol 45:1011-1022

Scatton B, Javoy-Agid F, Rouquier L, Dubois B, Agid Y 1983 Reduction of cortical dopamine, noradrenaline, serotonin and their metabolites in Parkinson's disease. Brain Res 275:321-328

Which motor disorder in Parkinson's disease indicates the true motor function of the basal ganglia?

C. D. MARSDEN

University Department of Neurology, Institute of Psychiatry, and King's College Hospital, Denmark Hill, London SE5 8AF, UK

Abstract. Parkinson's disease in its earlier stages is argued to be the best available model for human basal ganglia dysfunction. The negative motor symptoms of Parkinson's disease are considered to give the greatest clue to normal function of this region of the brain. Particular attention is given to disorders of movement. These include delayed initiation and slowed execution of simple fast movements, due to abnormal specification of initial agonist activity. This might compromise predictive motor action, but this is shown to be preserved in Parkinson's disease. Disorders of more complex movements, such as repetitive, concurrent and sequential motor actions, are also abnormal in Parkinson's disease. These various defects are discussed in terms of a motor strategy involving the selection and sequencing of motor programmes to form a motor plan, and the initiation and execution of that motor plan. On the evidence available, it is suggested that patients with Parkinson's disease are unable to automatically execute learnt motor plans.

1984 Functions of the basal ganglia. Pitman, London (Ciba Foundation symposium 107) p 225-241

Any theory of basal ganglia action must accommodate the observed effects of basal ganglia damage in humans. But there are major problems in interpreting this information. (1) The damage in humans is rarely confined to basal ganglia structures. (2) Basal ganglia diseases are generally progressive. As they progress, the damage they cause increasingly extends beyond the basal ganglia, making clinicopathological correlation in this region of the brain uncertain and, often, speculative. (3) Different basal ganglia diseases have their major impact on different regions of the basal ganglia.

These problems make it difficult to decide which human disease(s) can be considered a suitable model for disordered basal ganglia function in humans. This issue is discussed both in this volume (see p 221) and elsewhere (Marsden 1982, 1984a,b).

If one is forced to decide which human illness is closest to providing a picture of abnormal basal ganglia function in humans, I opt for Parkinson's disease, in its earlier stages before dementia supervenes. Although the putamen is affected more than other parts, there is still a profound depletion of dopamine in the caudate nucleus and in the ventral striatum. Furthermore, unlike Huntington's disease, the cortical changes in Parkinson's disease are likely to occur later in the illness. Parkinson's disease is not a perfect model, but it may be the most informative that we have.

Two crucial questions surround discussion of Parkinson's disease as a model of abnormal basal ganglia function.

(1) Are the basal ganglia concerned solely with the control of movement, or are they involved in other complex brain functions loosely described as cognitive? (2) If the basal ganglia are primarily involved in motor behaviour, which aspect of motor function do they influence?

The problem of whether the basal ganglia are involved in some cognitive processes as well as in movement has been a major theme of this conference, and my views are expressed elsewhere (see p 108 and Marsden 1982, 1984b). In this paper I will concentrate on the second issue, namely what is the critical motor defect of Parkinson's disease which illustrates the normal motor functions of the basal ganglia.

What is the critical motor disorder?

An analysis of the motor abnormalities produced by basal ganglia disease must start with Hughlings Jackson and his concept of dissolution of the nervous system. The effect of brain disease is to cause a reduction to a lower level of function, but to Hughlings Jackson (1881) this dissolution involves two processes. 'There is a negative state with each positive state, in each degree the patient's condition is duplex . . . The negative element in the symptomatology . . . . is the dissolution . . . Physiologically, it is so much loss of function of so many nervous arrangements of the least organised, the continually organising centres . . . The positive element is physically activity of lower nervous arrangements, which are, except for over-activity, healthy . . . healthy except, figuratively speaking, for "insubordination from loss of control".' Dewhurst (1982) succinctly summarizes this view: 'A double process is involved in all dissolutions. There is loss, or suspension of function, giving rise to negative symptoms combined with the emergence of the most obtrusive positive symptoms resulting from "letting go" of lower centres after removal of higher control. He (Hughlings Jackson) emphasised that a positive sign such as involuntary movement, cannot be caused by a brain lesion ("nothing cannot be the cause of something") but only through the removal of inhibitory processes.'

This concept is fundamental to the interpretation of the significance of the symptomatology of Parkinson's disease. In general, the 'positive' symptoms such as tremor and rigidity represent the abnormal activities of distant brain mechanisms, released by basal ganglia damage. As such, they do not tell of what is lost as a result of disordered basal ganglia function. What is lost in Parkinson's disease? Two major defects have been proposed: (1) a disruption of normal postural control, and (2) an inability to execute normal voluntary movement.

Disorders of posture in basal ganglia disease

Purdon Martin (1967) has argued the case that the basal ganglia are the centres responsible for control of postural organization. His conclusions are that 'A) The Basal Ganglia as a group are concerned with posture, other than the support of the body against gravity. B) The Pallidum is a centre for postural reflexes. C) The Putamen, Caudate Nucleus and Corpus Luysii have controlling functions over postural reflexes, each being concerned with its own group or groups of reflexes.'

Martin's observations leading to these conclusions were based on studies of the loss of postural fixation and equilibrium, and of righting reactions in response to unexpected body displacement, in 30 patients with postencephalitic parkinsonism, eight patients with Wilson's disease, and 20 patients with Huntington's disease. In addition, close examination of three patients with hemiballism suggested that the abnormal involuntary movements consisted of 'exaggerated reactions to instability'.

These superb clinical studies certainly highlighted the major defects of postural control that can be seen in these illnesses. However, two arguments can be marshalled to suggest that the basal ganglia are concerned with more than just postural control.

(1) Disorders of posture are not a major feature of the earlier stages of Parkinson's disease and other basal ganglia illnesses, when damage may be thought to be more closely restricted to these regions of the brain.

(2) Disorders of movement occur, even in the early stages of such diseases, that do not involve postural control. For instance, obvious delay in initiation and slow execution of movement of the top joint of the thumb in Parkinson's disease can be demonstrated when the rest of the arm is clamped in such a way as to abolish all need for postural fixation (A. Beradelli et al, unpublished work).

For these reasons, it is concluded that the critical motor defect resulting from basal ganglia damage is a disorder of voluntary movement. The 'negative' symptoms of loss of movement (akinesia) and slowness of movement (bradykinesia) in early Parkinson's disease are seen as the clue.

Motor programmes

Further discussion will be built around the concepts of 'motor programmes' (Brooks 1979) and the 'motor plan' (Marsden 1982) (Fig. 1).

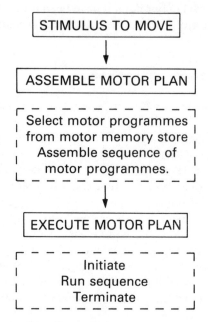

FIG. 1. Scheme for the execution of movements.

Motor programmes may be defined as 'a set of muscle commands that are structured before a movement sequence begins' (Keele 1968) which can be delivered without reference to external feedback. That humans are capable of calling up such motor programmes is well illustrated by the extensive range of both simple and complex motor acts that could be executed by a deafferented man (Rothwell et al 1982). This individual, without sensation from the arms or legs, could still drive a 'motor car'! Built within his nervous system were a wide range of learnt motor programmes capable of accomplishing many motor skills.

Each motor programme may be conceived of as consisting of a set of motor commands specifying the timing and amount of muscle contraction in agonists, antagonists, synergists and postural fixators required to execute a simple motor act. Such a simple motor programme might achieve movement of a limb from one position in space to another.

Normal movement, however, usually involves a sequence of such simple motor acts. Thus, the aim of moving the arm to the table might be to pick up the glass, bring it to the lips, drink, and then return the glass to the table. This sequence of motor acts, each produced by a motor programme, comprises the motor plan.

The execution of the complete motor act is seen as a series of steps (Fig. 1). The first step is to formulate the motor plan itself, and this requires selection and sequencing of the necessary subunit motor programmes. Then the motor plan must be initiated, after which the sequence of motor programmes must be run until the objective is attained, when the motor plan must be terminated.

How is this concept of voluntary movement distorted in Parkinson's disease?

Assembly of motor programmes in Parkinson's disease

A number of studies have used reaction-time tests to assess the ability of parkinsonian patients to prepare appropriate motor responses to external stimuli.

Evarts et al (1981) examined visual and kinaesthetic reaction times for forearm movement in response to a visual stimulus or to displacement of a handle, when a visual cue to the expected direction of movement was given 2–4 s before the stimulus. The times taken both to react and to complete the specified movement tended to be prolonged in 29 patients with Parkinson's disease compared to 21 age-matched elderly control subjects. Both visual and kinaesthetic simple reaction times were impaired. Choice visual reaction times were studied by providing the usual warning signal 2–4 s before the 'go' signal, but delaying the signal specifying the direction of movement until the 'go' signal. The median choice reaction time was only 61 ms longer than the simple reaction time in parkinsonian patients, which was less than that found in many previous studies of normal subjects. This extra time in a choice reaction test includes the period required to formulate the appropriate motor response to the combined 'go' and direction signal. Parkinsonian patients did not appear unable to formulate the correct response, nor did they appear to take any longer than normal to do so. In this test, at least, parkinsonian patients were not impaired in their ability to rapidly assemble the simple motor programmes required.

Heilman et al (1976) obtained similar results in a different experiment. They studied the ability of 10 parkinsonian patients to improve reaction times between a light cue and pressing a telegraph key when a warning auditory signal was given 0.5 or 1 s before the cue. Parkinsonian patients were slower to respond than normal subjects whether or not they were given the warning signal. However, the improvement in reaction time when the warning signal was provided was no different in parkinsonian patients than in controls. In other words, parkinsonian patients were able to assemble and deliver the appropriate motor response or programme normally in response to a warning cue.

Angel et al (1970) studied the capacity of parkinsonian patients to correct false moves. The subjects were asked to respond by appropriate arm movement to a visual target which jumped unexpectedly to a new position. In 50% of trials the visual display of the joystick movement was reversed, so that the subject moved the cursor in the wrong direction. Parkinsonian patients made no more false moves than control subjects. In other words, they had no difficulty in perceiving the visual stimulus or in selecting the correct motor response. They were slower to initiate and execute correct moves, as expected. Both normal subjects and parkinsonian patients took a shorter time to correct false moves than to initiate correct moves (see also Rabbitt 1966). This may have been because the response to correct moves represented a two-choice reaction whereas that to false moves was a simple reaction (see Angel 1976). The key observation, for the present purposes, was that the parkinsonian patients took longer to correct false moves than normal subjects. The delay in arresting false moves could not be attributed to rigidity or other peripheral motor disabilities, for these would have tended to hasten the end of the movement. The parkinsonian patients clearly had prepared the appropriate motor programmes for a correct move but were less able to switch to the opposite motor programme when they discovered they were moving in the wrong direction.

Rafal et al (1983) also used tests with false cues, but these were presented in advance of the stimulus to move the arm. They studied four parkinsonian patients both on and off medication. The patients were given a visual cue to move in a specified direction 50, 150, 350 or 550 ms before a target movement that they had to follow. Twenty per cent of cues were false, telling the subject to move in the opposite direction to the target. The reaction time to false cues was longer than that to valid cues, the extra time being that required to replace the initial motor programme, set up in the interval between the false cue and the target movement, with the appropriate response. Parkinsonian patients who were immobile when they were off drugs took as much extra time to correct invalid cues as when they were mobile during treatment. In other words, they had prepared the false motor programme irrespective of their state of mobility.

All these investigations point to the conclusion that patients with Parkinson's disease are quite capable of assembling motor programmes in response to warning signals. Their difficulty appears to be one of changing to a new motor programme if the cue proves false.

The content of motor programmes in Parkinson's disease

Simple motor acts are undoubtedly disrupted in Parkinson's disease. Movement is executed slowly and such patients have profound difficulty with fast ballistic movement (Wilson 1925). These commonplace clinical observations indicate that the simple motor programme is defective in Parkinson's disease.

Flowers (1976) has elegantly demonstrated the inability of parkinsonian patients to manually track step-changes in position when the changes are presented visually. The patients are unable to achieve the required velocity because of a breakdown of the normal control of ballistic activation of muscles. Hallett & Khoshbin (1980) have shown that the parkinsonian patient activates agonist and antagonist muscles in the correct time sequence but cannot deliver sufficient electromyographic activity in the initial ballistic agonist burst to achieve the required rate of force increase, or the required level of force to attain a large fast movement. The patient with Parkinson's disease achieves the point of aim by a series of small ballistic bursts which, summed together, jerkily drag the limb to the required position.

This basic defect clearly shows that the detailed content of the individual motor programme is specified abnormally in Parkinson's disease. The parkinsonian patient cannot deliver the correct force command to the agonist muscle in a fast movement. But the correct agonist is selected, and the normal pattern of antagonist command is delivered. In other words, parkinsonian patients, when they move, move in the right direction with the correct muscle(s) (see also Angel et al 1970). The parkinsonian patient also makes proper use of synergists. On the basis of extensive clinical observation, Wilson (1925) concluded that synergistic actions are 'executed normally by Parkinsonians, that is to say, the movements follow normal physiological lines and are neither lost nor perverted'.

The conclusion is that the overall form of motor programmes is preserved in Parkinson's disease, in that the selection of muscles and the relative timing of their activation is intact. However, the details of the number and frequency of motor neurons activated, at least in the first agonist burst, are inaccurate.

The failure to make fast movements has considerable implications for motor control in Parkinson's disease.

Predictive motor action in Parkinson's disease

Fast ballistic movements are required for predictive action. '(Normal) subjects operate with a repertoire of predetermined movements or movement sequences, predictively selected and, within limits, producing a predictable effect' (Flowers 1976). Since patients with Parkinson's disease cannot make accurate fast ballistic movements, Flowers concluded that they 'have a longer reaction time than normal, both in initiating and stopping (or correcting) movements, and they have less facility for producing appropriately the fast preprogrammed movements which in normals overcome the limitations of reaction time'. As a result, he suggested that 'A loss of prediction in the selection or performance of movements will make it difficult for these patients to do anything in the way of "skilled" movement, wherever skill involves the use of automatic predetermined actions and sequences of action to bypass the reaction time bottleneck performance, even where such movements are themselves relatively gross as far as dexterity is concerned. Where movements are unpredictable, also, large amplitude and fast movements will tend to go wildly astray within the reaction time period if the subject attempts them, and there will be a tendency to perform all action at a slow and steady pace so that a reasonable degree of control may be maintained.'

If this hypothesis is correct, it could be argued that the basal ganglia are responsible for predictive motor action.

In subsequent studies, Flowers (1978a,b) went on to examine this hypothesis in a series of pursuit tracking tasks using a manual joystick coupled to a visual display. He showed that parkinsonian patients were less accurate than control subjects at tracking a known sine wave, and they followed the target with a greater tracking lag. Parkinsonian patients showed little or no improvement when tracking a known sine wave compared with an irregular wave-form generated by a 'noise' signal (Flowers 1978a). He concluded that parkinsonian patients 'have difficulty in generating continuous regular movements spontaneously, and seem to lack a dynamic "internal model" of their own movements from which to control them predictively'. In a further series of experiments (Flowers 1978b), parkinsonian patients were found less accurate in following a known tracking pattern if the target was made to disappear temporarily, a strategy also used by Stern et al (1982). Flowers again interpreted this as evidence that parkinsonian patients have difficulty in utilizing prediction in their motor control. However, an alternative view is that these results indicate that patients with Parkinson's disease cannot accurately and continuously sequence motor commands, particularly if suddenly disturbed by unusual external stimuli such as disappearance of the tracking spot.

Day and I have also examined whether patients with Parkinson's disease are unable to utilize predictive motor control. We devised a method for examining the normal individual's capacity to use predictive action in a visually-guided motor task (Day & Marsden 1982). Twenty-two subjects performed a series of 150 visual tracking tasks each lasting 5 s. Each subject was asked to flex and extend the elbow, and the position of the elbow was sensed by a potentiometer whose output was displayed on a cathode ray oscilloscope. The subject was asked to move the elbow so as to track a second target spot displayed on the oscilloscope. The target-movement patterns used for the first 50 trials were all different, but for the remaining 100 trials they were identical. Subjects, however, were not informed of the repetition until the final 50 trials. When the task was made repetitive, even though the subjects were unaware of the repetition, some learning occurred, as evidenced by a small progressive reduction in tracking error, although tracking lag remained above the mean reaction time. Once subjects were aware of the repetition, tracking lags often fell to zero or even negative values and tracking error dropped even further. It was argued that the former learning was due to subconscious improvement in the intermittent response to visual inspection of tracking error, whereas the latter was achieved by adopting a truly predictive mode of tracking. Further experiments were devised to evaluate the role of visual information in movement control when the predictive strategy was used. The main finding was that even when the subject was moving predictively, visual information was used to regulate motor output, largely to modify the timing of the predictive response so that it synchronized with the stimulus.

We then adapted the task for use in patients with Parkinson's disease. This involved reducing the speed of change of the target movement to allow for the slower movement of parkinsonian patients. Twenty-five unselected patients with Parkinson's disease were examined, including patients with all degrees of duration and severity of the illness; no attempt was made to exclude individuals with cognitive defect. Nevertheless, 12 of the 25 were capable of adopting a predictive mode of tracking when they were aware of the repetitive nature of the final section of the task, as shown by a reduction of tracking lag well below visual reaction time (Marsden 1982).

In a further experiment (B. L. Day & C. D. Marsden, unpublished), five patients with moderately severe Parkinson's disease and intact general cognition were selected for study both when mobile on effective drug treatment and when akinetic after drugs were withdrawn. All five patients were capable of adopting the predictive motor control strategy, as shown by reduction of the tracking lag towards zero when the patients were either mobile or akinetic.

From these observations we conclude that a failure of predictive motor action is not an inevitable feature of Parkinson's disease. It is unlikely that the basal ganglia are concerned solely with this aspect of motor function.

However, in our studies and those of Flowers (1978a,b) parkinsonian patients could not make normal use of predictive action to reduce tracking error to the same extent as control subjects. Thus although they learnt an 'internal plan' of the movement required, and could deliver the correct actions more or less at the same time as the tracking spot changed course, this did not allow them to greatly improve accuracy. The simplest explanation of this failure is that the content of each motor programme was inaccurate, as discussed above. However, an alternative view is that parkinsonian patients have particular difficulty in automatically executing the motor plan.

Breakdown of the motor plan in Parkinson's disease

Having assembled the required sequence of motor programmes, each of which is correct in general form if not in exact detail, the parkinsonian patient then has to initiate, execute and terminate the whole plan. And the patients must be in a position to modify the plan at any time, in response to unexpected events. This they cannot do.

As indicated earlier, parkinsonian patients have particular difficulty in starting a movement. Motor reaction times, on the average, are delayed (Wilson 1925, Evarts et al 1981). The initiation of movement in Parkinson's disease is sometimes completely 'frozen'. 'Freezing' may affect gait, arm movement or speech, and often it is precipitated by sensory stimuli and unlocked by other sensory stimuli. All forms of simple movement are affected. Thus, the actions of both proximal and distal muscles are abnormal, both fast and slow movements may be disrupted, and few parts of the body are spared (see Marsden 1984a for review).

In addition, the parkinsonian patient has great difficulty in executing repetitive, concurrent or sequential motor actions.

Wilson (1925) and Schwab et al (1959) were impressed by the inability of patients with Parkinson's disease to sustain repetitive motor action, which is one of the best bedside indicators of Parkinsonian akinesia. The patient cannot continue to rapidly open or close the hand, or approximate the ball of the thumb to that of the index finger, or tap the foot on the ground. The movement progressively decreases in amplitude and slows in speed until it ceases. Schwab et al (1954) also demonstrated the inability of parkinsonian patients to undertake two simple motor acts at the same time, even though they could perform them individually. Talland & Schwab (1964) and Horne (1973) confirmed this observation with more complex simultaneous manual tasks. Schwab et al (1954) also remarked on the inability of parkinsonian patients to perform an uninterrupted sequence of motor actions, a defect subsequently explored experimentally by Flowers (1978a,b) and Stern et al (1982). Clearly, patients with Parkinson's disease are unable to execute either simple or complex motor behaviours. Indeed, the general conclusion is that the greater the complexity of the task, the more the parkinsonian patient is impaired.

Can the inability of the parkinsonian patient to execute complex motor acts be explained simply on the basis of the slowness in initiation of movement, and the inability to generate accurate motor programmes? The answer is probably not.

A failure to undertake repetitive motor actions might be explained by these simple defects. Each motor programme in itself is imprecise, so it is not altogether surprising that, when repeated, the whole sequence is in error. However, the details of the impairment of repetitive action suggests that there is more than this. The critical feature is that, as the sequence is continued, the individual movement or motor programme progressively degrades. The time to initiation gets slower and the size of the movement progressively fades. Thus, the final movement in a sequence is far worse than when that individual movement is performed by itself. Repetition of the movement unearths a greater deficit than is apparent in single movements.

Likewise, when two movements are made together, they are each performed much worse than if they are undertaken alone. This suggests that there is a fundamental breakdown in the capacity to run the sequence of movements that comprises a motor plan. In particular, there appears to be a major problem in switching from one motor programme to another. In other words, the sequence of the motor plan does not run smoothly in Parkinson's disease.

Do the basal ganglia sequence the motor plan?

Elsewhere, I have suggested that an essential defect in Parkinson's disease is the inability of patients to automatically execute learnt motor plans (Marsden 1982). The suggestion is that the motor function of the basal ganglia is to automatically and subconsciously run the sequence of motor programmes that comprise a motor plan. The programmes themselves may be learnt and stored elsewhere in the brain, and perhaps are assembled into the coherent plan in the premotor (and other frontal areas). But the initiation and automatic execution of the sequence of motor programmes required to complete the motor plan of a complex motor act may depend on the basal ganglia.

The anatomy and physiology of the basal ganglia allow such a notion. The multiple inputs from all areas of cerebral cortex into the striatum provide moment to moment information on the external environment and the wishes and feelings of the individual. At any moment, such inputs may alter the execution of the motor plan, in response to changes in the environment or aims of the subject. Since the workings of the basal ganglia are regarded as subconscious, such adjustment would be automatic. The internal workings of the striopallidal complex, with the subthalamus, could digest this mass of information and reduce it to appropriate signals relayed via pallidal output to thalamus and thence to premotor (and other frontal) cortex, so as to adjust the form and sequence of motor programmes.

Viewed in this light, the parkinsonian patient's motor deficit is reduced to an inability to specify the detailed accuracy of the motor programmes and to automatically run their sequence, switching from one programme to another as required.

This hypothesis also bears on the problem of learning motor skill. The capacity of the trained athlete, musician, or even the ordinary mortal, to automatically generate exquisitely accurate motor sequences in the face of varying external circumstances is marvellous. To begin with, motor performance is heavily dependent on checking the results against all modes of sensory feedback. With practice, however, the need to check becomes less and less, and the execution of the movement plan becomes more and more automatic. 'Ordinary singers do not always strike the note accurately at one jump, but must feel around a little after reaching the neighbourhood, guiding themselves by the sense of pitch. Probably the ordinary run of violinists, and certainly the beginner, find their notes in this groping way. The path to skill lies in increasing the accuracy of the initial adjustment, so that the later groping needs to be only within narrow limits; and through increasing the speed of the groping process, so that finally there seems to be no groping at all' (Woodworth 1899). The basal ganglia may be involved in generating skilled movement by the automatic execution of learnt motor plans.

REFERENCES

Angel RW 1976 Efference copy in the control of movement. Neurology 26:1164-1168

Angel RW, Alston W, Higgins JR 1970 Control of movement in Parkinson's disease. Brain 93:1-14

Brooks VB 1979 Motor programmes revisited. In: Talbot RE, Humphrey DR (eds) Posture and movement. Raven Press, New York, p 13-49

Day BL, Marsden CD 1982 Two strategies for learning a visually guided motor task. Percept Mot Skills 55:1003-1016

Dewhurst K 1982 Hughlings Jackson on psychiatry. Sandford Publications, Oxford, p 35

Evarts EV, Teravainen H, Calne DB 1981 Reaction time in Parkinson's disease. Brain 104:167-186

Flowers KA 1976 Visual 'closed-loop' and 'open-loop' characteristics of vountary movement in patients with parkinsonism and intention tremor. Brain 99:269-310

Flowers K 1978a Some frequency response characteristics of parkinsonism on pursuit tracking. Brain 101:19-34

Flowers K 1978b Lack of prediction in the motor behaviour of parkinsonism. Brain 101:35-52

Hallett M, Khoshbin S 1980 A physiological mechanism of bradykinesia. Brain 103:301-304

Heilman KM, Bowers D, Watson RT, Greer M 1976 Reaction times in Parkinson's disease. Arch Neurol 33:139-140

Horne DK de L 1973 Sensorimotor control in Parkinsonism. J Neurol Neurosurg Psychiatry 36:742-746
Jackson J Hughlings 1881 Remarks on dissolutions of the nervous system as exemplified by certain post-epileptic conditions. Med Press Circ 31:329-332, 339; 32:68-70, 380-389 In: Taylor J (ed) 1958 Selected writings of John Hughlings Jackson. Basic Books, New York vol 2:12-16
Keele SW 1968 Movement control in skilled motor performance. Psychol Bull 70:387-403
Marsden CD 1982 The mysterious motor function of the basal ganglia. Neurology 32:514-539
Marsden CD 1984a Motor disorders in basal ganglia disease. Hum Neurobiol 2:1-6
Marsden CD 1984b Functions of the basal ganglia as revealed by cognitive and motor disorders in Parkinson's disease. Can J Neurol Sci 11:129-135
Martin JP 1967 The basal ganglia and posture. Pitman Medical, London
Rabbitt PMA 1966 Error correction time without external error signals. Nature (Lond) 212:438
Rafal RD, Posner M, Walker J, Stebbins K 1983 Is slowness in initiating movement in Parkinson's disease or frontal lobe lesions due to slowness in mentally preparing to move? Neurology 33, Suppl 2:103
Rothwell JC, Traub MM, Day BL, Obeso JA, Thomas PK, Marsden CD 1982 Manual motor performance in a deafferented man. Brain 105:515-542
Schwab RS, Chafetz ME, Walker S 1954 Control of two simultaneous voluntary motor acts in normals and parkinsonism. Arch Neurol 72:591-598
Schwab RS, England AC, Peterson E 1959 Akinesia in Parkinson's disease. Neurology 9:65-72
Stern Y, Mayeux R, Rosen J, Ilson J 1982 Perceptual motor dysfunction in Parkinson's disease: a deficit in sequential and predictive voluntary movement. J Neurol Neurosurg Psychiatry 46:145-151
Talland GA, Schwab RS 1964 Performance with multiple sets in Parkinson's disease. Neuropsychologia 2:45-53
Wilson SAK 1925 Disorders of motility and muscle tone, with special reference to the striatum. Lancet 2:1-53, 169, 215, 268
Woodworth RS 1899 The accuracy of voluntary movement. Psychological Supplement 13. Macmillan, New York, p 1-114

DISCUSSION

Wing : In your experiment on predictive action in a visually guided motor task, 12 out of 25 patients with Parkinson's disease were able to perform the task successfully. Are the unsuccessful ones the demented patients? Is predictability as tested in that task a perceptual process?

Evarts: I have a similar question, going back to some of the observations on cognitive function. The patients who performed less well than controls on the tests of cognitive function were not necessarily the ones who had the most severe motor deficits: there appeared to be a dissociation of cognitive and motor performance. There are two ways of interpreting such a dissociation. You implied that the cognitive disturbance was not a necessary accompaniment of the parkinsonian motor deficit. But one can also say, given the wide range of functions of basal ganglia as a whole, that the patients really belonged in two different groups.

Marsden: The dilemma of this sort of work is that even if we are careful not to use those with dementia, we find that parkinsonian patients fall into two groups: non-demented parkinsonians who perform normally on tests of cognitive function or predictive tracking and so on, and those who, although not obviously demented, appear to be outside the normal range. The crucial question is whether we are looking in the former patients at effects of a relatively pure deficiency of dopamine in the motor striatum, with other functions remaining normal, and another abnormal group of patients with additional dopamine deficiency in the caudate and accumbens. Alternatively, are we looking at two groups of patients, both with a deficiency of dopamine in the basal ganglia but only one with additional deficiency of cortical dopamine or noradrenaline? This question is unanswerable at present, but in future we may be able to use radioactive fluoro-dopa, for example, to look at the extent and distribution of dopamine deficiency in the brains of patients with relatively early parkinsonism.

Calne: I agree with your eloquent analysis of the bradykinesia. But could you say more about features such as disturbance of posture, or flexion of the neck or shoulders? These are not due to weakness—patients can straighten up when asked. We can say that the loss of inhibition can give rigidity, and presumably some loss of control of oscillating circuits can give rise to tremor. Posture, rigidity and tremor are in some respects much more obvious than akinesia. Could you link these to your general hypothesis about akinesia? Or have you any other comments about the mechanism of these other clinical features?

Marsden: To distinguish the abnormal position that parkinsonian patients adopt from postural instability in the course of movement, which may be a separate issue, I shall use the word postural deformity to describe the former. There is the analogy of spasticity. Damage to the cortico-motor neuron pathway produces a characteristic posture caused by differential weakness and spasticity in different muscles. The posture is imposed by the spasticity. The spasticity occurs because the reflex machinery of the spinal cord works abnormally, due to the loss of cortico-motor neuron control. The positive symptom of spasticity is not telling us what the normal function of the cortico-motor neuron pathway is. The normal function is not to stop spasticity but to allow one to move. The posture is secondary to the positive symptom of spasticity in hemiplegia. In Parkinson's disease, the postural deformity may be imposed by the rigidity. The present concept is that parkinsonian rigidity is caused by enhanced long-latency stretch reflexes, which themselves are a release phenomenon. So the postural deformity may not indicate a true basal ganglia role in postural control.

Calne: But what is the link between that and striatal function?

Marsden: I suspect that the postural deformity of Parkinson's disease is a consequence of rigidity differentially affecting different muscles, so causing the head, trunk and arms to flex. The abnormality in the long-latency stretch reflex which causes rigidity has nothing to do with the normal function of the basal ganglia. Those reflexes do not go through the basal ganglia, but they seem to be released by basal ganglia lesions. When these long-latency stretch reflexes are enhanced, the patient becomes rigid and bent. That is how I would explain the postural deformities, and I would therefore dismiss them from the analysis of what the basal ganglia actually do.

The postural instabilities that occur in movement are a different issue. Every movement involves postural changes. If someone can't put a normal motor plan together in sequence, posture collapses just as much as the movement. The patient that Bob Schwab draws attention to falls over backwards when he puts his hand out, because he can't get the right postural motor programme to allow him to stand up at the same time as putting out his hand.

Calne: What about the tremor?

Marsden: I had an answer to that until NMPTP (*N*-methyl-4-phenyl-1,2,3,6-tetrahydropyridine) appeared! The answer was that everybody who wrote about experimental parkinsonian rest tremor claimed that to produce such a tremor one had to damage not only the nigrostriatal system but also the ascending cerebellar-fugal fibres. But now NMPTP is said to produce selective dopamine deficiency and it certainly can cause gross parkinsonian tremor. I am not aware that anybody has yet looked at what is happening to the cerebellar output fibres in human beings or monkeys given NMPTP.

Calne: Could rigidity, tremor and bradykinesia be as discrete in the kaleidoscope of basal ganglia deficiency in parkinsonism as the cognitive component? Perhaps we shouldn't lump the motor features together. Clinically they can occur separately.

Marsden: I am sure they have to be explained, but because of the Hughlings Jackson philosophy to which I am wedded, I think some of these features are not telling us the function of the part of the brain that is damaged.

Rolls: In analysing structures such as the head of the caudate nucleus and the limbic striatum we should look at which part of the cortex the major inputs come from. We should then test patients in whom we suspect deficits of that part of the striatum with tests appropriate to that part of the cortex. The Alice Heim type of test might not tax the prefrontal cortex particularly, but perhaps in your on–off study you used in addition the Wisconsin card-sorting task or fluency measures?

Marsden: We are doing that now, after starting by looking at global intellectual function. I agree that those selective tests are important.

Evarts: We can learn something from dopaminergic agonists and antagonists in relation to functions of the basal ganglia in controlling mood and emotional behaviour. Walle Nauta began this symposium with a description of links between the limbic system and basal ganglia, and David Marsden has just mentioned the research potential of studies to assess binding of fluoro-dopa in different parts of the basal ganglia in humans. If we were to discover that there is a disturbance of the binding of a ligand to the nucleus accumbens in a given psychiatric disorder, we would take this as an indication that the basal ganglia are involved in the disorder. How would we then interpret this, Professor Nauta? How would we integrate such a finding with what we know of the relation of basal ganglia dysfunction to motor disturbances?

Nauta: That is a most difficult question, Dr Evarts, on which I am afraid little more than speculation can be offered. It is embarrassing that, after a century of brain research, we are still unable to answer the naive question of where in the brain the impulse to a deliberate movement arises, and by what sequence of steps it is ultimately translated into patterns of activation and inhibition of motor neurons. Involvement of motor units is nearly the only attribute we can name that is specific to the motor system. Central to that level the system rapidly loses anatomical and physiological definition with each further synaptic remove from the motor neuron.

I believe there is reason to suspect that the central motor mechanism and the neural apparatus of ideation have things in common and may even be in partial overlap. Both somatic–motor and ideational processes require a motivational 'vis a tergo' to be initiated, as well as to be kept in progress. It requires an effort to change from a state of relaxed daydreaming—a sort of resting tonus perhaps—to one of goal-directed thinking, and it is something like an act of endurance to keep such a thought process going. Movement and thought share with each other a need for 'feedback' allowing for corrective adjustments of the process. Thought often spills over into the motor sphere, in the form of unintentional changes of body posture and facial expression. More similarities could be suggested, the most interesting perhaps for our present purpose being the suggestion that both movement and thought include in their physiological substrate a dopamine-dependent mechanism. The 'primary thought disorder' of schizophrenia may respond well to dopamine receptor-blocking agents, at least in the earlier stages of the disease, and the protracted use of such drugs has serious repercussions on the somatic motor system.

Our profound ignorance of the material substrates of movement and thought makes it difficult to suggest where in the brain the two might intersect. The evidence that not only the motor cortex but also all other cortical areas, and in addition the limbic system, send part of their output through the sequence striatum–pallidum could raise the question of whether it is reasonable to assume that all this abundant through-traffic of vastly different antecedents exclusively subserves somatic–motor function. This question is further encouraged by the consideration that so much of the striatopallidal outflow appears to be channelled back to the domain of its origin: to the cortex via various thalamic way-stations, and to the limbic system by way of various subcortical structures embedded in the limbic circuitry. Would it perhaps be more reasonable to suspect that the respective functions of the neocortex (including motor cortex) and limbic system share with each other a need for side inputs mediated by certain transmitter combinations (including dopamine) that are available only in the striatum–pallidum sequence? The basic question would thus become: do neocortex and limbic system, by passing part of their circuitry through the striatopallidal processing sequence, gain access to modulatory mechanisms crucial to their optimal functioning?

Hornykiewicz: We might then say that thought is movement confined to the brain.

Basal ganglia lesions and psychological analyses of the control of voluntary movement

ALAN M. WING and ED MILLER*

*MRC Applied Psychology Unit, 15 Chaucer Rd, Cambridge, CB2 2EF, and *Department of Clinical Psychology, Addenbrookes Hospital, Hills Road, Cambridge, UK*

Abstract. Psychological accounts of the voluntary control of movement recognize the contribution of perceptual and decision processes as well as processes such as motor memory and timing that are more obviously a part of motor control. A model including these and other components is outlined in relation to tracking, a task that has often been used to study human perceptual-motor performance. The need for experimental control in addressing such multi-process models is emphasized. A number of studies of perceptual-motor performance deficits in patients with Parkinson's disease are reviewed. The methods by which the studies relate the behavioural data to the hypothesized underlying processes may be broadly grouped according to whether they use a subtractive logic or take a decompositional approach. These studies suggest that the basal ganglia play a role in the activation of pre-planned movement.

1984 Functions of the basal ganglia. Pitman, London (Ciba Foundation symposium 107) p 242–257

Pursuit tracking

A neurological examination often includes a request to the patient to move his (or her) hand repeatedly between his (or her) nose and the examiner's finger. By moving the target finger the examiner can form an impression of the patient's ability to initiate, execute and modify aimed movement. In clinical contexts the finger–nose test is used primarily to test for cerebellar signs. However, the task of pursuit tracking, which may be considered as a laboratory analogue of the finger–nose test, has been used extensively in psychological studies of basal ganglia function. In a typical pursuit-tracking experiment the subject faces a visual display and holds a control lever that moves a cursor around the display (Fig. 1). The subject's task is to keep the cursor superimposed on a continuously shifting target which can be made to move with different degrees of predictability.

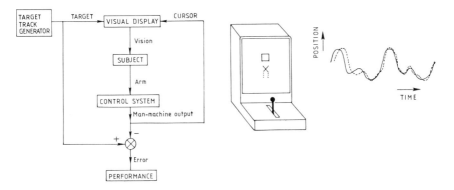

FIG. 1. Pursuit tracking.

Target predictability is a function of the range of frequencies present in the target track. If the high-frequency components in the target track are reduced, target predictability increases and performance improves. This is illustrated in Fig. 2 in which the performance measure is based on the area between the target track and the track of the subject-controlled cursor taken over trials of one minute duration.

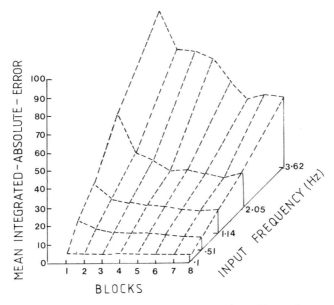

FIG. 2. The effects of target frequency and practice on pursuit tracking performance. (Redrawn from Pew 1974a.)

A special case of continuous tracking is step-tracking in which the target remains stationary for varying intervals of time before moving abruptly to a new position. This is similar to a choice reaction-time task with a variable interval between the previous response and the imperative signal that indicates which button to press. However, accuracy of movement is usually less critical in the reaction-time task. A performance measure in the continuous tracking case that parallels reaction time is phase lag. It is based on the time between movements of the target and the corresponding movements of the subject. A separate measure of performance is the amplitude ratio or gain. This refers to the size of the subject's movements relative to those of the target.

An information processing model of tracking (and reaction time)

Many papers have been published on models for describing human tracking performance; a readable introduction may be found in Pew (1974a). We now outline an account of movement control in tracking, emphasizing the psychological aspects or, in other words, the information-processing requirements of the task (Table 1). As our starting point we note that attention must

TABLE 1 Components for a functional model of pursuit tracking

(1) Attend display	
(2) Identify target-cursor error	(6) Identify target track regularities
(3) Select action	(7) Assemble sequence of actions
(4) Programme muscle commands	(8) Trigger successive actions
(5) Execute movement	

be directed to the visual input. With attention appropriately directed, the next stage is the detection of a discrepancy between the positions of the target and cursor. Given the inherent noise of the visual system, the subject must set up criteria for deciding whether a perceived discrepancy is 'real' and requires action. With information provided by the perceptual system about the direction of error, a response may be selected from a set of alternatives stored in memory. At this stage the response is probably represented in a symbolic form that allows a variety of different patterns of movement to be used to achieve the goal of reducing the perceived error. After this stage we assume that commands that will control specific muscles are programmed on the basis of specifications in motor memory. When the movement is completed, the newly available information about the discrepancy between the target and the subject's own position may call for further action. As psychologists professing an interest in voluntary movement control, our concern is with a cycle of perception and action taking place within a closed feedback loop.

With familiarity, subjects often begin to anticipate when signals will occur in reaction-time experiments and they reduce their so-called reaction time to such an extent that they may even respond in advance of the signal. In continuous tracking, if the pattern of target movements repeats in a simple manner, subjects begin to anticipate the target movements and reduce the phase lag, that is, the time between their position changes and those of the target. Reductions in phase lag probably underlie the practice effects evident in Fig. 2. Such improvement may take place without the subject being aware that the target track is being repeated (Pew 1974b). Thus to processes (1)–(5) in Table 1 we could add three more, as shown (6–8). The first of these (6), which could be associated with perceptual analysis of the input, identifies regularities in target movement. The second additional process (7), linked to output, is concerned with the sequencing of several separate movements. Where a sequence of several movements must be prepared, it is possible that the later elements are set up at the same time as the earlier elements are executed. The third process (8) provides a means of triggering movement elements at controlled times in order to space the elements and keep them in phase with movements of the target. These additions to the basic model allow anticipation, and also offer the possibility of opening the closed loop between action and perception. The benefit of opening the loop is that the subject can divert attention elsewhere for significant periods of time. Monitoring the progress of tracking may then take place intermittently (Craik 1947) so that, as skill develops, the balance changes from closed- to open-loop control.

Even with these revisions we do not claim that our model approaches completeness. However, it should be clear that, from an information-processing viewpoint, movement control entails more than the setting up of commands to the muscles. An important point in the present context is that a decline in tracking performance should not necessarily be attributed to a deficit in any one process. For example, an increased phase lag in tracking might arise from a deficiency in perception of the target track, from a lack of advance preparation, from a failure in timing, from slowness in the execution of movement, or some combination of these possibilities. We now review a number of behavioural studies of the control of voluntary movement in patients with Parkinson's disease, a degenerative disease of the basal ganglia associated with depletion of the dopaminergic neurons of the substantia nigra pars compacta in the basal ganglia.

In attempting to pinpoint which process or processes contribute to impaired performance it is helpful to distinguish between two methodological approaches. One is based on subtractive logic, the other employs analytic methods to decompose movement and reveal the contribution of separate underlying processes.

Subtractive logic approach

The approach based on subtractive logic has a history that dates back to Donders (1868) and extends beyond issues in the control of movement (for example, see Sanders 1980). In the present context the approach may be characterized as follows. Performance is contrasted on a pair of tasks, say T1 and T2, in which it is assumed that T1 draws on a set of distinct processes (P1, P2) while T2 draws on a superset (P1, P2, P3). Any performance differences, perhaps in the probability of being correct or in the reaction time, between T1 and T2 are attributed to the process P3. In this way it is argued that process P3 can be studied in isolation. In neuropsychology, the performance differences between tasks T1 and T2 are compared in two groups of people, G1 and G2. Suppose G1 consists of patients with a lesion in brain structure S and G2 consists of normal control subjects. If there are performance differences between G1 and G2 on the two tasks, the implication is that process P3 is associated with structure S. If, further, it can be demonstrated that no other lesion affects P3, it may be assumed that process P3 is localized at S.

An example of the subtractive logic approach is provided by Flowers' study (1978) in which subjects tracked a dot moving in the horizontal dimension of a visual display. Various waveforms were used for the target track, including predictable sinusoidal waveforms and unpredictable noise waveforms. Flowers reported that the tracking performance of normal control subjects was better with sinusoidal waveforms than with noise waveforms of the same average frequency. In contrast, parkinsonian patients with mild to moderate bradykinesia (slowing of voluntary movements) did no better with the sinusoids than with the noise (Fig. 3). A conclusion that might be drawn from this is that parkinsonian patients are deficient in using predictive, open-loop control of movement. This has been suggested, for example, by Marsden (1982).

One possible source of such a deficit might be difficulty in selecting individual responses when preparing sequences of movements. Evarts et al (1981) contrasted simple and choice reaction-time performance with the aim of evaluating the process of response selection in parkinsonian patients. In simple reaction-time tasks the subject is informed in advance what movement is required and an imperative signal indicates when the movement should be made. In this paradigm the subject can ready the required movement in advance and simply has to trigger it on receipt of the imperative signal. In choice reaction-time tasks the subject is not told which movement to prepare until the imperative signal to move is presented. Thus choice reaction time might be expected to exceed simple reaction time by the time taken to select the required response. Evarts et al (1981) found a difference between choice and simple reaction time in parkinsonian patients that was no greater than differences reported elsewhere for normal subjects. Apparently parkinsonian patients have no specific difficulty in response selection; indeed, when compared to the values for normal subjects, the findings of Evarts et al indicate that simple reaction time in parkinsonian patients is more slowed than choice reaction time.

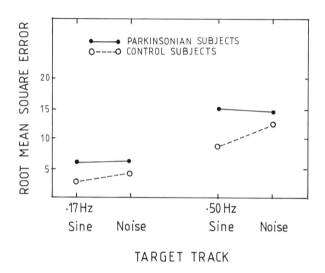

FIG. 3. The effect of target predictability on the manual tracking performance of parkinsonian patients compared with normal subjects. (Redrawn from Flowers 1978.)

In a study that included a direct comparison of parkinsonian patients with normal controls, Bloxham et al (1984) replicated the observation of Evarts et al that simple reaction time is differentially slowed in parkinsonian patients. Their results also showed that on those simple reaction-time trials with a longer interval between the warning signal that pre-cued the subject on which hand to move and the imperative signal, reaction time is reduced in both groups, but more in the normal controls (Fig. 4). This suggests that parkinsonian subjects prepare a response in advance on simple reaction-time trials but are slower in achieving a given level of preparation, with the result that, at any given time after the warning signal, the movement is less easily activated in response to the imperative signal.

Given these reaction-time findings, it is reasonable to question whether the tracking deficits observed by Flowers (1978) might stem from a primary failure in activating movement. The previously stated conclusion that prediction was defective was based on a coarse measure of the discrepancy between target and cursor tracks. However, it can be argued that reduction in phase lag for the sinusoidal target track relative to the noise track with the same average frequency is a more appropriate indication of subjects' use of predictability. It is thus significant that Flowers' results show that phase lag reductions in the parkinsonian group matched those in the control group. In

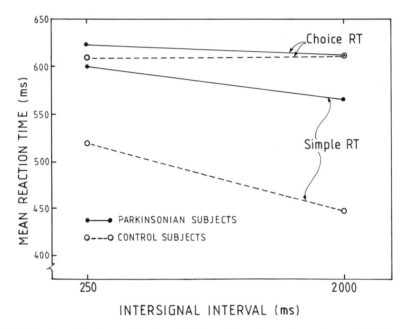

FIG. 4. Simple and choice reaction time (RT) in parkinsonian and normal control subjects as a function of the interval between the warning signal, which pre-cues the subject about which hand to move on simple reaction-time trials, and the imperative signal to move a particular hand. (Redrawn from Bloxham et al 1984.)

addition to their study of reaction time, Bloxham et al (1984) compared the tracking performance of parkinsonian patients with that of control subjects who were matched in terms of their noise-tracking performance. In going from noise to sinusoidal target movement Bloxham et al found the same degree of phase lag reduction in both groups. Lastly, in a study of smooth pursuit eye movements Melvill Jones & DeJong (1976) reported that the phase lag in parkinsonian patients was the same as that seen in normal control subjects. However, the performance of the parkinsonian patients was marked by reduced levels of gain in both saccadic and true smooth pursuit components of eye movement. What we should perhaps conclude from the tracking and reaction-time studies taken together is that parkinsonian patients have difficulty in activating movement either promptly or to a sufficient degree.

The decomposition approach

In the approach to behavioural data that we term decomposition, it is assumed that one or more of the processes underlying performance may be separately characterized on the basis of a set of distinct measures taken from each performance of a task. One example is that of separating reaction time (between imperative signal and movement onset) from movement duration (between movement onset and termination). In tasks requiring normal subjects to make accurate aiming movements, distance has effects on movement time rather than on reaction time. This has been taken to demonstrate that factors involved in movement execution are more important than those involved in movement preparation in determining the relation between speed and accuracy (Fitts & Peterson 1964).

Electromyography shows that the activity of biceps and triceps associated with rapid voluntary elbow flexion is in the form of a characteristic triphasic pattern. The pattern, which consists of a burst of activity first in biceps, then in triceps with a silent period in biceps, and finally in biceps again, has been observed in a patient with pansensory neuropathy (Hallett et al 1975a) indicating the pattern is centrally programmed independent of feedback. Patients with cerebellar deficits show abnormalities in the timing of the initial biceps activity and/or the activity of the triceps (Hallett et al 1975b). In parkinsonian patients, the timing of the first two phases of activity during fast flexion is within the normal range (Hallett et al 1977). However, Hallett & Khoshbin (1980) noted that when bradykinetic patients attempt rapid movements to a target there are many repeated cycles of alternating biceps and triceps activity. They suggested that this compensates for an overall downward scaling of activation in the initial centrally programmed triphasic pattern. Hallett and Khoshbin viewed the basal ganglia as serving an activational role for movements whereas timing and muscle selection functions are served by the cerebellum

The studies we have considered so far, in discussing possible deficits in the activation of a response prepared in advance, have all entailed the triggering of movement by, or in relation to, an external signal. It might therefore be asked whether activation deficits arise in relating the input signal to the stored movement rather than in the activation of movement itself. A study of controlled rate, repetitive finger tapping in a 44-year-old female hemiparkinsonian patient (Wing et al 1984) suggests this is not the case. The task required the patient to tap at regular intervals, using the index finger of either the more affected, dominant right hand or of the less affected left hand*. Two critical measures of performance on each trial, consisting of some 30 responses, were the variance and the lag one autocovariance of the interresponse intervals. (The lag one autocovariance is a measure of the statistical dependence between successive intervals; normalized by the variance it is the correlation between one interval and the next taken through the whole sequence). A simple model for performance of this task (Wing & Kristofferson 1973) assumes that departures from periodic responding arise from imprecision in a hypothetical timekeeper and from fluctuations in the time taken to execute responses triggered by the timekeeper (Fig. 5). Provided that the timekeeper intervals and response delays are independent it can be shown that the lag one autocovariance of the observed interresponse intervals is a measure of variability in motor delays, while the interresponse interval variance reflects both variability in motor delay and variance in a central timekeeper (for example, see Wing 1980).

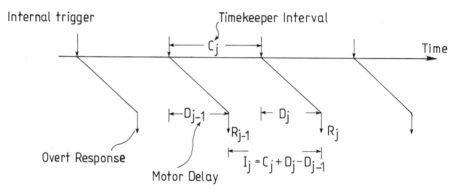

FIG. 5. Two-process model for repetitive tapping. The interval I_j between two successive overt responses, R_{j-1} and R_j, is given by the algebraic sum of the underlying timekeeper interval C_j, and the difference between the motor delay D_j associated with R_j and the delay D_{j-1} associated with R_{j-1}.

---

*This study involved the use of within-subject comparisons; the subject served as her own control. Within-subject experimental designs can increase the sensitivity of performance measures to differences resulting from a lesion of the central nervous system. The within-subject design is likely to minimize the number (and hence complexity) of differences between experimental and control results. Between-subject comparisons can suffer from the problem that differences between individuals in the many processes underlying performance may conceal small additional differences due to a lesion. In the exploration of parkinsonian behavioural deficits there are two broad possibilities: (a) evaluation of left/right assymetries in performance: such comparisons have the advantage that control trials can be nested to reduce the possibility that changes in strategy underlie differences in performance; (b) evaluation of changes in performance either as a result of drug therapy or, with longer time span, as a consequence of disease progression.

The results from two finger-tapping sessions separated by some seven months are summarized in Fig. 6. The estimates of standard deviation of the motor delay had the same value for the left and right hands in both sets of sessions. However, estimates of the standard deviation of the intervals of the

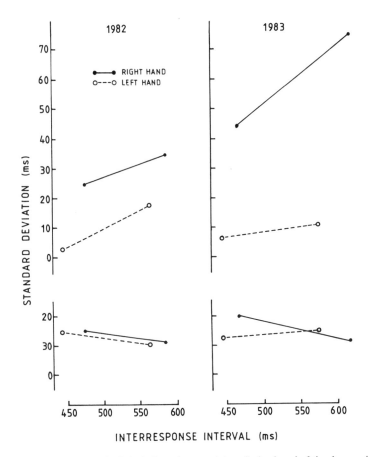

FIG. 6. Estimates of the standard deviation of motor delays (below) and of timekeeper intervals (above) for a hemiparkinsonian patient (data from Wing et al 1984).

timekeeper that triggers the responses show that the intervals associated with the left hand are much less variable than those for the right hand. Moreover, with progression of the disease over the seven months we see a deterioration affecting timekeeping of the right hand but not the left. This may be taken as a demonstration of a deficit in the activation of a preplanned movement when the trigger is provided internally.

A more complex task than finger tapping that requires sequences of internally triggered movements is handwriting. Small handwriting, or micrographia, can be associated with Parkinson's disease. In micrographia there is not only an overall reduction in amplitude relative to normal controls but also a progressive component, so that the written trace gets smaller. This occurs despite attempts to compensate for loss of force by increases in movement duration (Margolin & Wing 1983). The progressive aspect of micrographia indicates that a sequence of movements causes more difficulty than a single movement. However, recent research with bradykinetic parkinsonian patients at the Cognitive Neuropsychology Laboratory, Good Samaritan Hospital, Portland, Oregon has failed to show any differential lengthening of reaction time with longer movement sequences.

Conclusions

The studies of perceptual-motor control in parkinsonian patients that we have reviewed do not provide clear evidence for the idea that sequencing different movements is associated with any particular difficulty. Neither the prediction nor response-selection aspects of sequencing appear abnormal. Rather, the results suggest that these patients have difficulty in preparing and triggering voluntary movements. Thus, we offer the tentative conclusion that the basal ganglia play a role in the activation of previously prepared movement.

In concluding this review two serious limitations must be mentioned. The first is that the studies of voluntary movement control have been restricted to patients with Parkinson's disease. In this disease the lesions affect only a limited portion of the basal ganglia. Moreover, in many cases of the disease, projections from the basal ganglia to frontal cortex may also be disturbed, complicating the functional deficits. It is clearly desirable to extend studies of voluntary movement control to groups of patients with different lesions of the basal ganglia such as those associated with Huntington's chorea or hemiballismus. The other limitation to note is the failure of experimental psychologists to contrast performance on the above tasks of patients with lesions in brain structures other than in the basal ganglia. Without this we cannot hope to identify the particular contribution of the basal ganglia.

Despite these limitations, we feel that attempts to understand disturbances of perceptual-motor behaviour associated with Parkinson's disease have encouraged us, as experimental psychologists, to think about the functional components of the system in ways we might not have done if we had restricted consideration to normal performance. And, perhaps we can console ourselves with the thought that a good characterization of behaviour is a necessary first step in answering questions about what elements of behaviour are subserved by any given brain structure, such as the basal ganglia.

REFERENCES

Bloxham CA, Mindel TA, Frith CD 1984 Initiation and execution of predictable and unpredictable movements in Parkinson's disease. Brain, in press

Craik KJW 1947 Theory of the human operator in control systems: II. Man as an element in a control system. Br J Psychol 38:142-148

Donders FC 1868 Die Schnelligkeit psychischer Prozesse. Arch Anat Physiol (Leipz) 657-681

Evarts EV, Teravainen H, Calne DB 1981 Reaction time in Parkinson's disease. Brain 104:167-186

Fitts PM, Peterson JR 1964 Information capacity of discrete motor responses. J Exp Psychol 67:103-112

Flowers KA 1978 Some frequency response characteristics of parkinsonism on pursuit tracking. Brain 101:19-34

Hallett M, Khoshbin S 1980 A physiological mechanism of bradykinesia. Brain 103:301-314

Hallett M, Shahani BT, Young RR 1975a EMG analysis of stereotyped voluntary movements in man. J Neurol Neurosurg Psychiat 38:1154-1162

Hallett M, Shahani BT, Young RR 1975b EMG analysis of patients with cerebellar deficits. J Neurol Neurosurg Psychiatry 38:1163-1169

Hallett M, Shahani BT, Young RR 1977 Analysis of stereotyped voluntary movements at the elbow in patients with Parkinson's disease. J Neurol Neurosurg Psychiatry 40:1129-1135

Margolin DI, Wing AM 1983 Agraphia and micrographia: Clinical manifestations of motor programming and performance disorders. Acta Psychol 54:263-283

Marsden CD 1982 The mysterious function of the basal ganglia: the Robert Wartenburg lecture. Neurology 32:514-539

Melvill Jones G, DeJong D 1976 Visual tracking of sinusoidal target movement in Parkinson's disease. Defence Research Board, Ottawa, Canada, Report No DR225

Pew RW 1974a Human perceptual-motor performance. In: Kantowitz BH (ed) Human information processing: tutorials in performance and cognition. Erlbaum, Hillsdale, NJ, p 1-39

Pew RW 1974b Levels of analysis in motor control. Brain Res 71:393-400

Sanders AF 1980 Stage analysis of reaction processes. In: Stelmach GE, Requin J (eds) Tutorials in motor behavior, North Holland, Amsterdam, p 331-354

Wing AM 1980 The long and short of timing in response sequences. In: Stelmach GE, Requin J (eds) Tutorials in motor behavior. North Holland, Amsterdam, p 469-486

Wing AM, Kristofferson AB 1973 Response delays and the timing of discrete motor responses. Percept Psychophys 14:5-12

Wing AM, Keele S, Margolin DI 1984 Motor disorder and the timing of repetitive movement. Ann NY Acad Sci, in press

DISCUSSION

Porter: We should define what we mean by sequences of movements. Repetitive finger tapping seems to be quite different from the sequence of movements needed when somebody who is sitting down starts to stand up, especially if at the same time that person has to reach out to shake hands.

Evarts: P.J. Cordo and L.M. Nashner (personal communication) have done an experiment very similar to the one that Bob Porter has referred to. In their experiment subjects were asked to grasp a handle and pull it towards themselves. EMG recordings provided information on when the various sequences of muscle activities occurred, and though there was no defect in the time of initial onset of the EMG responses necessary to maintain the stability of the centre of gravity, there was an inappropriate selection of muscle groups. For the centre of gravity to be shifted properly as one pulls on an object, there is a certain *sequence* of activities in the various muscles of the legs and pelvic girdle. In parkinsonian patients there were abnormalities both of selection and of sequencing of muscle activities.

Marsden: I take issue with your statement that there is no defect of sequencing, however that is defined, Dr Wing. If you define it in terms of repetitive movements, one of the best ways of demonstrating the poverty of movement in Parkinson's disease is to look at repetitive movements which fade and disappear. We all use those movements every day to assess parkinsonian motor deficit, and they involve simple sequencing. Complex sequencing consists of putting together different sorts of movements and, again, parkinsonian patients show profound deficits. The tracking experiments illustrate this point.

The issue of whether one is using a predictive mode of motor behaviour depends on the time at which motor command signals are issued. If those signals are issued in advance of the target, i.e. if the phase lag from the target to response is reduced to zero, that is a predictive mode of motor behaviour in my terminology. It now seems from a number of studies that parkinsonian patients are able to adopt a predictive mode of motor behaviour. But we are all also agreed that their subsequent performance using that mode of motor behaviour is not as good as the performance of normal controls. The patients are unable to execute a sequence of movements and although they manage to move in advance of the target, they cannot use that phase advance to improve on performance. That is a defect of sequencing.

The question becomes: why is there such a defect? Does the defect lie in choosing the wrong muscles, in other words choosing the wrong motor programme, or in being unable to switch from one programme to another? The evidence suggests that the structure of an individual motor programme is, in fact, preserved in Parkinson's disease. The patients contract the right muscles and move in the right direction. The patterns of electromyographic firing and silence are as one would expect. The content of the programme seems to be intact. What is deficient is the first burst of the agonist. They can't get the right size of burst.

Wing: I agree that it would be wrong to say that parkinsonian patients have no difficulty with sequencing movements—clinical observation will often reveal a problem in producing rapid repetitive movement. But the question is whether their difficulty with sequences exceeds what one would expect, given that there are deficiencies in activating even a single preplanned movement?

What is their primary deficit? What is the essence of the motor deficit in bradykinesia? Does there have to be a sequence of movements before one sees a deficit or is it observable in a single activation? Hallet & Khoshbin's results (1980) indicate the latter is the case—even the first agonist burst is deficient. That does not mean that the later bursts are normal. They are also reduced in amplitude. This presumably underlies micrographia, in which there is reduced amplitude of handwriting movements. It must also contribute to the drop in gain in tracking tasks. If none of the components in a sequence are activated normally I would suggest that they may all be related to the first element of the sequence.

Marsden: I agree, but I don't think that is sufficient to explain the whole deficit of Parkinson's disease. One could put the two suggestions together in a single framework. That is, the parkinsonian patient, whether undertaking a simple single motor act requiring one motor programme or a motor plan requiring the sequence of motor programmes to be delivered to the appropriate point, is unable to respond to the right cue to produce either the single motor act or to run the sequence of motor acts.

DeLong: We are again talking about Parkinson's disease and also about the basal ganglia. We find supporting evidence for the importance of the basal ganglia in the scaling of the amplitude of movement in both cases. Parkinsonian patients clearly have great difficulty in moving quickly and that disturbance seems to result from a failure to scale the first agonist burst on the EMG. In virtually all the studies in primates that have looked carefully at the execution of movement and other movement parameters, the major effect of lesions of any kind has been on the execution of the movement. There is slowing and failure to energize the muscles, without any particular disruption of the sequence of activation of the muscles. This is very different from cerebellar lesions in similar experimental models. There is also a clear correlation between neural discharge in relation to the amplitude of movement in neurons of basal ganglia. I think everyone would agree that the basal ganglia play a role in the execution of simple movements. The controversial issue is the issue of sequencing of movement or more complex aspects of movement and whether basal ganglia do or do not play a role in the triggering or sequencing of those acts. I don't know of any experiments in animals that would help us to answer that for the basal ganglia.

Wing: Do parkinsonian patients find it more difficult to initiate a sequence of movements compared with a single movement? Does the initiation time get longer than in normal people? In work carried out in Portland, Oregon we have not found parkinsonian patients disproportionately slow in initiating movement sequences with more component responses (R.D. Rafal et al, unpublished).

Marsden: Another example of sequencing difficulties is seen in the Luria tests used to demonstrate frontal lobe problems. In these the subjects are asked to produce repetitive and different hand postures. This switching from one motor posture to another is grossly disturbed in Parkinson's disease.

Rizzolatti: Dr Wing, in the simple reaction-time experiments you cited (Bloxham et al 1984) the interval between the warning signal and the imperative signal was fixed at 250 ms. With a fixed interval of this kind, normal subjects can predict the occurrence of the second signal and respond on the basis of the prediction rather than on the actual stimulus presentation. Methodologically the experiment seems to me rather weak.

Wing: Those trials were presented in random order; on any trial the warning interval could be 250 or 2000 ms.

Rizzolatti: But the subjects should know from the warning signal when the imperative signal will appear. I am surprised that parkinsonian patients were unable to do so. It seems to be essentially an error in prediction rather than in responding to a signal.

Wing: When the warning interval was increased one might have expected time estimation would be harder for parkinsonian patients. Nonetheless, Bloxham and co-workers found that the reaction time of the patients improved with a longer warning interval.

Marsden: Another methodological problem is how the response is recorded. If the response is recorded in terms of a motor act and the person has to press a button, a patient may take an extra length of time to produce the necessary force to press the button. This would give a false impression of the delayed reaction time. So again one has problems in interpreting reaction times without knowing the exact nature of the response.

Evarts: Could you say more about the anticipatory, preparatory processes in movement?

Wing: Dr Deniau and I were discussing whether anyone has specifically explored spinal preparatory processes in animals with basal ganglia lesions.

Deniau: Various authors such as York (1972) have stimulated the basal ganglia to study the influence they exert on spinal reflexes. However most of these experiments are difficult to interpret since electrical stimulation can activate afferents of these structures and produce a spinal influence through axon reflexes.

Marsden: Some time ago we looked at what we call anticipatory postural responses. If a normal subject is standing up and one tugs their arm, they obviously contract various muscles to try and prevent themselves from falling over. On the old Sherringtonian concept, the calf muscles, for example, would contract as a result of the pull because of a local stretch reflex due to ankle movement. However, instead the person sets up a whole series of distant muscle contractions which come before the movement of the body part, so protecting against the fall. Hence the use of the term anticipatory postural reactions. These are defective in Parkinson's disease, though not in all patients who fall over spontaneously.

REFERENCES

Bloxham CA, Mindel TA, Frith CD 1984 Initiation and execution of predictable and unpredictable movements in Parkinson's disease. Brain, in press

Hallet M, Khoshbin S 1980 A physiological mechanism of bradykinesia. Brain 103:301-314

York DH 1972 Potentiation of lumbo-sacral monosynaptic reflexes by the substantia nigra. Exp Neurol 36:437-448

Final general discussion

Local regulation of transmitter release in the basal ganglia

Evarts: There are a number of major questions we could try to deal with in this final discussion. The first concerns the pharmacological, anatomical and behavioural aspects of local regulation.

Smith: It has become clear that there are drug receptors or transmitter receptors on nerve endings all over the body, including those in the brain, that are sensitive to drugs and transmitters and that can modulate the release of transmitters (Vizi 1979, Starke 1981). Jacques Glowinski was one of the first to describe these in the brain (Besson et al 1969). The general or simplified concept of regulation at the nerve ending and other presynaptic sites states that some actions of the transmitter released from the nerve not only trigger electrical events postsynaptically but also trigger the release of substances which can diffuse back and influence the nerve ending itself (see Fig. 1). The best-characterized of these substances are the prostaglandins. There is fairly good evidence that in peripheral systems, especially the sympathetic system, prostaglandins formed from postsynaptic sites can inhibit the release of transmitter (Hedqvist 1977).

Then there are a lot of receptors on the nerve ending or in the terminal network which are sensitive to drugs. Maybe a dozen drugs with different receptor properties influence the release of noradrenaline from peripheral sympathetic nerves (Starke 1981). This shows that nerve terminal networks separated from their cell bodies can be subject to extrinsic local chemical control. It doesn't prove whether this has any physiological significance, because one has to get the chemicals there in the first place. Possible sites of origin of endogenous chemicals are shown in Fig. 1, which could have functional significance in the heart.

Erich Muscholl and his group have done some elegant studies with the isolated perfused heart of the rabbit. They stimulate the sympathetic nerve and measure the release of noradrenaline into the perfusion fluid. If they simultaneously stimulate the vagus they can decrease the amount of noradrenaline released. This is entirely consistent with their pharmacological observations showing that if cardiac sympathetic nerves are perfused with acetylcholine this reduces the release of noradrenaline. This is a muscarinic effect, blocked by atropine. The inhibitory effect of stimulating the vagus on noradrenaline release is also blocked by atropine (Muscholl 1979). Muscholl and co-workers have now done the corollary experiment of stimulating the vagus and measuring acetylcholine release. They showed that perfusing the heart with noradrenaline reduces the release of acetylcholine and that stimulating the sympathetic at the same time as the vagus also inhibits the release of acetylcholine (Muscholl 1982). This peripheral organ provides an example of a possible physiological function for some of the many chemically sensitive sites on the nerve ending.

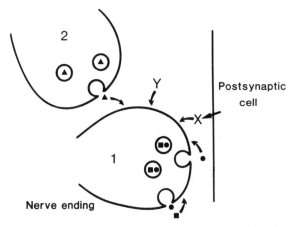

FIG. 1 (*Smith*). Possible sites of origin of chemicals that could modulate the release of a transmitter (solid circles) from nerve terminal 1. The transmitter itself can influence its own release; other chemicals that could influence the release are: a substance stored and released from the same nerve terminal (solid square), a substance (solid triangle) released from the nearby nerve terminal 2, a substance (X) released from the postsynaptic cell, a substance (Y) originating from more distant sites, such as a hormone.

Evarts: Something analogous may well be happening in the basal ganglia.

Smith: There is indeed a wealth of pharmacological evidence of presynaptic receptors in the basal ganglia (reviewed by Chesselet 1984), and Jacques Glowinski and his colleagues have made many contributions in this area.

Glowinski: As in the periphery, the presynaptic regulation observed in the basal ganglia should be demonstrated by specially designed experiments showing its physiological relevance.

Evarts: What are the effects of muscarinic receptor blockers on dopamine turnover as measured by its metabolites?

Glowinski: The existence of presynaptic regulation by cholinergic agents was first shown in *in vitro* experiments in which we estimated the effects of cholinergic drugs on the release of [^{3}H]dopamine continuously synthesized from [^{3}H]tyrosine (Giorguieff et al 1977).

Evarts: What are the effects of blocking muscarinic receptors in brain slices?

Glowinski: A muscarinic agonist increases the release of newly synthesized [^{3}H]dopamine; this effect is blocked by atropine.

Evarts: So cholinergic agents reduce the release of dopamine?

Glowinski: As I just said, muscarinic as well as nicotinic agents stimulate the spontaneous release of [^{3}H]dopamine and these effects are blocked by the respective antagonists. Inhibition of the release produced by muscarinic agents has been shown by Westfall (1974), who measured the effects of these agents on the potassium-evoked release of [^{3}H]dopamine.

Hornykiewicz: Doesn't the physiological role of the cholinergic influence on dopamine release in the striatum follow from the fact that if you use anticholinergic agents you get a change in dopamine release? This shows that under normal conditions there is apparently a tonic control of dopamine release by cholinergic neurons.

Glowinski: When a drug is injected at the periphery we cannot say what its precise site of action is. Cholinergic agents could act directly on dopaminergic cells in the substantia nigra, as well as acting directly on dopaminergic nerve terminals.

Hornykiewicz: You can perfuse the preparation, using a cannula.

Glowinski: Yes, but demonstrating the effect of acetylcholine or oxotremorine on dopamine release does not allow one to conclude that cholinergic interneurons are involved in the presynaptic regulation of dopamine release. This is a pharmacological but not a physiological demonstration.

Evarts: There is certainly circumstantial evidence that would be consistent with a central mechanism paralleling the one that Dr Smith described, given the overlap of the muscarinic receptors, the dopamine islands and the known clinical effects of anticholinergic agents in conditions in which there is a deficiency of dopamine. Can microstructural information cast light on these mechanisms?

Somogyi: It is not known where under normal conditions dopamine acts in the striatum. On the structural side the very nice studies of T.F. Freund and A.D. Smith (unpublished results) could shed some light on the mechanisms.

Smith: With Tamas Freund and John Powell I have studied the tyrosine hydroxylase-immunoreactive fibres in the striatum, as several people have done before (Pickel et al 1981). In the electron microscope one can see that tyrosine hydroxylase is present in immunoreactive structures that look like synaptic boutons. If this reflects dopamine, it indicates that dopamine can be released from synaptic boutons, but we wanted to know what information the synaptic boutons relay. One way of answering that is to identify the postsynaptic target neuron, which cannot be done with simple electron microscopy. The electron microscope reveals the minute detail of the postsynaptic target, such as whether it is a dendrite or part of the cell body or a spine or whatever, but it doesn't tell us much about the nature of that neuron—for example, whether it is a projection neuron or a local circuit neuron. So we used a method that Peter Somogyi and Tamas Freund developed, of Golgi impregnation combined with immunocytochemistry (Freund & Somogyi 1983, Somogyi et al 1983). We could then identify the Golgi type of neuron in the striatum that receives tyrosine hydroxylase-immunoreactive boutons. We found that the medium-sized spiny neuron receives these boutons. In addition to Golgi labelling Tamas Freund also managed to label the same cells from the substantia nigra with retrograde horseradish peroxidase and count the distribution of the tyrosine hydroxylase-immunoreactive boutons on the different parts of these identified striatonigral neurons. Almost all were on the dendrites and, surprisingly, 60% were at a particular position on the dendritic spine, forming a symmetrical synapse on the neck of the spine. The same spines also received another type of unlabelled bouton on the head of the spine which formed a classical type 1 asymmetrical synaptic specialization (T. Freund et al, unpublished work).

Studying the morphological specialization is one way of seeing how a transmitter can act. It indicates that one of the major actions of dopamine on striatonigral neurons is that it is likely to influence excitatory input from the head of the spine before it enters the dendrite. This illustrates what I said earlier about the new type of information that can be obtained by combining the ultrastructural study of synapses (some would call this 'reductionist') with the 'integrationist' approach (p 181).

DiFiglia: How does the acetylcholine neuron fit into this organization as a possible target for dopamine inputs?

Smith: We don't know yet whether tyrosine hydroxylase boutons form contacts with the cholinergic neuron. Much pharmacology implies that it might. If we block dopamine transmission, the acetylcholine level in the striatum falls and it has been a pharmacological concept that dopamine normally inhibits acetylcholine release in the striatum (Lehmann & Langer 1983). Tamas Freund didn't see a single tyrosine hydroxylase synapse on any identifiable postsynaptic structure that would be considered another synaptic bouton. So if we are looking for a morphological substrate for local control of acetylcholine release, all we can say is that we can't see it in terms of axo-axonic synaptic structures.

Glowinski: Did you see a non-dopaminergic bouton with the dopaminergic fibres or nerve terminals?

Smith: No.

The subthalamic nucleus

Evarts: The second question we might discuss is the subthalamic nucleus and its role in regulating the output of the basal ganglia.

Deniau: During this meeting the relationships between basal ganglia and motor cortex have been discussed. Dr Nauta described projections arising from the motor cortex to the striatum and showed us that they terminate in a rather restricted area of this structure.

Nauta: What I emphasized was the main area. Perhaps 80% of the projection has a dorsolateral distribution, and this is really quite a large field. In more caudal sections something that looks like a lion's tail extends from the main area through the whole length of the striatum in a ventrocaudal direction. I may have misled you a bit about the total size of the projection.

Deniau: In addition to the striatum, the subthalamic nucleus constitutes an important area through which the motor cortex can influence the activity of basal ganglia. It is striking that the efferent connections of the subthalamic nucleus seem to parallel those of the striatum. Like striatal efferents, single subthalamic neurons send bifurcated axons to innervate the internal segment of the globus pallidus and the pars reticulata of the substantia nigra. With Dr Kitai we have shown (Kitai & Deniau 1981) that stimulation of the motor cortex in the rat induces potent excitatory responses in the efferent subthalamo-nigral neurons. Thus the subthalamic nucleus should be regarded as an important component of the basal ganglia by which the motor cortex can regulate the activity of the internal segment of the globus pallidus and of the pars reticulata of the substantia nigra.

The synaptic influence exerted by the subthalamic nucleus is still being investigated. Dr Kitai may have some comments on this question.

Kitai: The subthalamic nucleus projects to the globus pallidus and to the substantia nigra. When Drs Kita, Chang and I made intracellular recordings of antidromic responses from subthalamic neurons after pallidal or nigral stimulation, we found that the latencies of pallidal and nigral activated antidromic spikes were similar, at 1.2–1.5 ms. Yet we know from anatomical studies that the substantia nigra is closer to the subthalamus than the globus pallidus. So why are the antidromic latencies the same? When we examined the recorded neurons labelled with HRP we found that a single neuron had bifurcating axons and the axon projecting towards the globus pallidus was much thicker than the one projecting towards the substantia nigra. This would be the morphological basis for the subthalamo-pallidal fibres having a greater conduction velocity than the subthalamo-nigral fibres. It is interesting to note that subthalamic target neurons are simultaneously activated by this mechanism.

DiFiglia: The intrinsic organization of the globus pallidus should be given more attention. Interneurons in the globus pallidus have been identified. Also complex synaptic arrangements involving triadic synapses and dendrites which are both presynaptic and postsynaptic have been observed (DiFiglia et al 1982). It is interesting that the small terminals with asymmetrical synapses that Steve Kitai describes are similar to those involved in the complex synaptic arrangements we have seen.

Yoshida: In 1971 we found that the caudate-nigral and caudate-pallidal pathways were inhibitory. Later the pallido-thalamic and nigro-thalamic pathways were also found to be inhibitory (Uno & Yoshida 1975, Ueki et al 1977, Uno et al 1978, Ueki 1983). There is therefore a double inhibitory neuronal chain. This chain does not work physiologically unless the second inhibitory neuron has powerful excitatory afferents or is a pacemaker cell.

Masao Ito and I studied the cerebellum for some years and a double inhibitory neuronal chain exists in the cerebellar cortex. The basket (inhibitory) cell connects with the Purkinje cell (also inhibitory). In the cerebellum, however, granule cell fibres (excitatory) have connections with both basket cells and Purkinje cells. Impulses arising from the granule cell fibre go first to the Purkinje cell, producing EPSPs, and then the Purkinje cell receives IPSPs produced via the basket cell. This makes sense physiologically.

I have tried to find this kind of excitatory input to either pallidal or nigral neurons. We could get powerful excitation or inhibition of nigral neurons by stimulating the peripheral nerves of the contralateral forelimb of the cat. Since the responses did not disappear after we cut the part just rostral to the substantia nigra, the peripheral nerve-evoked responses of the substantia nigra neurons must be produced by the ascending spino-nigral pathway, though not direct but via a polysynaptic pathway. There is an excitatory substance P-mediating caudato-nigral pathway (Kanazawa & Yoshida 1980). The pedunculopontine area produces excitation in pallidal neurons (Gonya-Magee & Anderson 1983). But these excitatory systems are probably all 'slow' systems. I really want to know which pathway produces responses in nigral neurons from the peripheral nerve. Interneurons might exist in perinigral structures.

Evarts: Do you think the source of the excitatory input might be the subthalamic nucleus?

Yoshida: Dr Kitai might think so, but Y. Nakamura (personal communication) recently found that the synaptic terminals of the subthalamo-pallidal pathway contain pleomorphic vesicles.

Kitai: One cannot say it is pleomorphic, therefore it is inhibitory, or that it is round and therefore excitatory. We can definitely say that subthalamo-nigral terminals do not resemble striatal or cerebellar GABAergic terminals.

DeLong: Steve Kitai and I have talked about the timing considerations. The pathway from the cortex through the subthalamic nucleus to GPi is a very fast pathway, with a conduction time of about 5 ms. By contrast, the conduction time from the cortex to the striatum alone may be 20 ms or more, with additional time required to the pallidum. It is striking that the subthalamic nucleus becomes so large in primates. A good part of this structure appears to be involved in limb control, from the anatomical and physiological studies. The arm and leg representations are quite large and of course the release of involuntary movements after lesions is a direct testimony to the role of this structure in controlling limb movements. Subthalamic cells are strikingly related to movement. Although the numbers are not large enough for us to make good comparisons between putamen and striatum, these neurons show striking and early changes in relation to movement, and they show clear parametric relations to direction and amplitude (Georgopoulos et al 1983).

As discussed earlier, when we look at the output neurons in the pallidum in relation to limb movement, the first change we see is often a sustained and powerful increase in discharge, rather than inhibition, so perhaps the subthalamic nucleus is excitatory on the pallidum.

Kitai: Our membrane study indicates that subthalamic neurons can fire at a very high frequency. Our study with Jean-Michel Deniau demonstrated that there are very powerful cortical excitatory inputs to the subthalamus. A small cortical stimulation always leads to large EPSPs accompanied by a barrage of spikes.

Carpenter: Many more subthalamic fibres project to the pallidum than to the substantia nigra in the primate. Even though the subthalamic nucleus consists of a single population of cells, I think there is a zone that is particularly the receptive zone, and this zone tends to be the rostral and lateral third of the nucleus (Carpenter et al 1981). We have never seen pallidal terminals in other parts of the nucleus. There is a caudal medial part that is always devoid of terminals; this portion of the nucleus is the main source of impulses that project to the medial pallidal segment. There are significant differences between rat and monkey and probably also between cat and monkey. This is probably why dyskinesia has not been produced in any animal other than the monkey.

If the subthalamo-pallidal fibres are not GABAergic, how is activity that is essentially the same as that produced by lesions produced by the GABA antagonist, bicuculline? I interpret this as suggesting that the neurotransmitter, at least to the pallidum, might be GABA.

Kitai: You are probably right about the species difference. Rats have very topographically organized cortical inputs into the entire subthalamic nucleus. But one possibility is that when one manipulates subthalamus with bicuculline or a GABA-related substance one might be manipulating the GABAergic inputs and so possibly the manipulation is affecting presynaptic inputs rather than postsynaptic events.

Marsden: I am still not clear about the anatomical or physiological relationship of the subthalamic nucleus to the second system going through the basal ganglia: the ventral striatum and ventral pallidum.

Nauta: In the rat the ventral (or 'limbic') pallidum projects to a very medial and also rather caudal part of the subthalamic nucleus. I have not looked at this in the monkey.

Carpenter: In 1966 you described a projection from substantia nigra which I have not seen, Professor Nauta (Nauta & Mehler 1966). How do you feel about that now?

Nauta: If I remember correctly the case in point was that of a monkey with an electrolytic lesion in the substantia innominata. That lesion may well have involved the ventral pallidum and thereby have caused at least part of the fibre degeneration we saw in the medial part of the subthalamic nucleus. But we must remember that results obtained by the lesion or fibre-degeneration method we used then are a good deal more difficult to interpret than autoradiographic findings.

Marsden: The subthalamic nucleus gates the output of the motor striatopallidal system. Is there any evidence of the subthalamic nucleus gating the output of the ventral striopallidum?

Nauta: There is no evidence I know of.

Evarts: It has been pointed out to me that I made a mistake in my discussion of Dr Hornykiewicz's findings on dopamine depletion in the putamen, compared to other parts of the basal ganglia (p 219). My mistake was in speaking of parkinsonism as if it were primarily a putamenal disorder. I was making the point that one could not assume uniformity of dopamine loss throughout the entire striatum, but I went too far.

Hornykiewicz: It is important to correct the possibly wrong impression arising from that discussion. Dopamine is reduced in all striatal areas in Parkinson's disease, including the ventral striatum and the limbic parts of the basal forebrain. The degree of decrease is different from area to area, with the putamen showing the largest loss of dopamine. In discussions about the relative importance of dopamine changes in the putamen and caudate, and in relation to symptomatology, it may be useful to consider results of a study in which we tried to correlate the severity of different symptoms with changes in dopamine in the different structures in the striatum and the concentration of homovanillic acid in the globus pallidus (Bernheimer et al 1973). At that time our analytical methods were not sensitive enough to detect the decreased dopamine levels in the parkinsonian globus pallidus, but the external pallidal segment does contain dopamine, which probably stems from collaterals of the nigrostriatal fibres terminating in the globus pallidus. Our results indicate that the decrease in dopamine in the caudate nucleus correlated best with akinesia. The changes in putamenal dopamine correlated better with rigidity than with akinesia. In neither of these two regions did the dopamine decrease correlate with tremor. However, the homovanillic acid decrease in globus pallidus correlated best with tremor but not so well with rigidity and akinesia. Admittedly these are only approximate correlations, performed with relatively small numbers of cases and large interindividual variations. Nevertheless, what this study shows is that probably one should avoid attributing a whole set of symptoms to changes in dopamine exclusively in one basal ganglia region.

Graybiel: That is very interesting information about the external segment. Is there any evidence from electron microscopy for dopamine-containing or tyrosine hydroxylase-immunoreactive terminals in the pallidum itself?

Bolam: I have seen tyrosine hydroxylase-immunoreactive fibres in the globus pallidus but haven't looked in the electron microscope for terminals.

DeLong: In the monkey the globus pallidus is a major pathway for fibres from the compacta, so the measurements may perhaps be contaminated by these. Does akinesia include bradykinesia?

Hornykiewicz: Akinesia in the patients we studied at that time lay mainly in the difficulty in initiating movement, not so much in the time taken to walk a certain distance.

You are quite right that the nigrostriatal dopamine fibres run through the globus pallidus. Of course we were aware of the fact that homovanillic acid was also contained in the fibres. Still, it is surprising that we didn't get a correlation between tremor and any of these biochemical changes in the caudate and putamen.

Carpenter: I am extremely interested that you found a correlation between the lateral pallidal segment and tremor. This is the only site, other than the subthalamic nucleus, where I have seen dyskinesia. This occurred spontaneously in a monkey that developed herpes encephalitis and a choreaform activity, essentially the same as we produced with subthalamic lesions (Carpenter & Strominger 1966). This animal had necrotizing lesions in a lot of places but the most crucial one was in the lateral pallidal segment. It involved the medial parts of the putamen and the lateral pallidal segment. This happened at a time when we had animals with dyskinesia due to lesions in the subthalamic nucleus. These animals looked exactly the same.

Porter: An aspect of control that hasn't received our attention is the anatomical substrate for potential control systems. We have now seen a large number of diagrams of circuits that connect regions of the cerebral cortex. Not one of those diagrams has yet shown that the receiving area of cerebral cortex also contains large numbers of corticothalamic fibres which project back to exactly the same zone of thalamus that feeds into that area of cortex. When we are talking about the potential for operation of controls, we have to think not only at the level of receptors on membranes of terminals, but also in terms of the potential that exists within these neuronal circuits. When we are trying to put together the picture of connectivity we should bear in mind that the supplementary motor area, if that is the major receiving area for the output from VA–VL, may be able to operate on the thalamus by selecting or controlling the input to itself as a result of all the activities that have gone on within the basal ganglia.

REFERENCES

Bernheimer H, Birkmayer W, Hornykiewicz O, Jellinger K, Seitelberger F 1973 Brain dopamine and the syndromes of Parkinson and Huntington. J Neurol Sci 20:415-455

Besson MJ, Cheramy A, Feltz P, Glowinski J 1969 Release of newly synthesized dopamine from dopamine-containing terminals in the striatum of the rat. Proc Natl Acad Sci USA 62:741-748

Carpenter MB, Strominger NL 1966 Corticostriate encephalitis and paraballism in the monkey. Evidence concerning distinctive functions of the globus pallidus. Arch Neurol 14:241-254

Carpenter MB, Batton RR, Carleton SC, Keller JT 1981 Interconnections and organization of pallidal and subthalamic nucleus neurons in the monkey. J Comp Neurol 197:579-603

Chesselet M-F 1984 Presynaptic regulation of neurotransmitter release in the brain: facts and hypotheses. Neuroscience, in press

DiFiglia M, Pasik P, Pasik T 1982 A Golgi and ultrastructural study of the monkey globus pallidus. J Comp Neurol 212:53-75

Freund TF, Somogyi P 1983 The section Golgi impregnation procedure. 1. Description of the method and its combination with histochemistry after intracellular iontophoresis or retrograde transport of horseradish peroxidase. Neuroscience 9:463-474

Georgopoulos A, DeLong M, Crutcher M 1983 Relations between parameters of step-tracking movements and single cell discharge in the globus pallidus and subthalamic nucleus of the behaving monkey. J Neurosci 3:1586-1598

Giorguieff MF, Le Floc'h ML, Glowinski J, Besson MJ 1977 Involvement of cholinergic presynaptic receptors of nicotinic and muscarinic types in the control of the spontaneous release of DA from striatal dopaminergic terminals in the rat. J Pharmacol Exp Ther 200:535-544

Gonya-Magee T, Anderson ME 1983 An electophysiological characterization of projections from the pedunculopontine area to entopeduncular nucleus and globus pallidus in the cat. Exp Brain Res 49:269-279

Hedqvist P 1977 Basic mechanisms of prostaglandin action on autonomic neurotransmission. Annu Rev Pharmacol Toxicol 17:259-279

Kanazawa I, Yoshida M 1980 Electrophysiological evidence for the existence of excitatory fibers in the caudato-nigral pathway in the cat. Neurosci Lett 20:301-306

Kitai ST, Deniau JM 1981 Cortical inputs to the subthalamus: intracellular analysis. Brain Res 214:411-415

Lehmann J, Langer SZ 1983 The striatal cholinergic interneuron: synaptic target of dopaminergic terminals? Neuroscience 10:1105-1120

Muscholl E 1979 Presynaptic muscarine receptors and inhibition of release. In: Paton WDM (ed) The release of catecholamines from adrenergic neurons. Pergamon, Oxford, p 87-110

Muscholl E 1982 Cholinergic-adrenergic interactions at the presynaptic level as studied in the heart. In: Yoshida H et al (eds) Advances in pharmacology and therapeutics II, Pergamon, Oxford, vol 2:93-102

Nauta WJH, Mehler WR 1966 Projections of the lentiform nucleus in the monkey. Brain Res 1:3-42

Pickel VM, Beckley SC, Joh TH, Reis DJ 1981 Ultrastructural immunocytochemical localization of tyrosine hydroxylase in the neostriatum. Brain Res 225:373-385

Somogyi P, Freund TF, Wu J-Y, Smith AD 1983 The section-Golgi procedure. 2. Immunocytochemical demonstration of glutamate decarboxylase in Golgi-impregnated neurons and in their afferent synaptic boutons in the visual cortex of the cat. Neuroscience 9:475-490

Starke K 1981 Presynaptic receptors. Annu Rev Pharmacol Toxicol 21:7-30

Ueki A 1983 The mode of nigro-thalamic transmission investigated with intracellular recording in the cat. Exp Brain Res 49:116-124

Ueki A, Uno M, Anderson ME, Yoshida M 1977 Monosynaptic inhibition of thalamic neurons produced by stimulation of the substantia nigra. Experientia 33:1480-1481

Uno M, Yoshida M 1975 Monosynaptic inhibition of thalamic neurons produced by stimulation of the pallidal nucleus in cats. Brain Res 99:377-380

Uno M, Ozawa N, Yoshida M 1978 The mode of pallido-thalamic transmission investigated with intracellular recording from cat thalamus. Exp Brain Res 33:493-507

Vizi ES 1979 Presynaptic modulation of neurochemical transmission. Prog Neurobiol 12:181-290

Westfall TC 1974 The effect of cholinergic agents on the release of ^{3}H-dopamine from the rat striatal slices by nicotine, potassium and electrical stimulation. Fed Proc 33:524

Closing remarks

E.V. EVARTS

Laboratory of Neurophysiology, National Institute of Mental Health, Bethesda, Maryland 20205, USA

1984 Functions of the basal ganglia. Pitman, London (Ciba Foundation symposium 107) p 269-272

I would like to start these remarks by reiterating Walle Nauta's statement about the enormous importance of the basal ganglia as a link between the limbic system and overt behaviour. It is easy to lose sight of this linkage, in part because the behaviours we study in laboratory animals are too elementary, or, put another way, are not sufficiently refined and demanding to bring out deficits in higher brain functions. At present we don't have experimental tools for producing sufficiently selective lesions or impairments of those parts of the basal ganglia whose malfunction will be expressed as cognitive deficits rather than as disturbances of movement. In the future there may be developments in genetics that will provide us with a disorder of basal ganglia dopamine that impairs higher brain functions in a way complementary to the way that motor function is impaired in the weaver mutant that Ann Graybiel has discussed. This complementary disorder might provide us with animals that have selective dopamine losses in those parts of the basal ganglia with linkages to the limbic system rather than to the motor system.

Another approach to discovering more about the limbic components of the basal ganglia was described by David Marsden. He pointed out that new information will emerge with the application of positron emission tomography and the use of ligands that bind selectively to different components of the basal ganglia. We know, for example, that the dopaminergic neurons innervating the different parts of the striatum are biochemically different. There have been a number of demonstrations that the various peptides contained by dopamine neurons are not uniformly distributed in all the components of the dopaminergic systems (e.g. A9 and A10 are different in terms of their 'unconventional' transmitters).

Another future approach will involve the use of immunological techniques to produce discrete lesions in one or another set of the dopaminergic neurons.

This too will provide us with a selectivity that we do not yet have. Paul Bolam talked about identified striatal neurons and it is impressive to note that there are now several very well characterized types of basal ganglia neurons in a structure that not long ago we knew so little about. Marian DiFiglia is one of those who has made tremendous contributions to our knowledge of the neuronal types in the striatum, and it is in large part due to work of the sort that she and her colleagues have pioneered that we are in a period of such rapid progress. Work in this area is now leapfrogging ahead faster than work on the cerebral cortex. We know more about identified neurons, their transmitters and their connections in the basal ganglia than in the cortex: the findings that Steve Kitai, Peter Somogyi and Paul Bolam have described at this symposium have been truly spectacular.

At a neurophysiological–biophysical level, Dr Deniau presented us with clear evidence for the functional meaning of successive inhibitory connections: increased discharge of striatal neurons will lead to reduced activity in substantia nigra pars reticulata (SNr) (caused by GABA release from active striatal terminals), and this in turn will disinhibit the neurons on which SNr terminals synapse and thereby give rise to increased activity in the tectum and other targets of SNr GABAergic outputs. Dr Deniau showed that local injections of GABA into SNr will inhibit SNr directly and thereby disinhibit the targets of SNr. As Peter Somogyi pointed out, things are finally making sense: Deniau's results fit nicely with the SNr single-cell recordings of Hikosaka and Wurtz, who found that pauses of SNr discharge precede saccades; such SNr pauses would lead to increased activity in the superior colliculus by disinhibition.

Mahlon DeLong's single-unit recordings in moving monkeys showed there are clear relations between the discharge properties of globus pallidus neurons and the parameters of movement. He also showed that the dopaminergic neurons of the substantia nigra pars compacta discharge tonically but that, unlike striatal and pallidal neurons, compacta neurons lack phasic relations to movement. This was the start of a set of presentations leading to the notion that the impulse frequencies of dopaminergic neurons cannot be thought of as the sole determinants of dopamine release. First, there is DeLong's finding that the amount of activity in dopaminergic neurons does not seem to be correlated with phasic changes of behaviour. Secondly, Jacques Glowinski made it clear that there are factors that can cause increased dopamine release at one locus and decreased release at another, even though the same dopaminergic neurons innervate both these loci. This means that dopamine release cannot be due merely to changes in the impulse frequencies of the dopaminergic neurons.

Ann Graybiel provided us with information about the ontogeny of basal ganglia biochemical anatomy. In a coffee break I asked her to say something in the general discussion about the way in which she was going to proceed with determining how environmental factors might bring about basal ganglia plasticity in adult life. 'That is the $64 000 question' she replied, and added that if she knew what experiments to do along these lines, she would be doing them right now. One can't help but believe that the basal ganglia, with inputs from parts of the cerebral cortex that mediate cognitive and mnemonic processes, must be a region that can link acquired patterns of activity in the cerebral cortex with structures that ultimately give rise to motor behaviour. The basal ganglia may provide a substrate for making new connections based on experience throughout life. Our problem is to get an experimental approach that parallels the Hubel–Wiesel experiments on effects of monocular deprivation. Their approach took advantage of the fact that normal animals have regularly interdigitating ocular dominance columns of equal width. In the basal ganglia, there are distinct patches of muscarinic receptors and dopamine islands, but there is no particular geometry that one can look at and immediately classify as normal or abnormal so as to infer the effect of an experience such as monocular deprivation. Within the next few years someone may hit upon a dramatic demonstration in which the properties of certain sets of striosomes, dopamine islands and muscarinic receptors will be changed by experience. That will be very exciting when it happens, and Ann might be the one to do it.

Jacques Glowinski demonstrated that there must be local factors that regulate dopamine release independent of dopamine neuron impulses, but the existence of these local factors does not exclude a possible complementary role of variations in the impulse frequencies of dopaminergic neurons. We know that there are feedback pathways from the striatum to the substantia nigra pars compacta, and we also know that impulse frequencies of compacta neurons can be regulated from 0–10 Hz as a result of pharmacological and behavioural manipulations. These changes of impulse frequency must have a purpose, and the coexistence of local regulation must not obscure the importance of changes of impulse frequencies. In this connection Dr Hornykiewicz mentioned that, especially under conditions in which dopamine neurons have been lost, as in Parkinson's disease, there may be mechanisms for up-regulation of the function of those neurons that remain, and it might be in situations such as this that the impulse frequencies of dopaminergic neurons are especially important.

Dr Jenner talked about GABA and behaviour. The use of techniques such as he described is becoming increasingly important, not only in behaviour but also (as Dr Deniau described for localized injections of GABA) in anatomical studies, and in studies involving measurement of brain metabolism using the deoxyglucose technique.

I was especially fascinated by Sue Iversen's description of the association and dissociation of defects in behaviour after selective lesions of different parts of the basal ganglia. The basal ganglia are made up of many different regions, and one can get a variety of selective disorders of behaviour from lesions of different parts of the basal ganglia.

Dr Divac has made a very strong case for his view that cognitive processes are mediated by the basal ganglia, and Walle Nauta spoke most eloquently along these lines in relation to basal ganglia function and mood: what is a thought except a movement that is not connected to a motor neuron? From the standpoint of our future work, we will find it useful to accept Dr Divac's point of view on the cognitive functions of the basal ganglia. Our problem, however, is that in experiments on animals it is difficult to address this point satisfactorily. Human studies of the sort that Teuber (1966) pioneered may help us: he thought of the frontal lobes and the caudate nucleus as highly linked and described a variety of perceptual and cognitive processes that would depend on this.

How should one formulate the nature of the impairment in Parkinson's disease? Alan Wing, David Marsden and Donald Calne have all contributed to our understanding of this issue and it has been noted that the motor impairments themselves often make it difficult to assess, in a teleological sense, why the basal ganglia have evolved. The kind of work that Calne, Marsden and Wing are currently doing is thus essential to our understanding of the role of the basal ganglia in particular and of the human brain in general.

REFERENCE

Teuber HL 1966 Alterations of perception after brain injury. In: Eccles JC (ed) The brain and conscious experience. Springer, New York, p 182-216

Index of contributors

Entries in **bold** *type indicate papers; other entries refer to discussion contributions*

*Alexander, G. E. **64**
Arbuthnott, G. W. 100, 177, 216

*Besson, M. J. **150**
Bolam, J. P. **30,** 42, 45, 46, 177, 266

Calne, D. B. 43, 59, 62, 80, 97, 99, 100, 101, 110, 174, 180, 199, 211, 212, 214, 218, 221, 223, 238, 239
Carpenter, M. B. 24, 26, 61, 97, 104, 210, 222, 264, 266
*Chéramy, A. **150**
*Chevalier, G. **48**
*Crutcher, M. D. **64**

DeLong, M. R. 27, 61, 62, **64,** 78, 79, 80, 81, 97, 98, 99, 100, 101, 111, 112, 148, 162, 199, 213, 214, 216, 255, 263, 266
Deniau, J. M. **48,** 60, 61, 62, 256, 262
Di Chiara, G. 28, 45, 60, 146, 160, 161, 171, 212, 222
DiFiglia, M. 43, 44, 262
Divac, I. 105, 108, **201,** 210, 211, 212, 213, 215, 218, 223
*Domesick, V. B. **3**

Evarts, E. V. **1,** 26, 27, 43, 58, 61, 80, **83,** 96, 97, 98, 99, 100, 101, 107, 146, 177, 212, 213, 216, 217, 218, 219, 220, 222, 237, 239, 253, 256, 258, 259, 263, 265, **269**

*Georgopoulos, A. P. **64**
Glowinski, J. 147, **150,** 160, 161, 162, 163, 175, 178, 180, 195, 259
Graybiel, A. M. 24, 26, 43, 44, 60, 62, 97, 106, **114,** 146, 147, 148, 177, 179, 180, 210, 222, 266

Hornykiewicz, O. 110, 144, 162, 179, 198, 199, 217, 218, 219, 220, 241, 265, 266

Iversen, S. D. 175, **183,** 195, 196, 197, 198, 199, 216, 217, 220

Jenner, P. G. **164,** 174, 175

Kitai, S. T. 24, 25, 42, 44, 45, 80, 108, 175, 179, 262, 263, 264

Marsden, C. D. 45, 60, 61, 62, 79, 81, 98, 101, 108, 110, 111, 112, 161, **164,** 196, 210, 212, 213, 214, 218, 219, 220, 221, 222, 223, **225,** 238, 239, 254, 255, 256, 264, 265
*Miller, E. **242**
*Mitchell, S. J. **64**

Nauta, W. J. H. **3,** 23, 61, 108, 109, 111, 180, 240, 262, 264, 265

Porter, R. 25, 81, 99, 103, 104, 105, 106, 253, 266

*Reavill, C. **164**
*Richardson, R. T. **64**
Rizzolatti, G. 60, 79, 97, 196, 256
Rolls, E. T. 27, 78, 79, 96, 98, 107, 111, 197, 198, 213, 214, 220, 239

Smith, A. D. 78, 110, 111, 148, 162, 172, 175, 181, 217, 258, 259
Somogyi, P. 24, 42, 43, 44, 45, 59, 60, 111, 146, 147, 178

Wing, A. M. 79, 96, 99, 176, 212, 237, **242,** 254, 255, 256
*Wise, S. P. **83**

Yoshida, M. 23, 42, 62, 80, 96, 107, 112, 174, 263

*Non-participating co-author.
Indexes compiled by John Rivers

Subject index

Acetylcholine
  caudate-putamen complex, in 1
  dopamine functions modulated by 193
  noradrenaline release and 258, 259
  striatum, in 184, 185, 186, 188, 261
Acetylcholinesterase 36, 37, 38, 69, 186
Akinesia, 48, 81, 86, 87, 207, 211, 212, 227, 238, 265, 266
Alternation, delayed 211, 212
γ-Aminobutyric acid *See under GABA*
Amphetamine 153, 176
Amygdala 5, 6, 24, 109
Angiotensin II, striatum, in 186
Ansa lenticularis 13, 19
Aphasia 213
Apomorphine syndrome 172
Arcuate premotor area 85
Arm movements, putamen neurons and 65, 67, 69, 70, 71, 72, 73, 76, 79, 80
Arousal, limbic systems and 86
Athetosis 48
Attentional factors 89, 97, 98, 106, 107, 110, 189, 190, 191, 196, 244
Autapse 37

Ballism 48, 222
Basal ganglia
  afferent and efferent relationships 3–29
  anatomical definitions 183
  cortical projections to 67, 81, 88
  cortical relations, segregation of pathways in 75, 76
  critical motor disorder in 226, 227
  disease 97, 207, 223, 227
  dysfunction, motor disorders and 48, 85, 86
  efferent pathways 19, 28
  functional organization 64–82
  GABA-containing neurons in 165, 166
  genetic disorders 222
  input pathways 115
  lesions 166, 167, 213, 214
    human model for 218, 219, 221–224, 225–241
  limbic and non-limbic projections 85
  loop circuits 133, 134, 137, 139, 179
  motor connections 25

  motor or cognitive functions 211–215, 222, 226, 240, 241, 271, 272
  neural circuitry 1, 2, 3–29, 30–47, 48–63
  neural interactions, time scale 136, 137
  neurochemical organization 123–131
  neuronal damage, dopamine loss and 100
  neuronal discharge 64, 65–70, 71, 72–75
  neurons, non-movement-related 78, 79
  neuropeptides 123, 136–139
  neurotransmission in, chemical regulation 115, 123, 129
  organization 214, 215
  output pathways 84–86, 94, 100
  predictive motor action and 232
  sensory input to, cortical pathway 81
  structure and function, species differences 26, 27
  transmitter release in, local regulation 258–261
  See also under anatomical components
Behaviour
  chemical manipulation, striatum and 187, 188, 197
  cortical lesions and 187, 188, 197
  deficits, complex 217
  GABA manipulation of basal ganglia outflows 168, 169
  manipulation, drugs and 175, 176
  Parkinson's disease, in 218
Bicuculline 52, 62
Bradykinesia 79, 89, 227, 238, 239, 246, 249, 252, 255
Bradykinin, striatum, in 186

Catalepsy 164, 167, 168, 169
Catatonia 167
Caudate nucleus
  basal ganglia output pathway through 100
  cholinergic manipulation 188
  cortical inputs to 79, 89
  dopamine levels in 145, 179
  dopamine deficiency in Parkinson's disease 218, 219, 220, 221, 226, 265
  dopamine release in 150, 151, 152, 153, 154, 155, 157, 160, 161
  efferent pathway 23

274

SUBJECT INDEX

GABA release, muscimol and 156, 157
innervation, Parkinson's disease, in 110
neurons 44
    discharge 65, 80, 91
    firing rate 96, 107
    movement initiation and 72
    non-movement-related 78
    set-related activity 94
    neurotransmitter patterns in 145
Caudatonigral fibres 42, 263
Caudatonigral synapses 174
Caudatopallidal pathway 263
Caudatopallidal synapses 174
Caudate-putamen 1, 4, 8, 122, 184, 197
Cerebellar cortex, double inhibitory neuronal pathways in 263
Cerebellothalamic/nigrothalamic projections, interactions 56, 57, 62, 163
Cerebellothalamocortical pathways 117
Cerebellothalamocortical projections 83, 85
Cerebral blood flow, regional 87, 88
Cerebral cortex
    connectivity within 26
    Huntington's chorea, in 223
    neurons, bilateral activation 179
Cholecystokinin 126, 179, 186, 193
Choline acetyltransferase 33, 34, 37, 144, 145
Cholinergic interneurons 37, 44, 45
Cholinergic neurons 33, 34, 37
Chorea 48, 75, 136
    See also Huntington's chorea
Choreoathetotic disease 134
Circling behaviour 164, 167, 168, 169
Cognitive disorders 197, 199, 211, 214, 217
    basal ganglia neurons and 79
    neostriatum and 207
    prefrontal lesions and 212, 213
Cognitive factors, frontal eye field neurons and 89
Cognitive processes, eye movements and 83
Cognitive state 94
Colliculus, superior *See under Superior colliculus*
Corpus callosum 153, 155, 157, 158, 161
Corpus striatum *See under Striatum*
Cortico-basal ganglia loops 76, 103
Cortico-basal ganglia relations, segregated pathways in 75, 76, 115
Cortico-cortical connections 76, 103, 104, 112
Cortico-neostriatal connections 203–205
Corticospinal dysfunction 114
Corticostriatal connections 5, 39, 46, 69, 104, 109, 115, 178, 186, 187, 210, 262
    functional aspects 201, 215, 235, 236

    lesions, behaviour and 202, 203
    species commonality 202, 203
    striatal neurons, to 32, 33, 39
Corticostriatal glutamatergic projections 155
Corticothalamic connections 61, 103
Corticothalamic pathway 84–86
Cortico-ventral pallidal connections 16

Diazepam, dopamine release and 155
Dopamine
    acetylcholine release inhibited by 261
    agonists, Parkinson's disease, in 217
    antagonists, neuroleptics, as 1
    basal ganglia, in 144, 145, 164–176, 181
    caudate-putamen complex, in 1
    cortical projections, loss of, Parkinson's disease, in 101
    depletion 1, 100, 191, 216–221
    histofluorescence 69, 179
    islands 122, 130, 138, 146, 147, 177, 179, 180, 260
    release in basal ganglia
        asymmetrical regulation 151–155, 157, 180, 217
        cholinergic regulation 259, 260
        local control 177–182, 271
        muscarinic agents enhancing 260
        nerve activity and 179
        nerve terminal interactions 178
        symmetrical regulation 151–157
    replacement 216–221
    striatum, in 184, 185, 188, 192, 193, 260, 261
Dopaminergic nigrostriatal neurons 32, 33, 46
    behaviour changes and 76, 270
    bilateral regulation, behaviour and 186, 216, 217
    cognitive and motor functions 69
    dopamine release from 150–163, 270
    firing rate 161, 162, 270, 271
    grafts, dorsal-striatal 216
    peptides 269
    ventral striatum 189–192
Dorsal raphe nucleus 39, 134
Dynorphin 124, 131, 147, 186
Dyskinesias 86, 164, 222, 266
Dystonia 48, 164, 214

Enkephalin(s) 35, 45, 69, 193
Enkephalin-positive neurons 9, 10, 15
    pallidal 124, 125
    striatal 31, 32, 33, 43, 128, 184, 186
    striato-pallidal 44, 124
Entopenduncular nucleus 23, 26, 49, 58, 59, 85

SUBJECT INDEX

Entopeduncular (contd)
  electrical stimulation 169
  GABA manipulation and 169
  GABAergic neurons 165, 166
  outflow, lesioning 169, 170
  projections 86
Extrapyramidal disorders
  GABA manipulation mimicking 164–176
  reaction time in 136, 137, 139
Extrapyramidal pathways 3, 49, 114, 115, 116, 117, 132, 133, 134
Eye fields, frontal 61, 85, 89, 92, 97, 98
Eye movements 83, 88, 89, 92, 97, 116

Face movements, neurons related to 65, 67, 69, 76
Fluorodeoxyglucose 59
Forebrain 122, 205, 207
Frontal cortex, dopamine depletion in Parkinson's disease 220
Frontal lobe damage 86, 213

GABA 31
  antisera to 43
  basal ganglia outflow pathways and 164–176
  infusions, dopamine release and 158, 159, 161, 162, 172
  input to basal ganglia, opiates modulating 45
  manipulation 166–176
  presynaptic activity 173
  release, muscimol and 156, 157, 163
  striatum in 184, 185
GABAergic
  drugs, dopamine release regulated by 151–153, 157, 158
  inhibitory effects of nigral efferents 51–53
  interneurons 34, 36, 37, 42, 43, 44, 45, 163
  neurons 31, 32, 33, 43, 44, 45, 123
  pathway, striatonigral 50, 59
  processes, muscimol and 156, 157
  synapses 172–174
Glial cells, basal ganglia, in 180
Globus pallidus 8, 9, 12
  afferent connections 12
  border cells, nucleus basalis of Meynert and 65
  efferent connections 13, 27, 115, 123
  external segment 67, 72, 165, 270
  homovanillic acid in 265
  internal segment
    anatomical relationships to SNr 65
    cooling of 87
    cortical pathway 84–86
    efferent connections 49

GABAergic neurons 165, 166
neuronal discharge pattern, movement and 72
neurons, somatotopic organization 67
substance P-positive striatal efferents in 23
subthalamic nucleus regulating 262
thalamic route to SMA 85
intrinsic organization 262
lateral segment, tremor and 266
limbic 107–113, 116
loop systems of 133
neurochemical subdivisions 123, 124, 126, 132
neuronal activity 73, 80, 98, 107
neurons
  behaviour and 79
  discharge rates 65
  proprioceptive control 81
  set-related response 92
  striatal inhibition of 80
premotor pathways 122
segments 12, 13, 23
somatosensory stimulation, pathways 81
striatal spiny neurons in 31, 35, 44
tyrosine hydroxylase-containing fibres in 266
ventral 15, 116
  neurons, synapses 35
  neuropeptides 124
  nucleus basalis of Meynert and 112
  pathways through, segregation 78, 79
  projections 15, 16, 19, 20, 109, 110
  striatofugal efferents in 23
  -thalamo-cortical connections 117
Glutamate 184, 185, 204, 205
Glutamate decarboxylase (GAD) 32
  antibodies to 43, 45
  antisera to 35, 36
  depletion, Huntington's chorea, in 174
  immunoreactive terminals 32, 33, 36, 37, 172, 173
  -like immunoreactivity, striosomes, in 128
Golgi impregnation, striatal neurons, of 30–39, 43, 44

Habenula 59, 86, 115, 166
Head turning, compulsive 48, 53, 167, 169, 170
Hemiballismus 75, 164, 166, 214, 227, 252
Hippocampus 5, 216
Huntington's chorea 28, 164
  cerebral cortex in 227
  cortico-neostriatal transmission and 205
  delayed alternation in 212
  dopamine transmission in 192

SUBJECT INDEX

GAD depletion in 174
genetic aspects 222, 223
model of basal ganglia function, as 218, 219, 223
nucleus accumbens in 222
oculomotor control in 89
postural disorders in 227
voluntary movement control in 252
6-Hydroxydopamine, dopamine depletion after 187, 188, 189, 190, 220
5-Hydroxytryptamine (5-HT) 33, 34, 39, 184, 185

Immunocytochemistry
  GABAergic neurons and 42, 43, 45
  striatal neuron studies, in 30, 32, 33, 37, 38
Interneurons
  GABAergic 36, 37, 38, 45
  nigrostriatal, dopaminergic 161
  pallidal 262
  striosomal 148

Kainic acid 187, 188, 202

L-dopa, 1, 217, 219
Leg movements, neurons related to 65, 67, 69, 76
Limbic system 24, 111, 116, 269
Limbic–extrapyramidal interface 109
Limbic–striatal connections 5, 6, 20
Locomotion 101, 167, 168, 169
Locomotor hyperactivity 164, 169
Locus caeruleus, noradrenergic pathways from, Parkinson's disease, in 101

Macaca fascicularis 97, 106
Macaca fuscata 112
Macaca irus 97
Macaca mulatta, basal ganglia function in 64–82
Melanocyte-stimulating hormone, striatum, in 186
Memory
  basal ganglia and 1, 88, 97, 98
  -contingent visual response 88, 89, 94, 98, 107
  deficits, neuronal grafts and 216
  spatial 88, 89, 96
Mesencephalon, pallidal projection to 15
Mesocortical pathways 122
Mesolimbic pathways 122
N-Methyl-4-phenyl-1, 2, 3, 6-tetrahydropyridine (NMPTP) 27, 221, 239
α-Methyl-p-tyrosine 153

Micrographia, Parkinson's disease, in 252, 255
Midbrain, dopamine neurons 116
Mnemonic processes, 88, 89, 96, 97, 98, 137
Motivation to movement 86, 98, 100, 101, 108
Motor centres 85
Motor control 84, 87, 94, 111
Motor cortex 85, 103, 104, 154–156
Motor set, supplementary motor area and 87, 94
Motor plan 228, 229, 233, 234, 235, 236, 254
Motor programmes 228, 229, 230, 231, 234, 244, 254
Movement
  activation 247, 249, 252
  amplitude and direction, basal ganglia neuronal discharge and 72, 75, 136, 137
  analysis
    decomposition approach 249–252
    subtractive approach 246–249
  anticipation 245, 256
  cerebral control 87, 88, 98
  closed-loop control 244, 245
  complex, control 87, 88, 245
  disorders
    GABA function and 164, 165
    Parkinson's disease, in 227–241
  execution 101, 228–241, 249, 255
  initiation 97, 101, 234, 235, 245, 255
  motivation to 86, 98, 100, 101
  open-loop control 245, 246
  planning 87 *See also Motor plan, Motor programmes*
  preparation for 87, 246, 247, 249, 252, 256
  reaction times 244, 245, 246–252, 256
  sequences 87, 88, 101, 110, 228, 229, 232, 234, 235, 245, 252, 253, 254, 255
  striatal control 108
  supplementary motor area and 105, 106
Muscimol
  behaviour and 167
  dopamine release and 155, 156, 157, 172
  GABA functions and 165, 167, 168
  GABAergic processes in basal ganglia and thalamus 156, 157

Neocortex 186
Neostriatum
  cerebral cortex and 201–215
  cognition and 207
  cortical inputs to 108, 109
  cortical stimulation, functional specificity 204, 205
  general features 38, 39, 108

Neostriatum (*contd*)
  heterogeneity 204, 205
  information processing in 207
  lesions, behaviour and 202, 203, 204, 207
  motor and association aspects 211–215
  neurons, synapses 30–47
  sensory functions and 203
  See also Striatum
Neuroleptics 1, 176
Neuron grafts 188, 189, 195, 196, 216, 217
Neurotensin 128, 186, 193
Neurotransmitters, target specificity of 178
Nigrocollicular relationships 53–56, 60
Nigrocortical projections 18
Nigroreticular efferents 50, 51, 53–57
Nigrostriatal
  connections 7, 8, 17, 35, 39, 45
  dopamine system, Parkinson's disease, in 221
  GABAergic pathway 59
  neurons, dopaminergic 33, 50, 122, 147, 150–163
    modulatory effect on striatum 67
    monosynaptic connections 110, 111
    postsynaptic targets 111
  pathway, autoregulation 135
Nigrotectal
  neurons 50, 51, 54, 56
  pathway 53, 116
  projection 19
Nigrotelencephalic projection, non-dopaminergic 60
Nigrothalamic
  -cerebellothalamic interactions 56, 57, 62
  -cortical pathways 117
  neurons 50, 51, 53–57, 153, 154
  -nigrocollicular pathways 49, 50, 51
  -pallidothalamic projections 61
  pathways 174, 263
  projections 18, 35, 56
Noradrenaline 144, 145, 179, 184, 185
Nucleus accumbens 1, 5, 8, 10, 11, 12, 19, 20, 24, 184
  dopamine deficiency in 217, 219, 220, 222
  dopamine innervation and 20, 189
  GABAergic neurons in 166
  neurons, excitation of 108
  neurotransmitter patterns in 145
  Parkinson's disease, in 110
  peptide distribution in 130
Nucleus basalis of Meynert 65, 112, 210

Ocular and cephalic motor activity 53, 54, 56, 60
Oculomotor control 88, 89
Olfactory tubercle 8, 107, 116, 130, 184

Opiate peptides, modulating GABA input to basal ganglia 45

Pallidofugal projections 13, 115, 124
Pallidohabenular connections 14, 15, 24, 26, 86
Pallido-habenulo-raphe pathway 134
Pallidolimbic confluence 15
Pallidomesencephalic projection 15
Pallidonigral neurons 35
Pallido-pedunculopontine nucleus connections 133, 134
Pallidospinal pathway 94
Pallidosubthalamic connections 12, 13, 80, 109, 110, 175
Pallido-subthalamo-pallidal loop 124, 133
Pallidotegmental projections 49
Pallidothalamic pathway 174, 263
Pallidothalamic projections 13–15, 26, 49
Pallidothalamocortical pathways 117, 124
Pallidothalamocortical projections 83, 85, 86, 115
Pallidum *See under Globus pallidus*
Parkinson's disease 28, 164
  akinesia 211, 218
  basal ganglia in 101
  behavioural disorders in 218
  cognitive functions and 199, 211, 213, 233, 237, 238
  deficits in 96, 97, 101, 217
  demented and non-demented patients 233, 238
  dopamine deficiency in 1, 198, 217, 218, 238
  dopamine transmission in 192
  emotional state in 110
  experimental 86, 100
  eye deficits in 97
  frontal cortex dysfunction in 213, 226
  micrographia in 252, 255
  model of basal ganglia disease, as 226, 227
  motor defect in 254, 255
  motor disorders in 225–241
  motor plan in 228, 229, 233, 234
    breakdown 234, 235
  motor programme in
    assembly 229, 230
    content 231, 234, 236, 254
    fast movement performance 231, 232, 255
    switching 230, 235, 236, 254, 256
  movement in
    amplitude 74, 75
    control 97, 110
    execution 101, 228, 229, 234
    initiation 97, 101

SUBJECT INDEX

sequencing 245, 252, 253, 254
oculomotor control in 89
pathological changes in 218
postencephalitic 174, 218, 223
postural deformity in 227, 238, 239
postural instability in 238, 239
predictive motor action in 231–234, 237, 246, 254
reaction time in 96, 101, 136, 229, 230, 232, 246, 247, 248
striatal abnormalities and 211
substantia nigra in 1, 96, 97
surgical thalamic lesion in 99
unilateral 59
voluntary movement control in 242–257
Pedunculo-pallidal connection 133, 134
Pedunculopontine nucleus 19, 116
  descending connections 26, 27, 28
  GABAergic neurons 166
  loop pathways 133, 134
  neuropeptides 128
  projections 127, 134
Perception, basal ganglia and 1
Perceptual-motor skill 212, 213
Picrotoxin, GABA function and 165, 167
Postural deviation 164, 167, 169, 238
Postural disorders, basal ganglia disease, in 48, 227
Postural responses, anticipatory 256
Potassium, dopamine release and 155, 157
Prefrontal cortex 84, 88, 89, 96, 212, 213
Prosencephalic systems 205, 207, 212
Prostaglandins, local regulation of transmitter release and 258
Psychological approaches, to basal ganglia function 242–257
Pursuit-tracking 242–244
Putamen
  basal ganglia output pathways through 100
  cortical projections to 69, 73
  dopamine deficiency, Parkinson's disease, in 217, 218, 219, 220, 221, 226, 265
  lesions 214
  neurons 44
    caudate neuronal response and 96
    movement related 72, 73, 78
    organizaton 65, 67, 69, 70, 71
    response to somatosensory stimulation 70, 71, 81
    set-related response 83, 90, 92, 94
    tonically active 90, 92
  neurotransmitter patterns in 145
  Parkinson's disease, in 110
  ventral 79, 107, 108
Putameno-nigral fibres 44
Pyramidal system 114

Raphe, dorsal, input to striatal neurons 39
Reserpine
  L-dopa and 85
  metabolic activity and 58, 59
  motivational state in monkeys and 100
  motor effects, species differences 86
  striatal activity and 80
Rigidity 48, 86, 99, 207, 227, 238, 239, 265

Schizophrenia 176, 192, 240
Secretin, striatum, in 186
Sensorimotor, definition 197, 198
Sensorimotor neglect 217
Sensory guidance of movement and speech 87
Sensory neglect 189, 196
Single-cell recording 64–82, 101, 162, 270
Sleep states, neural activity in 134
Somatostatin, striatum, in 186, 193
Somatostatin-like immunoreactivity 128, 131
Somatostatin-positive fibres, substantia nigra pars lateralis, in 124
Somatostatin-positive neurons 34, 35, 37, 38
Spatial reversal 187, 197
Spinal reflexes, basal ganglia and 256
Stereotypy 164, 167, 168, 169, 176
Striatal neurons
  afferent, segregation of pathways and 79
  aspiny 35, 42, 43, 44, 45
  discharge rates 65
  dopamine 189–192
  electron microscopy of 30, 31, 34, 35, 36, 43, 44
  enkephalin-positive 31, 32, 33, 43
  function, selective testing 239
  GABAergic 123, 165, 166
  GAD-immunoreactive terminals in 173
  Golgi impregnation 30–39, 43, 44
  grafting 188, 189, 195, 196
  input to 32–34, 38, 39
  intrinsic 184
  output 34, 35, 39
  projection neurons 31, 35, 43, 44, 45
  spiny 31–35, 36, 37, 39, 42, 43, 44, 45
  synapses 30–47
  types 270
Striatofugal fibres 8, 9, 10
Striatonigral
  connections 9–12, 17
  dopaminergic projections 10
  GABAergic fibres 42, 165
  neurons 31, 33, 148
  return loop 10
Striato-nigro-striatal circuit 134, 135
Striatopallidal projection 8, 9, 12, 20, 44, 109, 112, 186

Striatum 4–12
  afferent connections 4–8, 27, 79, 130, 131
  behaviour switching and 197, 198
  cellular islands 69
  chemical morphology 184–186
  dopamine in 1, 176–182, 226, 260, 261
  dopamine functions modulated by 193
  dopamine innervation 122, 123, 132, 134,
    135, 136, 138, 147, 184, 185, 192, 193,
    198, 199, 217
  dopamine islands 122, 130, 138, 146, 147,
    177
  dorsal 184, 190, 191
  dorsolateral 7, 20
  efferent connections 8–12, 19, 23, 25, 130,
    131
  enkephalin-positive fibres 128
  interneurons 34, 36, 37, 38, 39, 42, 43, 44,
    45, 123, 131, 146
  lesions, unilateral 53
  limbic 5, 6, 7, 20, 24, 26, 107–113
  limbic and non-limbic 108, 110–112
  loop nuclei 116, 127, 133
  muscarinic receptors 128, 146
  neurochemical subdivisions 123, 128–131,
    132
  neurons *See under Striatal neurons*
  neuropeptides 177, 184, 185, 186
  neurotransmitter functions 186–189
  non-limbic 7, 20, 24, 26, 109
  opiate binding sites 128
  role in basal ganglia 49
  sensorimotor interface, as 188, 189, 197
  tyrosine hydroxylase in 260, 261
  ventral 116, 184
    connections to ventral pallidum 108, 116
    dopamine neurons 189–192
    emotional response in 107
    Parkinson's disease, in 226
    visual response in 78, 107
  ventromedial 7, 20
  *See also under Caudate nucleus,
    Neostriatum, Nucleus accumbens,
    Putamen*
Strionigral pathways 122, 133
Striopallidal pathways 122, 123, 124, 133,
  166, 235
Striosomes 128, 130, 131, 132, 138, 139, 146,
  147, 148, 177, 180
Subcortical dopamine innervation 122
Substance P-positive fibres 9, 10, 15, 23, 109
  dopamine regulation and 193
  nigral 124
  striatal 128, 186
  striatopallidal 124
  striosomal 131

Substance P-positive striatal neurons 31, 32,
  33
Substantia nigra
  afferent connections 17
  cell groups 16, 17
  descending projections 19
  dopamine neurons 7, 16, 17, 18, 19, 24, 25,
    35, 39, 123
  dopamine release, drug-induced 150–163
  efferent connections 17, 19, 48–63
  GABA release in, muscimol and 156, 157
  lesions 53, 188
  limbic connections 17, 18
  mesencephalic projections 101
  neuronal discharge, movement parameters
    and 72
  nigrostriatal projections 7, 8, 17, 25
  non-dopamine neurons 16, 19, 25
  Parkinson's disease, and 1
  pars compacta *See below under Substantia
    nigra, pars compacta*
  pars reticulata *See below under Substantia
    nigra, pars reticulata*
  striatal projections to 23, 36
  striatonigral connections 9, 17, 25, 44, 50,
    112, 186
  subcortical projections 17
  ventral pallidal connections 15
  ventral tegmental area *See under Ventral
    tegmental area*
Substantia nigra, pars compacta (SNc) 11,
  12, 24
  dopamine-containing neurons 126, 127,
    147, 270
  neurochemical subdivisions 126, 127
  neuronal discharge 65, 67, 78
  neurons, dopamine receptors of 135
  Parkinson's disease, in 96, 97
  striatonigral input to 134
Substantia nigra, pars reticulata (SNr) 10,
  11, 19
  anatomical relationship to globus pallidus,
    internal segment 65
  cortical pathways 84–86
  dynorphin activity 124, 147
  efferents 49, 50, 51–57, 123, 167, 270
  frontal eye field and 61, 81, 88, 89
  GABA manipulation 167
  muscimol and 167, 168
  neurochemical subdivisions 123, 124, 125,
    132
  neurons
    behaviour-related activity 85
    discharge latencies, initiation of
      movement and 98, 107, 270
    discharge rates 65, 96, 99

SUBJECT INDEX

GABAergic 166, 167
non-dopaminergic 50
single neuron activity 88, 89
somatotopic organization 67
neuropeptides 124, 125
nigrothalamic and nigrocollicular
pathways 49, 50, 51, 85, 88, 116
output system of basal ganglia, as 49, 50
picrotoxin and 167
premotor pathways 122
subthalamic nucleus regulating 262
thalamic route to supplementary motor
area 85
Subthalamic nucleus 16, 20, 116
basal ganglia output regulated by 261–268
connections of 12, 13, 15, 16, 25, 264, 265
cortical projections to 16, 133
efferents 262
excitatory functions 263, 264
GABA manipulation 175
lesions 166, 167, 213, 214
locomotor activity and 169
movement control by 263, 264
neurons
discharge, movement parameters and 72
GABAergic 166, 264
somatotopic organization 67
neuropeptides 127
picrotoxin, effects 166, 167
projections 50
species differences 264
Subthalamonigral fibres 262
Subthalamopallidal projections 12, 127, 262, 264
Superior colliculus
cell bursts, GABA-induced 55, 56, 62
connections 28
frontal eye fields and 61, 85
intermediate layers 60
SNr projection to 85
Supplementary motor area (SMA) 83
ablations 86, 87
attentional response and 106
basal ganglia and 86–88, 266
cortical inputs to 99, 104
damage, movement and 87
motor set and 87, 94
movement and 105, 106
neuronal discharges, time relations 105, 106, 107
output pathways, motor 94, 99, 100
pallidal input to 94, 115
projections 104, 105
thalamic inputs to 85, 99, 115
thalamus affecting 266

Taurine, striatal efferent neurons, in 31, 32
Tectospinal neurons, eye and head
movement and 53, 54
Tectospinal/tectodiencephalic neurons,
disinhibitory mechanisms in 54, 55, 56
Tectospinal-thalamic projections 60, 61
Temporal cortex, projections to basal
ganglia 84, 88
Tests of function 239
Thalamus
basal ganglia and cerebellar inputs to 13, 14, 18, 83, 84, 85
basal ganglia bilateral regulation and 150–163
cortical projections 61, 103, 115
frontal eye field connections 97
GABA infusion into, dopamine release in
basal ganglia and 157, 158, 161
GABAergic processes in, muscimol and 156, 157
nigropallidal connections, overlap 61
nuclei, intralaminar and motor 116, 159, 161, 163
nucleus ventralis lateralis pars ovalis
(VPL$_o$), connections of 100, 103
pallidal connections 13–15, 83, 115
pallido-recipient zones, innervation of 124
projections to cerebral cortex 103
striatal connections 7, 46, 69, 115
surgical lesions, Parkinson's disease, in 99
VA-VL complex 13, 14, 16, 25
VM region, GABAergic neurons 166, 167
VM-VL stimulation 153, 154, 156, 157, 158
Thyrotropin-releasing hormone, striatum, in 186, 193
Tics 136
Tremor 48, 86, 99, 207, 227, 238, 239, 265, 266
Tyrosine hydroxylase-containing fibres 32, 33, 35, 39, 127, 260, 261, 266

Vasoactive intestinal polypeptide, striatum,
in 186
Vasopressin, striatum, in 186
Ventral pallidum *See under Pallidum*
Ventral tegmental area (VTA) 11, 12, 17, 18, 19, 24, 118–121, 122, 147, 189, 190, 191, 196
Visual response 78, 79
Visually-guided behaviour 87, 89

Weaver mouse mutant, 147, 222, 269
Wilson's disease, postural disorders in 227